Advanced Techniques in Biological Electron Microscopy

Edited by

J. K. Koehler

With Contributions by

S. Bullivant · J. Frank · K. Hama · T. L. Hayes
J. H. Luft · F. A. McHenry · D. C. Pease
M. M. Salpeter

With 108 Figures

Springer-Verlag New York Heidelberg Berlin 1973

James K. Koehler, Ph. D.
University of Washington
Department of Biological Structures
School of Medicine
Seattle, WA (USA)

The cover illustrations show examples of frog red blood cells prepared in five different ways for electron microscopy. Included are critical point dried cells viewed with the scanning electron microscope, shadowed cell "ghosts", negatively stained "ghosts", frozen etched cells and a thin section preparation. The micrographs were prepared by J. K. KOEHLER and T. L. HAYES.

ISBN 0-387-06049-9 Springer-Verlag New York Heidelberg Berlin
ISBN 3-540-06049-9 Springer-Verlag Berlin Heidelberg New York

Preface

The past decade has seen a remarkable increase in the use of electron microscopy as a research tool in biology and medicine. Thus, most institutions of higher learning now boast several electron optical laboratories having various levels of sophistication. Training in the routine use of electron optical equipment and interpretation of results is no longer restricted to a few prestigious centers. On the other hand, techniques utilized by research workers in the ultrastructural domain have become extremely diverse and complex. Although a large number of quite excellent volumes dedicated to the basic elements of electron microscopic technique are now available which allow the novice to acquire a reasonable introduction to the field, relatively few books have been devoted to a discussion of more advanced technical aspects of the art. It was with this view that the present volume was conceived as a handy reference for workers already having some background in the field, as an information source for those wishing to shift efforts into more promising techniques, or for use as an advanced course or seminar guide.

Subject matter has been chosen particularly on the basis of pertinence to present research activities in biological electron microscopy and emphasis has been given those areas which seem destined to greatly expand in usefulness in the near future. It would be impossible to adequately cover the myriad technical developments available to cell biologists interested in fine structure within this modest volume, and the knowledgeable reader could cite many worthy areas that have not been touched upon. Nevertheless, the subject matter included will be seen to cover a number of subdisciplines within the ultrastructural area and is not restricted to a narrow range of interests.

Not only have considerable improvements been made in the design and use of conventional transmission electron microscopes, but entirely new instruments have appeared on the scene such as the scanning and high voltage electron microscopes which provide quite different kinds of information and require modified or new methodologies of specimen preparation and manipulation. The potentialities of these tools are explored in chapters by T. L. HAYES and K. HAMA, respectively. Alternatives to chemical fixation and staining have been devised, particularly those employing freezing methods of preservation. The freeze-fracture or freeze-etching approach has been particularly fruitful in this regard and is the subject of S. BULLIVANT's contribution. Substitution methods, either with or without freezing involvement, have also turned out to be successful alternatives and complementary approaches to traditional chemical methods of tissue preservation. These techniques and the unique inert dehydration method form the subject of the chapter by D. PEASE. Even the subject of embedding materials for fine structural studies (a field that many probably considered closed after the advent of epoxy resins) receives an imaginative reappraisal and infusion of new ideas in the article by J. LUFT. Although autoradiography has for many years been included in the repertoire of electron microscopy, the urge to incorporate more specificity and dynamic information into ultrastructural studies has recently caused many more laboratories to turn to this technique. Consequently, it may seem to be considered an almost routine part of the technology of fine structure workers. M. SALPETER and F. A. McHENRY, however, indicate in their contribution the great care that one must exercise in optimizing considerations of resolution, specificity and morphology in addition to the many pitfalls which stand in the way of accurate inter-pretation of electron microscopic autoradiography. Finally, at the physical end of the spectrum, J. FRANK illustrates the possibilities opened up by computer processing of electron optical data and details the mechanisms by which such information can be obtained.

I wish to express my sincere appreciation to these authors for the time and effort which they have expended on these contributions and for their patience and understanding during the editorial processing. Dr. KONRAD F. SPRINGER and the Springer-Verlag have been most cooperative in this venture and have striven to achieve the highest quality possible in word and picture.

Seattle, February 1973   J. K. KOEHLER

Contents

Embedding Media — Old and New. J. H. Luft

A. Introduction 1
B. Early Embedding Media 1
    I. Gelatin 2
    II. Celloidin 2
    III. Paraffin 2
    IV. Methacrylate 3
C. Conventional Embedding Media 6
    I. Polyester Resins 6
    II. Epoxy Resins 7
    III. Water-soluble Embedding Media 10
D. Advantages and Disadvantages of Conventional Embedding Media   12
    I. Polymerization Damage 12
    II. Beam Damage 13
    III. Erratic Polymerization 15
    IV. Shrinkage 16
    V. Viscosity 17
    VI. Osmotic Damage 18
    VII. Toxicity 20
    VIII. Miscellaneous Defects 22
E. New Embedding Media 23
    I. Low Viscosity Epoxy Resins 24
    II. Exotic Embedding Materials 26
        1. Hydrophilic Gels 26
        2. Polyurethane Resins 28
F. Conclusion 28

References 31

Substitution Techniques. D. C. Pease

A. Introduction 35
B. Inert-dehydration 36
    I. Method 38
    II. Artifacts 40
    III. Fine Structure 41

C. Freeze-substitution 42
     I. Experiments of this Author 44
     II. The Work of FERNÁNDEZ-MORÁN and BULLIVANT 53
     III. The Experiments of REBHUN and Associates 57
     IV. Experiments of VAN HARREVELD, CROWELL and MALHOTRA 58
     V. Pertinent Findings of Other Investigators 59
D. Conclusions 60

References 63

Freeze-Etching and Freeze-Fracturing. S. BULLIVANT

A. Introduction 67
B. Freezing of Biological Systems 67
C. Methods and Instrumentation 70
     I. Historical Development 70
     II. Physical Basis of Technique 71
         1. Fracturing 71
         2. Etching 73
         3. Replicating 73
         4. Cleaning 73
     III. A Simple Freeze-Fracture Device 74
         1. Pre-Treatment 74
         2. Freezing 74
         3. Fracturing 75
         4. Replication 76
         5. Cleaning of Replica 78
     IV. A Microtome Freeze-etch Device 79
         1. Freezing 80
         2. Fracturing 80
         3. Etching 80
         4. Replication 80
         5. Cleaning of Replica 81
     V. Other Simple Devices 81
         1. GEYMEYER 81
         2. WINKELMANN 81
         3. WEINSTEIN 82
         4. McALEAR 84
     VI. Other Microtome Devices 84
         1. KOEHLER 84
         2. STEERE 84
         3. PRESTON 84
         4. EDWARDS 85

VII. Complementary Replicas 85
    1. CHALCROFT 85
    2. STEERE 87
    3. WEHRLI 87
    4. SLEYTR 88
    5. WINKELMANN 88
VIII. Technical Variations 88
    1. Pretreatment 88
    2. Freezing 88
    3. Replication 89
D. Interpretation 90
    I. The Membrane Fracture Face 90
      1. Complementary Replicas 94
      2. Surface Labelling 94
      3. Thin Sectioning 95
    II. Particles in Membranes 97
      1. Lack of B Face Pits 98
      2. The Nature of the Particles 100
    III. Contamination 102
      1. Particulate Contamination 102
      2. Plaque Contamination 105
E. Conclusions 106
    I. Choice of Equipment 106
    II. Future 107
References 107

Electron Microscope Autoradiography. M. M. SALPETER and F. A. McHENRY

A. Introduction 113
B. Distribution of Developed Grains Around Radioactive Sources . 115
C. Analysis of Autoradiograms 123
    I. Qualitative Assessment 125
    II. Quantitative Analysis 129
      1. "Simple Grain Density" Analyses 129
      2. "Per cent" Analysis 134
      3. "Probability Circle" Analysis 136
      4. "Density Distribution" Analysis 143
D. Conversion of Developed Grain Data to Information on Radio-
activity 149
References 151

Scanning Electron Microscope Techniques in Biology. T. L. HAYES

A. Introduction 153
    I. General Principles of Operation 153
    II. A Comparison of Resolution 156
    III. Comparison of Information Transfer 161
        1. Analytic Information 161
        2. Subjective or Experiential Information Transfer . . . 163

B. Specimen Preparation 163
    I. Selection of Tissue 164
        1. Natural Surfaces 164
        2. Dissected Material 166
        3. Sectioned Tissue 167
        4. Living Specimens 167
        5. Ion Etching 169
        6. Freeze-Etching Techniques 169
    II. Fixation 170
        1. Ultrastructure Fixatives 170
        2. Light Microscope Fixatives 170
    III. Dehydration and Drying 171
        1. Freeze-Drying 171
        2. Critical Point Drying 174
        3. Air Drying 174
    IV. Improving Conductivity 175
        1. Metal Evaporation 175
        2. Conducting Sprays and Solutions 177

C. Viewing Techniques 177
    I. Standard Specimens 178
    II. Signal Monitor 179
    III. Accelerating Voltage 179
    IV. Specimen Current 180
    V. Contrast; Photo-multiplier 180
    VI. Scan Rate 180
    VII. Astigmatism Correction 183
    VIII. Final Aperture Size 184
    IX. Viewing Aspect 184
    X. Micromanipulation 185

D. Signal Processing 185
    I. Differentiation 185
    II. Deflection Modulation 187

III. Color Modulation 188
IV. Computer Processing 188
E. Recording Techniques 189

  I. Photographic Integration 189
    1. Polaroid Film 189
    2. 35 Millimeter Standard Roll Film 189

  II. Stereo-Pairs 190
    1. Resolution of Analytic Ambiguities 190
    2. Enhancement of Experiential Contact 190
    3. Methods of Stereo-Pair Presentation 190

  III. TV Tape 192
F. Information Assimilation by the Observer 193

  I. Analytic Information Processing 193
    1. Geometric Information 193
      a) Metric Geometry 193
      b) Topologic Geometry 194
    2. Chemical Information 196
      a) Characteristic X-Ray Elemental Analysis 196
      b) Auger Spectra 196
      c) Cathodoluminescene Analysis 197
      d) Enery-loss Spectra 197
    3. Electrical Properties and Charging 198

  II. Experiential or Subjective Information Processing . . . 198
    1. Models of Perception 198
    2. Limits of Analytical Information Processing 199
    3. Possibilities of Complementary Subjective and Analytic
      Investigations 199
G. Conclusion 200

  I. Questions Regarding a Scanning Electron Microscope
  Program for Biological Study 200
    1. Is the SEM Really Necessary? 200
    2. Which Instrument? 201
    3. What Auxilliary Equipment Might be Needed? . . . 201

  II. Prospects for the Future 202
H. Appendices 203

  I. Optical Aids for the Viewing of Vertically Mounted Stereo-
  Pairs 203
  II. Projection of Stereo-Pairs by Means of a Superimposed
  Color-Coded Transparency 205

References 206

Computer Processing of Electron Micrographs. J. Frank

A. Introduction 215
B. Linear Systems and Fourier Processing 217
   I. The Concept of Linear Systems 217
   II. Fourier Integrals and Theorems 218
   III. Implementation on the Computer 221
C. Digitizing of Electron Micrographs 221
   I. Photographic Recording 221
   II. The Densitometer 222
   III. Sampling 223
   IV. The Effect of the Scanning Aperture 225
   V. The Effect of the Image Boundary 226
D. Noise Filtering 227
   I. Noise Sources 227
   II. Noise Filtering in the Case of Periodic Objects 228
   III. Noise Filtering in the Case of Aperiodic Objects 231
E. The Cross Correlation Function and its Use for Image Alignment 232
   I. Two Electron Micrographs with Identical Defocus Value . 233
   II. Two Electron Micrographs with Different Defocus Values . 234
   III. A Technical Note 235
F. Two-Dimensional Restoration 235
   I. Restoration of Phase Objects from a Single Phase Contrast
     Image 235
   II. Restoration of Phase Objects from a Focus Series 243
   III. Restoration of the Complex Object 244
   IV. Restoration from Dark Field Images 247
G. Object/Support Separation 248
   I. Optimal Filtering 248
   II. Matched Filtering 249
   III. Separation Based on Knowledge of the Film Structure . . 249
   IV. Separation Based on the Z-dependence of the Imaginary
     Scattering 250
H. Three-Dimensional Reconstruction 250
   I. The Fourier Method 251
     1. Principle of the Three Dimensional Fourier Reconstruc-
       tion 251
     2. The Interpolation Problem 254
     3. The Use of Symmetries 256
     4. Implementation 257
     5. A Two-dimensional Fourier Reconstruction Scheme . . 262

　　II. Real Space Methods 262
　　　　1. Exact Solution 262
　　　　2. Superposition Method 264
　　　　3. Iterative Approximation 265
References 269

High Voltage Electron Microscopy. K. HAMA

A. Introduction 275
B. Merits of the High Voltage Electron Microscope 275
　　I. Specimen Penetration 276
　　II. Resolving Power 278
　　III. Beam Damage 279
C. Biological Applications 281
　　I. Specimen Preparation 281
　　II. High Resolution Observation 281
　　III. Observation of Thick Specimens 287
　　IV. High Voltage Stereoscopy 289
　　V. Observation of Undehydrated Specimens 292
　　　　1. Ultracryotome Method 292
　　　　2. Wet Cell Method 292

D. Design and Construction of High Voltage Electron Microscopes . 294

References 296

Subject Index 299

Contributors

BULLIVANT, STANLEY, Professor, Department of Cell Biology, University of Auckland, Auckland (New Zealand)

FRANK, JOACHIM, Dr., Max-Planck-Institut für Eiweiß- und Lederforschung, Abteilung Röntgenstrukturforschung, München (W. Germany)

HAMA, KIYOSHI, Professor, Institute of Medical Science, University of Tokyo, Tokyo (Japan)

HAYES, THOMAS L., Professor, Donner Laboratory, University of California, Berkeley, CA (USA)

LUFT, JOHN H., Professor, Department of Biological Structure, School of Medicine, University of Washington, Seattle, WA (USA)

MCHENRY, FRANCES A., School of Applied and Engineering Physics, Cornell University, Ithaca, NY (USA)

PEASE, DANIEL C., Professor, Department of Anatomy, University of California, Los Angeles, CA (USA)

SALPETER, MIRIAM M., Professor, School of Applied and Engineering Physics, Cornell University, Ithaca, NY (USA)

Embedding Media — Old and New

John H. Luft

A. Introduction

Embedding media for microscopy have no other value than that of a convenient means to achieve a particular end, namely, to enable the object of interest to be cut sufficiently thin for the microscope to develop its full resolution. The embedding does not contribute to the staining of the object nor to the resolving power of the microscope. The best embedding medium permits thin sectioning with the least damage during specimen preparation and gives the least interference during microscopy. This is not to say that embedding is a trivial part of specimen preparation; the cutting of the tissue in the embedding matrix is a mechanochemical event which can be interpreted only in terms of sophisticated concepts of the properties of materials. The most fundamental approach to the problem in the biological literature is that of WACHTEL, GETTNER and ORNSTEIN (1966). Useful mechanical concepts are developed in various texts, such as NIELSEN (1962) or McCLINTOCK and ARGON (1966) and mechanochemical concepts in the paper by WATSON (1961). Despite the advanced state of materials science, it has had little impact in improving our understanding of the mechanism of the cutting of embedded tissue, beyond what is intuitively obvious to biologists. It is clear that the embedding medium "supports and holds together" the tissue, but this phenomenon seldom is encountered in industrial processes where a detailed analysis is sufficiently important to engender research. It is possible that embedding material can be compared usefully to the matrix in composite materials, but that it functions in embedded tissue to produce an effect opposite to that intended for industrial laminates and composites.

B. Early Embedding Media

Before discussing the newer embedding materials, it would be profitable to examine the advantages of the older media as well as whatever faults may have prompted the search for substitutes. A glance at any edition of Lee's Microtomist's Vade-Mecum (LEE, 1928) suggests the variety of materials which have been explored for embedding. Included are "fusion masses" such as paraffin, soap and gelatin; and "cold masses" such as celloidin (nitrocellulose), lead-gum, gum arabic-glycerin and shellac, as well as gums and sugar syrups to ptotect tissues during freezing. Of these, gelatin, celloidin and paraffin are worth examining in some detail along with the more recent methyl and butyl methacrylates.

I. Gelatin

Gelatin appears to have been the first embedding medium with a reference in Lee dated 1802. Later procedures in the 1880's consisted of soaking the fixed and washed tissue in warm gelatin solutions over a period of days to a week in increasing concentrations up to 25% gelatin and 10% glycerin, which after cooling, was hardened with formaldehyde and the gel cut wet. The defects were that the gelatin was a large molecule which gave viscous solutions requiring prolonged infiltration times, and that the gel was not strong enough to permit even moderately thin sections to be cut. However, it avoided temperatures above 35 °C and did not extract or displace lipids, since the tissue was never dehydrated. Shrinkage could be kept very low.

II. Celloidin

Celloidin embedding was introduced in 1879 and consisted of dehydrating the fixed tissue with absolute ethanol and then gradually infiltrating the tissue with increasing concentrations of nitrocellulose dissolved in an ether-alcohol mixture from 1—2% to 10—15% over a prolonged period. A small piece of tissue might take 2—3 days whereas an entire human embryo could require months. The most concentrated nitrocellulose solution was allowed to thicken further by very slow evaporation, and the nitrocellulose was converted to a firm gel by precipitation with chloroform, in which the nitrocellulose is insoluble. The tissue blocks were cut wet. The defects, again, were the high molecular weight of the nitrocellulose which gave very viscous solutions, and prolonged infiltration times. The ether-alcohol must have extracted most of the lipid. Somewhat thinner sections could be cut than from gelatin, down to 25—50 μ, and if the infiltration was done sufficiently slowly, large objects could be infiltrated completely and sectioned with remarkably little shrinkage.

III. Paraffin

Paraffin was introduced as an embedding agent in 1881. The tissue was fixed and dehydrated, and then "cleared" with some agent which was miscible both with the dehydrating alcohol and paraffin, such as benzene, chloroform, cedar wood oil, etc. (Some of these liquids by accident had a refractive index close to that of the tissue, so that the tissue became relatively transparent or "clear" when the alcohol was replaced by them.) The tissue then was transferred to a warm mixture of the clearing agent with paraffin, and then into pure molten paraffin, each for an hour or so, whereupon the paraffin, together with the tissue was solidified by quick cooling. Paraffin had the advantages of speed — tissue requiring 3 days in celloidin could be embedded in an hour in paraffin (LEE, 1928) — and sections could be cut as thin as a few microns. Speed resulted from the low viscosity of the molten paraffin, but the heat produced shrinkage. The paraffin blocks could be cut dry, and the sections would adhere to form ribbons, which was a great simplification in preparing serial sections. The advantages of paraffin so far outweighed its disadvantages that it became the routine method for histology. With the introduction of electron microscopy in the late 1940's it became apparent that even paraffin could not deliver sections in the 1/10 micron range which was required for good electron imaging (PEASE and BAKER, 1948), and the search was launched for a substitute.

IV. Methacrylate

Methacrylate, and particularly butyl methacrylate, was introduced as an embedding medium by NEWMAN, BORYSKO and SWERDLOW (1949a, b). The procedure was almost too good to believe in terms of providing nearly ideal solutions to the difficulties of earlier media. For the first time a low molecular weight monomer was employed to infiltrate the dehydrated tis-

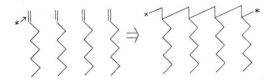

Fig. 1. Polymerization (curing) mechanism of methacrylate ($\vee\!\wedge\!\vee$ =) with a free radical (*) as initiator

sue, with rapid penetration resulting from the low viscosity and rapid diffusion from the low molecular weight, all at room temperature or even at 0 °C if necessary. After infiltration, the monomer was polymerized to the final solid plastic by gentle heating, or by alternative methods polymerization could be accomplished at room temperature or in the cold. The mechanism of methacrylate polymerization is shown schematically in Fig. 1. Best of all, the resulting plastic was strong enough to be cut to the necessary thinness. The most serious fault was occasional injury to the tissues during polymerization which the authors identified from the first, suggesting that such damaged tissues be discarded since such blocks were easily recognized. The chemicals required were relatively inexpensive, nontoxic, easily obtained, and the hardness of the final plastic block could be varied continuously and predictably by using different proportions of methyl (hard) and n-butyl (soft) methacrylates. In retrospect, it was obvious that the methacrylate blocks were easy to cut, and the tissue in sections one micron thick could be stained for superb light microscopic cytology. Although the plastic could be dissolved away from the tissue easily, it was soon apparent that the resulting loss of support of the tissue components by the resin seriously distorted the ultrastructure, so that the fine structure was better preserved if the plastic was left with the sections for electron microscopy. During observation in the electron microscope, the sections "cleared" and became more transparent as some of the plastic sublimed from the section, and the improved contrast which resulted enabled ultrastructural details to be recognized without special staining of the sections.

The disadvantages which led to the replacement of methacrylate were two: the "polymerization damage" to the tissue block, and the susceptibility of the plastic in the section to damage by the electron beam. The latter problem was inherent in the methacrylates; they were unusually sensitive to depolymerization by heat, so much so that, for example, a high yield of methyl methacrylate can be obtained merely by distilling scrap pieces of polymethyl methacrylate above 300 °C (RIDDLE, 1954, p. 39). Although attempts were made to reduce damage by protecting the section by layers of evaporated carbon, the final solution lay in adopting different plastics.

The damage during polymerization appeared as a swelling of the tissue block, in the worst cases to double its original dimensions (8-fold increase in volume) and sometimes was referred to as "explosion damage" because in the electron microscope the tissue elements appeared separated from each other as if blown apart. Methacrylate was accused of uneven and erratic polymerization and a variety of empirical procedures, some verging on witchcraft, were proposed to reduce or eliminate polymerization damage. In fact, methacrylate polymerization usually is smooth and predictable, and it was more valuable to understand the mechanism underlying polymerization damage than to curse it. The problem was first identified by BIRBECK and MERCER (1956) and by WATSON (1963), who suggested that the "explosion" was due to osmotic swelling due to unpolymerized monomer dissolving in the polymer which somehow had been formed rapidly within the tissue block. At the same time WATSON (1963) proposed the use of cross-linking additives to the methacrylate so that the polymer within the tissues would swell less. The reason that polymerization was accelerated within the tissue lay in the well-documented evidence that the polymerization rate in methacrylate esters is highly dependent upon the viscosity of the mixture. The maximum rate of polymerization, which occurs at about 20% conversion in methyl methacrylate, may be of the order of 10 times the initial rate (BAMFORD et al., 1958). The fine texture of the fixed tissue itself apparently is sufficient to restrict the motion of the growing polymer chains, and thus to simulate the effect of viscosity in reducing chain terminations, which is the mechanism of the accelerated polymerization of the gel effect (BEVINGTON, 1961). The finer the texture of the fixed tissue block, the faster the methacrylate could polymerize within it with respect to the surrounding methacrylate, and the greater the osmotic swelling which could result. In retrospect, it is clear why viscous, prepolymerized methacrylate was useful to control polymerization damage (BORYSKO and SAPRA-

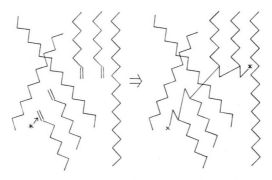

Fig. 2. Curling of prepolymerized methacrylate syrup, with completely formed polymer chains (⌇⌇⌇⌇⌇) dissolved in monomeric butyl methacrylate (⌇⌇⌇=). Note that polymer chains (⌇⌇⌇⌇⌇) strictly should look like right side of Fig. 1. Free radical initiator at (*)

NAUSKAS, 1954): it more or less matched the rates of polymerization outside and inside the block by the increased viscosity outside, so that the osmotic gradient inside and outside the block was minimized. Polymerization of prepolymerized methacrylate is shown schematically in Fig. 2. Low temperature polymerization was no help because it gave more time for the osmotic damage to occur. High temperature polymerization, which was studiously avoided by NEWMAN et al. (1949b) for fear of damage, paradoxically gave better results over-all (BORYSKO, 1956): the tissue was stronger than the nearly molten plastic, and solution rather than swelling was the result. Polymerization by ultraviolet light, in ideal cases, should have produced rapid polymerization in the surrounding monomer but slower rates in the opaque black (osmium-fixed) tissue block, thus equalizing the rates in another manner, but the ideal was seldom realized.

Despite the heroic efforts which were expended in attempting to counteract these two defects of the methacrylates, the results were only partially successful at best. When the first electron micrographs of tissue embedded in epoxy or polyester resins appeared in 1956 (Maaløe and Birch-Andersen, 1956; Kellenberger, Schwab and Ryter, 1956), it was clear that methacrylate embedding left much to be desired. The newer resins retained much more delicate structure in the cytoplasm of cells as well as in membranes. The methacrylates reappeared later as special variants in attempts to avoid lipid extraction by using water-soluble methacrylate (Cope and Williams, 1968) or to permit better enzyme accessibility to the tissue for histochemical purposes (Leduc and Holt, 1965; Leduc and Bernhard, 1967; Bernhard, 1969), or to obtain a highly ionic embedding medium (McLean and Singer, 1964). The methacrylates are a highly versatile group of plastics, all polymerizing by the same mechanism, but with a choice of substituents limited only by the ingenuity of organic chemists and the experimenters.

C. Conventional Embedding Media

The methacrylates have been all but replaced by the epoxy and polyester resins for general ultrastructural use, where the optimum preservation of cell structure is the prime consideration. Where other properties of the tissue are important or essential as in the field of histochemistry or immunochemistry, other embedding materials and particularly the "water-soluble" resins may be necessary. Some of the more popular embedding media in these groups are discussed below.

I. Polyester Resins

This class of resins can be defined as "the polycondensation products of dicarboxylic acids with dihydroxy alcohols[1]" (Bjorksten, 1956). This definition likewise includes the fiber-forming polyesters so common in modern textiles. The resins which concern the electron microscopist are the *unsaturated* polyesters (made with a small percentage of unsaturated acids, such as maleic, or less commonly an unsaturated alcohol) and particularly those using styrene as a cross-linking agent, and polymerized by a free-radical mechanism. This resin system was described first in 1940 (Rust, 1940) and became available in the late 1940's. The resin is relatively cheap

---

1 This is more a concise description of the method of manufacture in terms of the initial ingredients than a unique description of the final resin molecule. The latter might be represented as short segments of substituted hydrocarbon chains linked together primarily by ester groups, but this, in fact, is equally true of the epoxy resins cured by anhydrides which are used widely in electron microscopy. The significant difference between the polyester and epoxy resins lies in the nature of the prepolymerized resin molecules and particularly in the atomic arrangement of the strained group which carries the energy to drive polymerization, and hence determines the mechanism of polymerization or curing. Thus, the definition quoted above is terse, but correct.

and easy to manufacture and experimental batches can even be made on a laboratory scale. The resin before polymerization usually is quite viscous, from 300—5000 centipoises. The resin cures through a free-radical mechanism similar to the methacrylates; in fact, methacrylates can be used as the cross-linking agent in place of styrene (RUST, 1940; BJORKSTEN, 1956). The mechanism of polymerization is diagrammed in Fig. 3. There are similarities to the prepolymerized methacrylates (Fig. 2). The polymerization is most conveniently accomplished by using small amounts of an organic peroxide, together with a metal salt as an activator (cobalt naphthenate, popular in the paint industry). The first description of polyester embedding

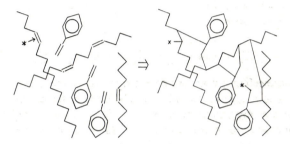

Fig. 3. Curing mechanism of polyester resin with styrene (⬡— =) as cross-linking agent. Free radical initiator at (*)

for electron microscopy was that of KELLENBERGER, SCHWAB and RYTER (1956) using Vinox K3, a proprietary resin of Swiss manufacture. In their more detailed paper (RYTER and KELLENBERGER, 1958) they had chosen another resin, Vestopal W. A different polyester resin system using Japanese resins (Rigolac) has been described by KUSHIDA (1960). The details of the use of these resins is summarized in the book by HAYAT (1970). All of these polyester resins are rather nonpolar and are not miscible with alcohol, so that acetone is usually used for dehydration. Styrene has been used as a "clearing" agent (KURTZ, 1961) for the polyester resins in which it serves a function similar to that of propylene oxide for epoxy resins; propylene oxide is, of course, of no value for polyester resins, nor is styrene of any use for epoxy embedding, since the curing mechanism of the two resins are completely independent of each other.

II. Epoxy Resins

This group of resins is characterized by the presence of the epoxy group, a strained 3-member ring of two carbon atoms and one oxygen, which upon rupture, provides the energy (~ 22 kcal/mol) (LEE and NEVILLE,

1967) to drive polymer formation. The epoxy ring (epoxide, oxirane) is uncommon among organic compounds and, although the basis for epoxy resin manufacture was laid in Switzerland (Ciba) in 1936, resins were available commercially only in the late 1940's (LEE and NEVILLE, 1967). The methacrylates and polyester resins polymerize exclusively through their ethylenic double bonds by a free-radical chain reaction which is highly selective and ignores most other common chemical groups in the vicinity. The epoxy ring, on the contrary, reacts avidly with any hydrogen atom which can be pried loose as a proton from any source, so that, in addition to reacting as intended with a second molecule in the resin mixture to form the polymer, the epoxy rings may also react with their container or other available surface as well as with the tissue block being embedded. In the case of the methacrylates and polyesters, the polymer chain which is initiated by a single free-radical, is completed within a few milliseconds (BAWN, 1948). The epoxy rings, however, react essentially at random as the rare, favorable collision breaks a ring open here and there, so that the polymer molecules independently grow larger very slowly. Whereas the methacrylates or polyesters are able to polymerize upon themselves with only a free-radical source as a trigger, the epoxies usually require large quantities of a second, nonepoxy molecule with which to condense, called the "hardener". Although potentially there are a great many different compounds which can be mixed with the epoxy-containing molecules as "hardeners" to form a solid polymer, in practice relatively few satisfy the necessary chemical restrictions and at the same time are cheap and convenient to handle. The first restriction is that for polymer formation, the participating molecules must be polyfunctional, i.e., each molecule must have at least two sites which are reactive. Thus, all of the compounds used to make epoxy polymers are di-compounds. The epoxy molecules themselves are di-epoxides in chemical terminology (although they usually are given trade names and numbers for humanitarian reasons). The second restriction is more a rule of thumb, in the sense that the successful compounds which give useful resins are four: organic acids, organic primary amines, sulfhydryl and alcoholic hydroxyl groups. Historically, the organic acids (which had to be dicarboxylic acids) were used first, and the most convenient were the anhydrides; these still are the "hardeners" used in most formulations for electron microscopy. The organic amines came later, typified by ethylene diamine, but replaced by others which were easier to live with, such as diethylene triamine; it was this "hardener" which MAALØE and BIRCH-ANDERSEN (1956) employed in their first publication. Sulfhydryl groups are available commercially in the "Thiokol" compounds (Thiokol Chemical Corp.), and a formulation using a Thiokol hardener has been published by KUSHIDA (1965). Alcoholic hydroxyl groups are available in the epoxy molecules themselves in sufficient amount to form a solid resin without additional "hardener". Even if another hardener is added, the hydroxyl reaction cannot be completely suppressed, and it results in branching of the polymer chain with consequent formation of cross-links, whether wanted or not. This propensity to form cross-links results in a very strong, infusible resin, properties which are valuable for industrial purposes, but not for cutting thin sections for electron microscopy (LUFT, 1961). The different hardeners each react with the epoxy groups to give different bonds which hold the monomer units together as the polymer. The acids (from anhydrides) react with epoxy groups to form ester bonds, so that the epoxy-anhydride polymer chemically could be called a "polyester" resin. It is the presence of these alkali-sensitive ester linkages which makes it possible to

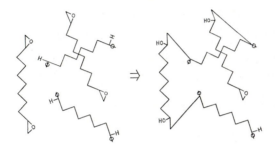

Fig. 4. Curing mechanism of epoxy resins (△〜〜△) with dicarboxylic acids (〜〜〜)

dissolve the epoxy resin from one-micron sections for light microscopy using alcoholic sodium hydroxide (LANE and EUROPA, 1965). The primary amine hardeners react to form secondary amines, the alcoholic hydroxyl groups react to form ethers and the sulfhydryl groups form thio-ethers, all of which are resistant to alkaline attack. In addition to the hardener itself, it is common to add an "accelerator" to the mixture of the epoxy and the hardener, particularly with the anhydrides. The accelerator usually is a strong organic base — often a tertiary amine — such as benzyldimethyl-amine (BDMA), dimethylamino ethanol (DMAE) or DMP-30 (tri[dimethyl-aminomethyl]phenol). The primary amines are sufficiently strong bases to be self-catalyzing. Fig. 4 shows diagrammatically the polymerization of an epoxy with an acid (anhydride) hardener.

The first epoxy resins which became available were phenolic derivatives, and these still constitute the bulk of epoxy resins in commercial use because of their strength, chemical resistance and lower cost. MAALØE and BIRCH-ANDERSEN (1956) probably used an epoxy resin of this type, and the first practical method for electron microscopic embedding (GLAUERT, ROGERS and GLAUERT, 1956; GLAUERT and GLAUERT, 1958) used "Araldite

Casting Resin M" (Ciba) which certainly was phenolic. The latter authors also contributed a particularly useful anhydride hardener (DDSA, or dodecenyl succinic anhydride) which has remained, by itself or as close relatives, a standard ingredient in epoxy embedding mixtures for electron microscopy. This final Araldite embedding mixture was about as viscous as the polyester resins, and equally unpleasant to use in comparison with the methacrylates, but the tissue preservation which it provided was so good that it, along with the polyester resins, quickly replaced the methacrylates (see for example, Birbeck and Mercer, 1956). A few nonphenolic epoxy resins also were available commercially, in particular one based on glycerol known as Epon 562 and later Epon 812 (Shell). Various recipes for embedding resins based on this epoxy were proposed (Kushida, 1959; Finck, 1960; Luft, 1961). These resin mixtures are sufficiently polar to be soluble in alcohol, and the original procedures recommended that the tissue be dehydrated in alcohol and then taken directly into the resin through an intermediate alcohol-resin mixture. However, replacing the alcohol with resin thoroughly enough to give reliable embedding was found to be a very slow process. This could be greatly accelerated by the use of a very low viscosity epoxy liquid, propylene oxide (epoxy propane), as a "clearing" agent between the alcohol and the resin to ensure uniform infiltration and embedding (Luft, 1961).

III. Water-soluble Embedding Media

Ever since the replacement of gelatin gels with celloidin and paraffin, concern has been voiced regarding the extraction of various substances during dehydration, and particularly the extraction of lipids. It has also been concluded that dehydration with "organic solvents" such as ethanol or acetone is harsh treatment for tissues. Certain resins, either by design (glycol methacrylate, Durcupan) or by accident (Aquon) have enough polar residues in their structure to endow them with moderate water solubility. It has occurred to various people (see Hayat, 1970, p. 126 for details) that less harm would be done to the tissue, and perhaps less extraction would take place, if the tissue were dehydrated and embedded in the same material. After fixation the tissue is taken through ascending concentrations of the particular water-soluble resin in water, instead of through graded alcohols, and into the final embedding mixture which usually (but not always) is anhydrous. The fallacy underlying this argument is the failure to recognize that the liquid resin itself is an organic solvent; it may be an unusual and expensive one, but it is no less a solvent for all that. Perhaps some workers thought that the resin, by virtue of its water solubility, would extract less from the tissue. Again the logic is faulty, because some of the most valuable solvents are water soluble, and if it were not for the expense, glycol methacrylate, for example, might be one of them. An illustration by analogy may be illuminating. Several decades ago, nitrocellulose was a valuable plastic; it was soluble in a mixture of alcohol and ether, but in neither alone. The chemical industry produced "Cellosolve", a glycol ether which contained both the alcohol and ether group in one molecule, was soluble in all proportions in water, and was an excellent solvent for many purposes. Glycol methacrylate is a glycol ester which is very similar to the glycol ether in the Cellosolves.

It was thought that these water-soluble resins might dissolve less lipid from the tissue, but it is clear (COPE and WILLIAMS, 1968) that the benefits are small in comparison with conventional resins. What many fail to recognize is that it is not so much the solvent or resin that extracts the lipid or allows it to migrate, but rather the *water* that keeps the lipid in place in the tissue. The following experiment may clarify the last statement. Take a wet piece of filter paper and place on it about a gram of butter; the butter remains in one place and does not penetrate the filter paper. Now, merely let the paper dry and observe the results; the butter begins to bleed into the paper and gradually produces a typical grease spot. The paper has not changed in any way and no solvents are involved; only the water has left. Most water-soluble organic solvents are poor solvents for lipids as long as they contain substantial percentages (20—30%) of water, so that the use of hydrous resin gels as embedding media is the most rational procedure to retain lipids (LEDUC and HOLT, 1965). Even if the tissue is embedded in a resin which does not extract lipid, in the absence of water the lipid is very likely to migrate unless it can be converted to a crystalline solid, for example by freezing, or by digitonin (STERZING, SCALETTI and NAPOLITANO, 1970).

There is another and more rational function which the water-soluble embedding media can serve, and which was alluded to previously. The monomers are soluble because they have sufficient polarity to associate strongly with water. When they polymerize, the molecular weight increases greatly, but most of the polarity remains. They are so hydrophilic that the sections tend to swell in water and addition of a small percentage of conventional monomer may be necessary to prevent the sections from disintegrating on the water in the knife trough. Their hydrophilic properties and controlled swelling permit relatively easy access of reagent solutions to the tissue in the section and hence they should be useful for histochemistry and particularly so for enzymatic reactions (ROSENBERG, BARTL and LEŠKO, 1960; LEDUC and HOLT, 1965), where the large enzyme molecule may be excluded sterically from the tissue by the hydrophobic resins. There are water-soluble epoxy resins such as Aquon (GIBBONS, 1959) and Durcupan (HAYAT, 1970, pp. 167—170) which have been recommended to avoid dehydration by conventional solvents. They do not seem to be as suitable as glycol methacrylate from the standpoint of the sensitivity of the tissue to enzymatic hydrolysis, partly because of the reactivity of the epoxide for the same chemical groups on the tissue which are necessary for enzyme function (Leduc and Bernhard, 1961 and 1962). Chemical reactivity aside, however, swelling and porosity of the resin gel also must be important, because merely the addition of 1% of a cross-linking agent to glycol methacrylate considerably reduced the activity of ribonuclease (Leduc and Holt, 1965). This is a special problem with the epoxies since cross-linking inevitably is present.

D. Advantages and Disadvantages of Conventional Embedding Media

The polyester and epoxy resins have replaced butyl/methyl methacrylate for good reasons, but they have, in turn, created problems of their own which should be recognized. The water-soluble resins confer certain other valuable properties upon the embedding, but these benefits also are bought at a price. If the facts are available, however the well informed microscopist can select whatever feature he wishes to optimize in his research and will be prepared to work around the limitations which accompany his choice.

I. Polymerization Damage

The methacrylates produced a form of damage in tissue characterized by expansion and tearing apart of tissue components, probably due to the osmotic swelling mechanism discussed previously (p. 4). Damage by this mechanism should be inherent in the polymerization of vinyl compounds generally, and should not be unique to methacrylates. Although seldom mentioned in papers concerning glycol methacrylate embedding, the fact that authors employ viscous, prepolymerized glycol methacrylate, rather than the more fluid monomer (Leduc and Bernhard, 1967) suggests that it still is a problem. The observation that tissue preservation is much better in either the polyester resins or the epoxies suggests that "polymerization damage" is much less in them. From a theoretical standpoint, the mechanism implicated in polymerization damage with the methacrylates should be less intense in the polyester resins and nonexistent in the epoxies. On the other hand, it is much easier to identify tissue damage if one has "undamaged" tissue to compare; i.e., the methacrylates look particularly bad in retrospect by comparison with epoxy-embedded tissue. The damage that the polyester or epoxy resins may produce is less easy to assess. "Fixation damage" traditionally has been offered to account for suspected tissue alterations, and there is no doubt that it is real (Strangeways and Canti,

1927), so that it came as a shock when Borysko and Sapranauskas (1954) identified polymerization damage as the culprit in their preparations. Until a "superior" resin becomes available there is nothing with which to compare for an assessment of morphology, except perhaps for the freeze-etching pictures, which have their own uncertainties. At the molecular level the detection of damage must be indirect. In the case of fixation, for example, Lenard and Singer (1968) have demonstrated drastic alterations in the conformation of isolated proteins after several common fixation procedures. Periodic structures in cells, such as the myelin sheath of nerve, lend themselves to nondestructive measurement of the layer spacings by X-ray diffraction, and the measurement can, in principle, be made on the same nerve from the freshly dissected living state through fixation, dehydration and embedding. Fernández-Morán and Finean (1957) and Finean (1958) have shown with rat sciatic nerve that the 178 Å spacing measured in fresh nerve falls to 164 Å with OsO_4 fixation and decreases further to a minimum of 128 Å with alcohol dehydration. With methacrylate embedding, the layer spacing expands back to 155 Å, but with epoxy embedding (Araldite) the expansion does not take place (Joy and Finean, 1963). Thus emerges the paradox that "polymerization damage" in methacrylate produces an expansion in myelin which partially restores the shrinkage which took place during processing, whereas the less destructive Araldite retains the shrunken periodicity in the polymerized block. The moral is that indirect measurements must be interpreted with care, and with as much knowledge about the various factors involved as possible.

II. Beam Damage

Since the electron microscope uses a highly energetic stream of electrons as its probing radiation, it is intuitively obvious that the object being examined must be reasonably resistant to the electron beam. A corollary of this statement is that the resin, which supports the object in the section, must be equally resistant, since early experiments with sections clearly showed that ultrastructure suffered when the plastic was dissolved away. It is equally apparent that the methacrylates are one of the worst choices among the various plastics in regard to radiation sensitivity (Little, 1956). The epoxy resins (phenolic) are very resistant to radiation and in this respect, are much superior to the polyester resins (Lee and Neville, 1967, pp. 6—45). The same reference shows that polyurethane, silicone and furane resins are also highly radiation resistant. The proteins of tissue are not particularly radiation resistant; silk and keratin lose strength at moderate doses of radiation and collagen is still more sensitive. This provides a further reason for using a radiation-resistant embedding resin to reinforce the tissue proteins in the section.

One obvious manifestation of radiation damage in electron microscopy is the increased transparency of the section which occurs during the first few seconds of beam exposure. This phenomenon of clearing has been examined carefully by several quantitative methods under conditions applicable to electron microscopy, and various resins have been compared. REIMER (1959) measured the intensity of the electron beam directly in the microscope with a Faraday cage, and reported a mass loss before and after beam exposure of 40—50% for methacrylate, 20% for Araldite and 13% for Vestopal W. ZELANDER and EKHOLM (1960) measured changes in optical path length with an interference microscope in thin methacrylate sections. They detected a 10% loss of mass after an hour's exposure to vacuum only, and a 56% mass loss after beam exposure. COSSLETT (1960) also used an interference microscope, but on thick (1500—2500 Å) sections at high beam intensity and with various beam diameters. She measured a 40—60% loss in methacrylate, 20—50% in Araldite and Aquon, and 30—50% in Vestopal. The subject of radiation damage has been reviewed in depth by REIMER (1965).

These experiments make it clear that methacrylate suffers the greatest mass loss from the electron beam, but it is puzzling why the loss is still so high in the resins which allegedly are superior. Some other factor must be operating as well against the methacrylates. It is obvious that the mechanical strength of the resin in the section is not directly related to the mass (a liquid and a solid both can have the same density), and this is clearly depicted by REIMER (1959, Fig. 1) illustrating melting of the resin and dislocation of tissue elements. PEASE (1964) likewise suggests specimen collapse at liquid interfaces as the resin melts and decomposes in the beam. The important consideration is to separate the melting (loss of strength) from the loss of mass during electron bombardment. There is no reason why radiation sensitivity necessarily should be bad in itself, and in some respects, it may be beneficial. The clearing of methacrylate sections increased tissue contrast substantially and, during the early years of biological microscopy, permitted sections to be examined without the use of stains. If it were possible to maintain the strength of the resin in the section during irradiation, so that the various elements of the tissue would remain at the same distance from each other, mass loss from radiation damage would increase tissue contrast and so would be an advantage rather than a defect. It was precisely this line of argument that led LOW and CLEVENGER (1962) to the use of a mixed polyester-methacrylate embedding mixture. They added up to 50% n-butyl methacrylate to a polyester resin, hoping that the polyester resin would form an insoluble and thermostable reticular matrix in which the methacrylate resin would be interspersed. Then, after embedding the tissue and cutting a section, the methacrylate could be selectively removed from the section by heat or solvents and still leave enough polyester behind to support the tissue with reduced mass and consequently higher contrast. It was a very clever idea, but it does not seem to have become popular. One serious defect may be that it is improbable that the polyester and the methacrylate will polymerize as separate resin phases. It is more likely that the whole will copolymerize together, since RUST (1940) initially demonstrated that the methacrylates function as cross-linking agents for the polyester resins as well as styrene. Thus the methacrylate will become part of the polyester framework and will weaken the whole to the extent that the methacrylate is decomposed or dissolved.

The legacy of Low and CLEVENGER's work reappears at another level, however. As will be pointed out later, with regard to the low viscosity resins, it is possible to reduce the density (gms/cm³) of the resin by choosing the appropriate organic components such as nonaromatic, noncyclic, hydrocarbon derivatives. This provides an embedding resin of low density initially. If, in addition, mass can be removed from the resin by radiation "damage" at the atomic level without significant mechanical displacement at the level of EM resolution, then the end envisaged by Low and CLEVENGER may be attained by different means.

III. Erratic Polymerization

All of the monomers which are in common use among electron microscopists polymerize smoothly and in an easily controlled manner in the small quantities necessary for embedding tissues. It is true of the methacrylates in particular, and the vinyl resins in general, that there are a number of factors which can affect the rate of polymerization. The effect of viscosity has already been noted in the discussion of the methacrylates, with the suggestion that the tissue itself may mimic high viscosity. The decomposition of the peroxide "catalyst" is very sensitive to the presence of certain types of reducing agents; the addition of N,N-dimethyl aniline to methacrylate containing 1% benzoyl peroxide will reliably produce in a few hours at room temperature polymer blocks which are no worse — or no better — than those polymerized at 50 °C. It is possible that some of the charges of "erratic" polymerization leveled against the methacrylates have been due to reducing agents in the tissue block itself which could accelerate polymerization locally. This is particularly likely to occur if aldehydes are used as fixatives and if the postfixation in osmium tetroxide was brief, or worse still, omitted. It is less likely to occur with nonoxidizing catalysts, such as α,α-azodiisobutyronitrile. Thus the polymerization may appear to be erratic or capricious to the extent that polymerization is influenced by factors of which one is ignorant. When these factors are taken into consideration, vinyl polymerization is smooth and predictable.

In the case of polyesters, the polymerization also takes place through the vinyl type of free-radical mechanism. However, the difficulties which were encountered with the methacrylates seem not to be significant, although they might be detectable if looked for carefully. The epoxy resins are cured through mechanisms which are more or less independent of the presence of tissue. There is at least one instance, however where an epoxy resin may fail to polymerize within the tissue block but hardens properly outside. If the tissue has been exposed to ionic silver (silver nitrate solution), the polymerization fails unless the excess silver has been eliminated from the tissue by 15—30 min exposure to sodium thiosulfate before dehydration. The rationale for this failure may be that residual ionic silver (not metallic silver) complexes with, and hence inactivates, the amine accelerator (BDMA, DMP-30, etc.) so that in the tissue the accelerator concentration, which is low in the resin formula to begin with, is too small to permit normal polymerization within the tissue (LUFT, unpublished data).

IV. Shrinkage

All of the plastics used by electron microscopists for embedding shrink to a certain extent during polymerization as a consequence of the chemical reactions which drive polymerization. This is particularly clear with the alkyl methacrylates, where polymerization can be carried out with a series of related and pure monomers, resulting in a contraction of 22.5 cc/mole (RIDDLE, 1954). This has the consequence of reducing the percentage shrinkage as the molecular weight of the monomer increases; methyl methacrylate shrinks about 21% whereas n-butyl methacrylate shrinks about 15% in volume in going from monomer to polymer (RIDDLE, 1954). The polyesters shrink less, and the epoxies the least of all; Vestopal W shrinks 7—10% during cure and Araldite about 2% (REIMER, 1959). While no one suggests that shrinkage is a valuable attribute for an embedding resin, it is a relatively minor defect even in the methacrylates. The 15% shrinkage in volume for n-butyl methacrylate corresponds to only about 5% linear shrinkage and is small in comparison to other dimensional changes which the tissue can suffer during processing (TOOZE, 1964; also section D.VI). The magnitude of the shrinkage as felt by the tissue is less still because the majority of the shrinkage occurs in the liquid resin before gelation even in the cross-linking epoxy resins (LEE and NEVILLE, 1967). That fraction of the volume contraction which takes place while the resin is still liquid can be relieved easily by fluid flow in the interstices of the solid, fixed tissue, and only when the resin becomes a gel will volume changes in the resin transfer stress to the tissue, as the tissue-resin system takes on the mechanical characteristics of composite materials (KELLY, 1967). The resulting strain in the embedded tissue, expressed as a volume contraction, for these reasons must always be substantially less than the values given for the resin itself.

V. Viscosity

The polyester resins and the older epoxy resins were very viscous liquids which were thoroughly unpleasant to work with. PEASE (1964, p. 89) describes the epoxies as "the stickiest and messiest substances that one is ever apt to handle." The viscosities of several resins are listed in Table 1.

Table 1

Resin System	Viscosity (cps)	Density
Vestopal W	900	1.170
Araldite 502/DDSA (LUFT, 1961)	2500	1.081
Epon 812, A : B = 1 : 1 (LUFT, 1961)	1400	1.158
Maraglas, complete (SPURLOCK, KATTINE and FREEMAN, 1963)	565	1.121

Viscosities in centipoises at 25 °C and densities for several complete resin systems (containing hardener and/or accelerator, mixed and ready for embedding). From SZU-BINSKA and LUFT, 1973.

In retrospect, the low viscosity of the methacrylates made them a delight to use. The high viscosity of the unpolymerized (liquid) polyester and epoxy resins, in addition to ethetic faults, caused technical difficulties. The most serious was the failure to achieve complete mixing of the components of the resin, and particularly, of the small amounts of accelerator. The techniques appropriate for aqueous chemistry, such as shaking to mix, rely on generating turbulence, and turbulent mixing is useless in highly viscous fluids. The problems are more akin to those of mixing a cake from flour, eggs, butter, etc., rather than of dissolving sugar in coffee; slow but positive displacement of the viscous material from the wall of the container with a flexible spatula is much more efficient than a few back and forth strokes with a spoon. Mechanical mixers often operate at high speeds and entrain bubbles of air into the resin mixture which, in addition to their obvious interference with embedding, may oxidize and hence inactivate the accelerator in the epoxies. The several resin components have a remarkable tendency to stratify against the walls of the mixing container and to remain there unless deliberately displaced. A thin (2—3 mm diameter) nylon stirring rod has been found to be very useful for mixing small quantities in a glass conical (pharmaceutical) graduate; glass rods can chip tiny fragments of glass from both the rod and the graduate, wooden sticks can quickly absorb some of the accelerator and release sawdust as well, and metal rods are too stiff. The liquid epoxy resin can appear to be completely mixed and yet show refractive index striae when the stirring rod is rotated or wiped against the wall of the graduate. The absence of these striae is good evidence of complete mixing.

A less damaging aspect of the high viscosity of the polyester and epoxy resins, namely slow infiltration, is reduced by the very small samples of tissue normally used in electron microscopy. It can be made still less objectionable by the technique of using after dehydration a clearing agent which can react without harm with the resin itself. In the case of the polyester resins which use styrene as the cross-linking agent, styrene is the logical clearing agent (KURTZ, 1961). For epoxy resins, propylene oxide, has been recommended (LUFT, 1961). Most tissues are adequately infiltrated with the resin in a few hours when these reactive clearing agents are employed. Very small objects (a single amoeba for example) do not require a clearing agent, but can be embedded directly from the dehydrating liquid. Conversely, organisms which have unusually resistant cell walls or a cuticle may require very long infiltration times or a mechanical breaching of the barrier in some way.

VI. Osmotic Damage

There is another source of damage to the tissue during embedding which seems generally to be unrecognized in the electron microscopic literature, although it was known to the light microscopists. It is related indirectly to viscosity in the sense that it is the high molecular weight (actual or effective) of the resin components which results in the high viscosity and at the same time, but by a different mechanism, which generates substantial hydraulic pressure on the tissue itself (SZUBINSKA and LUFT, 1973). The light microscopists who used celloidin embedding were careful to begin with very low concentrations of nitrocellulose and to build up the concentration over days or weeks to attain adequate infiltration, and at the same time, to avoid shrinkage. The phenomenon was noticed again when prepolymerized methacrylates were introduced to control polymerization damage; Pacinian corpuscles, which retained their shape in life when embedded in n-butyl methacrylate monomer, frequently were flattened severely when embedded in prepolymerized, viscous methacrylate (SZUBINSKA and LUFT, 1973). A common factor is present for both the nitrocellulose solution and the prepolymerized methacrylate, namely, that each contains a high molecular weight polymer dissolved in a solvent. An hydraulic pressure can be set up across a membrane which has semipermeable properties, i.e., is freely permeable to the solvent but restricts the large polymer molecule (BULL, 1971). The more delicate the fixation events, the finer will be the porosity of membranes in the cells making up the tissue block, and the more likely it will be that the solvent molecules will pass freely and the larger molecules be held back, thus generating greater osmotic pressure. This is very nearly the *inverse* of the argument invoked to explain the mechanism of polymerization damage, with the polymer/solvent gradient, and hence the hydraulic forces, acting in the opposite direction. Whether or not there are noticeable dimensional changes in the tissue depends not only on the osmotic gradient of polymer concentration and on the degree of semipermeability of the various tissue membranes, but also on the mechanical rigidity of the tissue itself. Thus the firm tissues such as kidney or liver, with a high protein content and many organelles, are strong enough to survive embedding in prepolymerized methacrylate without noticeable distortion, whereas the delicate onion skin structure of the Pacinian corpuscle is crushed under the same circumstances.

Table 2

Fixative/Buffer	Embedding Resin	Shrinkage Average	Range
OsO_4/PO_4	Epon (1 : 1)	45%	—
OsO_4/PO_4	Araldite	51%	—
OsO_4/PO_4	Maraglas	41%	35 — 55%
$Acr/OsO_4/PO_4$	Vestopal	27%	6 — 47%
OsO_4/PO_4	DGE	6%	0 — 13%
$Acr/OsO_4/PO_4$	DGE	13%	7 — 23%

Shrinkage of individual *Amoeba proteus* fixed in various ways, dehydrated in ethanol (or acetone for Vestopal), and infiltrated in several epoxy or polyester resins. Shrinkage is the percentage decrease in length of the amoeba in the resin compared with the same dimension in the fixative. Acr = Acrolein; DGE = diglycidyl ether. From SZUBINSKA and LUFT, 1973.

All of this discussion would be ancient history were it not for the fact that similar effects have been shown to occur with both the polyester and epoxy resins, and that the magnitude of the effect seems to be more closely correlated with viscosity of the resin than with any other factor (SZUBINSKA and LUFT, 1973). In the course of experiments involving single *Amoeba proteus* (SZUBINSKA, 1964a, b; SZUBINSKA and LUFT, 1971), it was noted that dramatic, but uniform, shrinkage of the amoeba took place when the amoeba was transferred from absolute ethanol into liquid Epon (LUFT, 1961) resin. In order to quantitate the observations, the length of a particular amoeba was measured from a light micrograph taken when it was in the fixative and the same amoeba was measured again from another micrograph taken after an hour's exposure to the pure, catalysed resin. The shrinkage listed in Table 2 is the percent reduction in a linear dimension in amoebae while in the resin, as compared with the same dimension in the fixative. Thus, an average amoeba embedded in Epon shrank 45% in length compared to its length in the fixative. If, as seems likely, the same percentage of shrinkage occurred in the other two dimensions, the volume of the amoeba would have been reduced to only about 17% of that in the fixative. Although this amoeba was shrunken to only 1/6 of the volume it had when it was in the fixative, the ultrastructural preservation appeared to be very good by the usual criteria, i.e., intact membranes, continuous cytoplasmic density and crisp definition of ribosomes, filaments and vesicles (SZUBINSKA and LUFT, 1973).

From visual observations as well as some measurements of single amoebae, it was apparent that there was little if any shrinkage during the dehydration steps and that virtually all of the shrinkage took place quickly when the amoeba encountered the resin. Even when the resin was diluted serially with alcohol in 10 steps of 10% increments, from 10% resin in absolute ethanol through 90% resin with 10% ethanol and into 100% resin, spending 10—15 min in each dilution, the total shrinkage was essentially the same; it merely took place in 10 smaller increments rather than shriveling once magnificently. An experimental epoxy resin based on diglycidyl ether (SZUBINSKA and LUFT, 1973) had been under investigation in the laboratory and since this mixture had a particularly low viscosity, it was tested with amoebae under the same conditions as the other resins. It resulted in the least linear shrinkage of any of the resins tested at that time (average 6—13%) and has been used for this purpose subsequently. It has certain drawbacks which do not recommend it for general purpose embedding (see Section D.VII), but low viscosity resins certainly may serve a useful purpose in embedding delicate structures. Strong and solid tissues do not require special attention; for example, KUSHIDA (1962) was unable to measure any shrinkage in sea urchin eggs during infiltration in a variety of embedding media, although shrinkage took place at other stages.

VII. Toxicity

The toxicity of embedding materials is worth considering separately since it has become a problem with the epoxy resins. The older, viscous epoxy resins, and their curing agents particularly, were capable of producing serious dermatitis (LEE and NEVILLE, 1967, pp. 25—25) but this could be avoided by scrupulous cleanliness. The resins had such a low vapor pressure by virtue of their high molecular weight that intoxication by inhalation was remote when used in small batches and cured at low temperatures as in electron microscopy applications. Some of the newer, low viscosity epoxy resins, however, have properties such that they can no longer be ignored safely.

Paraffin and gelatin are virtually nontoxic. The solvents used in celloidin embedding have a substantial vapor pressure as do the methacrylates, and the maximum permissible concentrations of these organic chemicals can be found in various publications (e.g., SUNSHINE, 1969). It would be more useful to have some measure of the risk of these various liquids under conditions found in the laboratory, i.e., small amounts of the materials exposed at room temperature with normal ventilation. When this is done[2], alcohol (for celloidin embedding) has a low risk, while the methacrylates have a moderate risk along with ether, ethyl acetate and trichloroethylene. Styrene falls into the same category. As will be seen in the next section, the low viscosity epoxy resins employ low molecular weight bifunctional epoxides (diepoxides) which have low but significant vapor pressures at room temperature. They are very toxic but their low vapor pressure gives them a "risk factor"[2] in the same range as the methacrylates and styrene. Despite this appearance of safety, however, extra caution is recommended, for there is good reason to suspect that they may be carcinogenic. This danger may be present with liquid epoxies in which the epoxy groups are separated by four or fewer carbon atoms (LEE and NEVILLE, 1967, pp. 25—32). The antimitotic and carcinogenic activity seems to increase as the epoxy groups are closer together, the limit being reached with the compound 1,2,3,4-diepoxybutane (butadiene dioxide) with the formula:

$$\begin{array}{cc} \text{H} & \text{H} \\ \text{H}_2\text{C}-\text{C}-\text{C}-\text{CH}_2 \\ \diagdown\!\diagup \quad \diagdown\!\diagup \\ \text{O} \qquad \text{O} \end{array}$$

This liquid is in the same class as the nitrogen mustards in its antimitotic properties (ROSE, HENDRY and WALPOLE, 1950; ROSS, 1962) and is frequently used in mutation research (WATSON, 1972). The bifunctionality (two epoxy groups per molecule) greatly increases its toxicity in comparison

---

2 It is useful for this purpose to combine a measure of the toxicity of a substance (the TLV or threshold limit value) with its volatility. This has been done (SZUBINSKA and LUFT, 1973) to arrive at a "risk factor". Data from SUNSHINE (1969).

$$\text{Risk Factor} = \frac{\text{(vapor pressure at room temperature in mm Hg)}}{\text{(TLV in ppm)}}$$

Substance	Risk Factor		Substance	Risk Factor	
Acrolein	214/0.1	= 2140	Methyl methacrylate	68/100	= 0.7
Osmium tetroxide	5/0.005	= 1000	Ethyl acetate	100/400	= 0.25
Butadiene dioxide	9/0.5	= 18	Diglycidyl ether	0.09/0.5	= 0.2
Propylene oxide	445/100	= 4.5	Butyl acetate	15/150	= 0.1
Iodine (I$_2$)	0.2/0.1	= 2	Styrene	10/100	= 0.1
Ethyl ether	439/400	= 1.1	Ethyl alcohol	50/1000	= 0.05
Trichloroethylene	77/100	= 0.8	Mercury	0.002/0.1	= 0.02

with the monofunctional compounds (HENDRY et al., 1951) but the relation-ship to carcinogenicity is more complex (SZUBINSKA and LUFT, 1973). However, it is precisely this same bifunctionality which is essential for resin polymerization. Thus, there is no hope of finding a low viscosity, nontoxic epoxy resin. The best that can be done is to use the low viscosity resins only when it can be demonstrated that they are necessary, and then to be very careful to work under conditions of good ventilation and clean-liness, and to place all waste in a closed container to control evaporation. Skin contact also should be meticulously avoided.

VIII. Miscellaneous Defects

There have been various problems with the epoxy resins which emerged after the recipes were published. Perhaps the first of these materialized when workers outside the United Kingdom tried to follow the Araldite procedure of GLAUERT, ROGERS and GLAUERT (1956). The Araldite M which they specified was not available in other countries because of the franchise restrictions imposed by Ciba, and the "equivalent" resin manu-factured by the Ciba licensees in other countries under the Araldite name did not behave in the same way. Since the epoxy resins are not pure compounds, and since the actual constitution of any batch of resin depends not only on the amount of the components of the batch, but also on the sequence and rate at which they are combined, differences should be the rule rather than the exception.

When the Epon procedure was published (LUFT, 1961), it was hoped that the quality would be more uniform, since a single manufacturer (Shell) with world-wide distribution facilities would supply the product. Although uniformity of one sort was achieved in this way, another form of variability took its place, namely, a gradual drift upward over the years of the epoxy equivalent of the Epon 812. The original specifications for Epon 812 (earlier Epon 562) set limits of 140—160 for the epoxy equivalent. When the embedding recipe was being developed, three presumably independent samples of Epon 562 and 812 were titrated as having epoxy equivalents of 140, 141 and 142, and the recipe was published assuming the average of the three, i.e., 141. Over the years the epoxy equivalent of the Epon 812 which we have used has risen so that it is now about 170, with one sample at 178. This drift changes the recipe substantially if it is desired to maintain the A:E ratio (anhydride to epoxy ratio) at 0.7. If the old recipe is used with the new, high equivalent Epon 812, the effect is to raise the A:E ratio into a region where sectioning is more difficult (LUFT, 1961). We have corrected the recipe according to Table 3. The epoxy equivalent of any particular batch of Epon 812 can be obtained from the Shell Chemical Co., or titrated (LEE and NEVILLE, 1967, p. 4—17).

Another unfortunate feature of Epon concerns the rapidity with which its viscosity increases, even at room temperature, once the accelerator has been added. Table 1 lists its viscosity at 1400 centipoises when mixed and ready to use. The viscosity of this particular resin mixture (refrigerated 2 months) without accelerator was 950 centipoises, and the viscosity of the resin plus accelerator increased about 40% during the 45 minutes that the viscosity was being measured. Mixed Araldite 502 increases only about 10% over the same length of time. The problem with Epon 812 may be that it contains some of the triglycidyl ether of glycerol along with the major components of the di- and mono-glycidyl ethers of glycerol (LEE and NEVILLE, 1967, pp. 4—66). The trifuctional component should cross-link the forming resin with concomitant rapid increase in viscosity. On the other hand, the keeping qualities of Epon 812 are very good. One particular batch of Epon 812 was titrated for its epoxy equivalent upon arrival, and after 3 years of storage at room temperature with occasional opening for use, the epoxy equivalent was within one percent of its original value.

Table 3

		Epoxy Equivalent of Epon 812							
Table 3a		140*	150	155	160	165	170	175	180
A : E ratio	0.70*	62*	66	68	70	73	75	77	79
desired in	0.65	66	71	74	76	78	80	83	85
mixture "A"	0.60	72	77	80	82	84	87	90	92
Table 3b	0.70*	90*	84	81	78	76	73	71	69
A : E ratio	0.65	84	78	75	73	70	68	66	64
desired in	0.60	77	72	70	67	65	63	61	60
mixture "B"									

Tables of proportions of Epon 812, DDSA and NMA required to maintain a specific A : E (anhydride : epoxy) ratio in mixtures "A" and "B", using Epon 812 with a variable epoxy equivalent. Table 3a lists the *ml of Epon 812* to be mixed with *100 ml of DDSA* to give corrected mixture "A". Table 3b lists the *ml of NMA* to be mixed with *100 ml of Epon 812* to give corrected mixture "B". The volumes are listed in order to make the volume of the larger component 100 ml, for easier measuring with a graduate. This accounts for the odd pattern of values in the two tables. The values indicated by (*) are close to the proportions listed in the original paper (LUFT, 1961) since the equivalent of the first Epon samples was about 141.

E. New Embedding Media

Although organic chemistry provides an embarrassment of riches with respect to possibilities for producing synthetic resins, many of them can be eliminated from consideration for convenient embedding of biological tissues under the usual laboratory conditions. Polymerizations requiring emulsions, solid catalysts, high temperatures or pressures of a thousand atmospheres, or those releasing corrosive by-products are not competitive with the embedding systems in current use. There are, however, some newer epoxy formulations which offer viscosities nearly as low as butyl methacrylate, as well as some more exotic aldehyde-based resins which form hydrated gels, and which may be useful in avoiding solvent extraction of tissues.

I. Low Viscosity Epoxy Resins

The first of these to be published was the formula of SPURR (1969). This embedding mixture employed a relatively pure diepoxy compound, vinylcyclohexene dioxide, which has a molecular weight of 140, viscosity of 8—10 centipoises, boiling point of 227 °C and a vapor pressure of about 0.055 mm at room temperature. The two epoxy groups are separated by two carbon atoms. It is available from Union Carbide as ERL-4206 or UNOX-206, aad from Ciba as Araldite RD-4. The formula of vinylcyclohexene dioxide is:

It polymerizes readily with various anhydrides, and nonenyl succinic anhydride (NSA) was chosen for its low viscosity. The result is a very hard resin, but this could be plasticized conveniently with DER-736 (a diglycidyl ether of polypropylene glycol available from Dow Chemical Co.). It was cured with a low viscosity tertiary amine, dimethylaminoethanol (available from standard organic chemical sources). The formula was given as follows (SPURR, 1969):

VCD (vinylcyclohexene dioxide)	10.0 gm
DER-736	6.0 gm
NSA (nonenyl succinic anhydride)	26.0 gm
dimethylaminoethanol	0.4 gm

The viscosity is 60 centipoises (Table 4) and the quality of the embedding and the ease with which the resin can be cut is at least equal to the other epoxy resins. The anhydride : epoxy (A:E) ratio of this recipe is 0.71, at the level previously suggested to be valuable for good cutting qualities (LUFT, 1961).

In 1962 the author obtained as a gift from the Shell Development Co., a sample of diglycidyl ether which was used as the basis for an epoxy resin using nonenyl succinic anhydride (NSA) with an A:E ratio of 0.6:

DGE (diglycidyl ether)	15.5 ml
NSA (nonenyl succinic anhydride)	34.5 ml
Total	50.0 ml

This was cured with 1.0—1.5% DMP-30 for 24 hours at 60 °C. This mixture had a viscosity of 23 centipoises and cut well. However, the toxic properties of diglycidyl ether were known at that time, and its advantages did not seem to outweigh its disadvantages, so the details were not published. Later (SZUBINSKA, 1964b, 1971), its value became apparent to reduce the shrinkage of amoebae. Diglycidyl ether, however, was not available commercially, and correspondence with Shell in 1967 revealed that they no longer had diglycidyl ether under development, principally because of its high toxicity. Both vinyl cyclohexene dioxide and diglycidyl ether (di-[2,3-epoxypropyl]ether) have been known for their carcinogenic

Table 4

Resin System	Viscosity (cps)	Density
VDC/736/NSA (SPURR, 1969)	60	1.06
DGE/NSA (SZUBINSKA and LUFT, 1973)	23	1.07
DEO/NSA (SZUBINSKA and LUFT, 1973)	20.5	1.02

Viscosities in centipoises at 25 °C and densities for several complete resin systems (containing hardener and accelerator and ready for embedding). These resins are between 1 and 2 orders of magnitude less viscous than resins listed in Table 1. From SZUBINSKA and LUFT, 1973.

or cytoxic properties (HENDRY et al., 1951; SZUBINSKA and LUFT, 1973). A search for a substitute that might be less toxic, be available commercially, and still provide low viscosity culminated in the compound 1,2,7,8-diepoxy octane[3] (SZUBINSKA and LUFT, 1973). The two epoxy groups in diglycidyl ether are separated by two carbon atoms and one oxygen, and in diepoxy octane they are separated by 4 carbon atoms. Their formulae are:

$H_2C — CH—CH_2—O—CH_2—CH—CH_2$

diglycidyl ether, M.W. 130, B.P. 260 °C, Vapor pressure 0.065 mm at 20 °C

$H_2C — CH—CH_2—CH_2—CH_2—CH_2—CH—CH_2$

1,2,7,8-diepoxy octane, M.W. 142, B.P. 240 °C, Vapor pressure 0.045 mm at 20 °C.

---

3 Diepoxy octane is available from Aldrich Chemical Co., Inc., 940 St. Paul Ave., Milwaukee, Wisc. 53233 under their stock number 13, 956—4. The price is $ 4.00 for 25 gm or $ 14.00 for 100 gm.

No data are available on the toxicity of diepoxy octane, but it should be somewhat less toxic than either diglycidyl ether or vinylcyclohexene dioxide (Szubinska and Luft, 1973). When diepoxy octane is substituted for diglycidyl ether and the proportions are adjusted to maintain the A:E ratio at 0.60, the recipe is:

DEO (diepoxy octane)	18 ml
NSA (nonenyl succinic anhydride)	32 ml
Total	50 ml

The resin also is cured with 1.0—1.5% DMP-30 for 24 hours at 60 °C. The viscosity is even lower than that for diglycidyl ether, and all three resins are compared in Table 4. The cutting properties of these resins are no worse (nor much better) than the other epoxy resins. Thin sections for electron microscopy stain in the usual manner and are capable of high resolution pictures with minimum granularity (Szubinska, 1971), in common with all epoxy resins. These low viscosity resins also have low density (compare Table 4 with Table 1) which should make them useful for high voltage (1 mev) stereo microscopy of thick (1—2 μ) sections, since the lower the specific gravity of the resin, the less interfering electron density there will be between tissue elements in the image.

II. Exotic Embedding Materials

The literature records a modest amount of effort to examine some new embedding materials, or new variations on older themes.

1. Hydrophilic Gels

Fernández-Morán and Finean (1957) and Gilëv (1958) embedded osmium-fixed tissue in concentrated solutions of gelatin, permitted them to harden by drying and then were able to section the blocks thin enough to examine them in the electron microscope. This procedure had the obvious advantage of dehydration and embedding in a hydrophilic, polymeric substance which should have very poor, if any, solvent-like properties. However, gelatin blocks appear to be difficult to cut (Casley-Smith, 1967) and the search for similar, but better, material has continued.

Casley-Smith (1963) first explored the use of urea-formaldehyde synthetic resins, with further details later (Casley-Smith, 1967). After aldehyde fixation, the tissue blocks were placed in increasing concentrations of the water-soluble, urea-formaldehyde prepolymer (Monsanto) up to the maximum concentration of 60% to infiltrate them. Polymerization was initiated by the addition of 2% by volume of 10% ammonium chloride solution, whereupon the mixture set to a firm gel at room temperature.

Polymerization continued for several days, and the blocks were air-dried (with shrinkage) for sectioning using a water-filled trough. Lipid droplets were retained by the procedure, but sections were not obtained easily.

A variation on the theme was introduced by ROBERTSON and PARSONS (1969) using, this time, a resorcinol-formaldehyde prepolymer sold under the name Cascophen (Borden Chemical Co.) as a solution of 65.5% resin and 34.5% water. Myelinated nerve was fixed and taken through increasing concentrations of the aqueous resin and finally polymerized by the addition of 6—7% formaldehyde (ROBERTSON and PARSONS, 1970). The hydrated gel was then dried with a 30% volume loss and 40% weight loss and sections were cut on water and stained with lead citrate. They noted that the resin penetrated poorly into the axons, but that the myelin period was retained at 160 Å, rather than the 130 Å spacing found in Epon.

Another variation was the method of PETERSON and PEASE (1970a, b; 1971) in which they generated their resin directly around and within the tissue using glutaraldehyde and urea solutions. Various rat tissues were fixed in aldehydes and then quickly infiltrated with a concentrated solution of urea and glutaraldehyde. Although polymerization began promptly near neutral pH, it was necessary to lower the pH to about 4 to form a water-insoluble polymer; gelation allowed only 30—60 minutes for infiltration. Again, the hydrated gel was allowed to dry, becoming brittle with cracking, and sections were cut from dried blocks on a water surface. They were stained with uranyl acetate, lead or phosphotungstic acid. Myelin was stated to retain its true 180 Å periodicity.

A similar procedure, yet different in detail, was described by FARRANT and McLEAN (1969), in which they substituted serum or egg albumin for the earlier gelatin procedures, and cross-linked it to a hydrated gel with glutaraldehyde. As before, the cross-linked protein was dried, and the dried material was cut onto water and stained with phosphotungstate.

All of these methods have in common the formation of a highly hydrated gel of polymeric material around, and hopefully within, the tissue, after which the gel is dried with substantial shrinkage and cracking. Presumably it is impossible to cut the blocks thin enough in the hydrated condition, although LEDUC and HOLT (1965) describe cutting hydroxypropyl methacrylate blocks containing 3 to 20% water. All of the materials employed in these procedures are so thoroughly hydrated and later, during dehydration, are so polymeric, that solvent extraction should be negligible. It is puzzling that FINEAN (1958) reported that when myelinated nerves were embedded in gelatin, the X-ray diffraction pattern changed to that of an air-dried specimen, whereas several of the workers mentioned above found that the myelin spacing, as measured by electron microscopy, changed little or not at all despite the shrinkage and cracking of the block during drying. Another element common to all of these reports is an element of heroism among the authors in obtaining their results; none of them suggest these methods as a replacement for conventional embedding.

2. Polyurethane Resins

For the sake of completeness, it is worth mentioning some earlier, unpublished results with these plastics. Polyurethane resins result when organic diisocyanates react with diols (di-alcohols) (SAUNDERS and FRISCH, 1962). The resins can be made either cross-linked or thermoplastic, and they have good radiation resistance. Some resins based on hexamethylene diisocyanate and propanediols were described by BREWSTER (1951), and from this was developed a resin[4] which cured to a clear transparent solid at room temperature. The main difficulty is that the diisocyanates are exceedingly sensitive to the presence of water, with which they react to form carbon dioxide. No more than 5% water is necessary to produce polyurethane foams as used for pillows. This sensitivity to water requires such vigorous dehydration of the tissue to obtain bubble-free blocks that the resin has not been pursued further.

F. Conclusion

Embedding media were discovered intuitively and have been developed empirically to the stage where microtomy operates at macromolecular thicknesses. It seems remarkable that it should have been so successful; the progress of the art must have been aided by the availability of a great variety of materials to play with, by ancilliary skills in producing sharp edges, and by the ease with which experiments in embedding and sectioning can be done and a good result recognized. It is also likely that intelligence as well as blind luck played their parts. Even though a rational approach cannot yet provide a new embedding medium that is as superior to our polyesters and epoxies as the methacrylates were to paraffin, at least we might enjoy the satisfaction of realizing why things turned out so well. This, as well as learning from past mistakes, suggested the review of the older embedding materials. Another aspiration has been to sort out those features which seem to be the necessary, or at least useful, attributes of an ideal embedding medium. These are listed as follows.

---

4 The resin was a polymer of an equimolar mixture of two diols with the stoichiometric amount of hexamethylene diisocyanate.

2-ethyl-1,3-hexanediol	7.3 gm
2,2-dimethyl-1,3-propanediol	5.2 gm

These dissolve to form a liquid stable at room temperature, density 0.95, avg. MW = 125. The polymerization mixture consists of 1.57 ml of the diol mixture with 1.90 ml of hexamethylene diisocyanate plus 1 drop of triethyl amine as an accelerator. The mixture will cure in several days at room temperature, and in shorter times at 50 °C but with more likelihood of bubbles.

1) The medium should be generated by conversion of a monomer to a polymer. The monomer should be of low molecular weight, have low viscosity, be nontoxic and stable, and should polymerize smoothly near room temperature without release of by-products and with minimum shrinkage.

2) The polymer should be transparent. This assures that the "grain" size of the structural units of the polymer is small, i.e., well below the wavelength of light, admits "glassiness" and excludes microcrystallinity.

3) The polymer should be mechanically stable to radiation. It may be radiation-resistant, but if not, the mass lost during irradiation should not weaken the embedding material enough to displace the tissue elements within the section. Certainly it should not melt under the beam.

4) The density should be low. The specific gravity of the resin should be as low as possible to provide the greatest contrast for the tissue. Mass loss during radiation may contribute to this provided that strength is retained.

5) The resin should bond tightly to tissue. There should be good adhesion between the resin and tissue so that sectioning will not tear tissue from resin. Covalent bonding between the resin and the tissue carries this to the ultimate extent.

6) The resin should not chemically alter the tissue. For histochemical purposes, the chemical groups, and perhaps the conformation, characteristic of the tissue macromolecules should not be altered by the resin. As a corollary, the resin should not block access of reagents to the tissue in the section, nor should it extract or dissolve tissue components. Items 5 and 6 probably are mutually exclusive.

7) The resin should cut well. This is the least well understood condition. At least the coefficient of friction of the resin against the knife should be as low as possible, and lubrication by water should occur.

None of the embedding media available today satisfy all of the conditions listed above, but various media emphasize one or more at the expense of others. The well informed microscopist will be able to pick that embedding medium which best satisfies his requirements from the abundance of recipes available, and need not apologize for the sacrifices inherent in the choice. Prudence, however, requires the microscopist to be aware of the compromises he has made, since all of his electron micrographs of sections inevitably will be modulated by the embedding material. In morphology, as well as under the law, ignorance is no excuse.

What of the future? Resin chemistry is reasonably well understood and the cream has been skimmed in that field. The development of new resins, even more now than in the past, requires effort and capital investment on the industrial scale. New embedding materials for electron microscopy in the future will be based on the crumbs left over from commercial-industrial banquets. There is always the possibility of new and ingenious uses to which older polymers may be put, such as PETERSON and PEASE' urea-glutaraldehyde aqueous gels (PETERSON and PEASE, 1971).

It seems to the author that the area of weakest knowledge, in which important returns may reward further investigation, is contained in condition (7) of the list above. Friction appears to be a simple concept and obeys a simple equation only at the gross level. The microscopic events implicated in frictional phenomena involve sophisticated chemistry and physics which encounter the greatest difficulty in the macromolecular domain — exactly where ultrathin sectioning must function (RABINOWICZ, 1965). Another missing link is a conceptual model of the mechanism by which the embedding medium assists cutting. It is clear that it does help, because "cutting" in the absence of an embedding medium is more properly described as mashing, squeezing and tearing. An attempt was made along these lines by WACHTEL, GETTNER and ORNSTEIN (1966), but it is not adequate. Viscosity enters importantly into their discussion, and while all embedding media are viscoelastic, the time-dependent viscous component should contribute minimally during the very brief times (1—2 sec) involved in section cutting. Viscosity enters into consideration primarily when one invokes the concept of melting of the polymer at the knife edge from the heat generated by the work done on the resin as the knife advances. This model seems to fit the results obtained with the thermoplastic methacrylates (LUFT, 1961). Something else is needed for an explanation which can include the thermoset resins such as the polyesters and epoxies.

A fruitful approach may be to view the tissue embedded in the resin as a composite material, using the concepts developed to describe fiber-reinforced plastics (KELLY, 1966, 1967). It is known that very strong composite materials may result when strong filaments (e.g., fiberglass) are embedded in a plastic. For best strength, the filaments should be strong and stiff (have a large elastic modulus), and the plastic matrix should adhere tightly to the filaments and be more elastic (have a much smaller elastic modulus). The essence of the model lies in the analysis of composite materials which shows that when a load is applied to the composite, both fibers and matrix experience the same *strain* since they are bonded together, but most of the *stress* is taken by the filaments and very little by the plastic, in proportion to the ratio of their respective elastic moduli (KELLY, 1967). On the other hand, if the composite carries a lateral defect or notch, then under certain conditions, "stress will be effectively concentrated in the composite because, due to the concentration of *strain* in the matrix, fibres close to the end of the notch will be more highly stressed than remoter ones" (KELLY, 1966, pp. 160—161). It would seem that the wedge effect of the knife advancing into the embedded tissue would fit nicely into the general model of notch failure, and the embedding matrix would function to concentrate the stress induced by the knife on the "filaments" of the tissue just ahead of the knife edge (see Fig. 5.15, p. 161, KELLY, 1966). The result would be stress sufficient to exceed the strength of the tissue components which would be concentrated or "focussed" by the matrix onto a line just ahead of the knife and which would advance as a plane through the tissue block to produce the "cut". Theory indicates which factors should be controlled to *maximize* the work required to break a composite (KELLY, 1966); a reversal of these factors should permit one to *minimize* the work in cutting. If the mechanical properties of the various tissues and resins are known, it should be possible to improve the cutting qualities of embedded tissues in general. It might even be possible to predict the mechanical requirements of an embedding resin to provide optimum cutting quality for any specified tissue fixed in a specified way. The great variety of embedding resins available today, and particularly the endless variations possible in the epoxy formulations, should enable these hopes to be realized, if the model for composite materials can be applied quantitatively to embedded tissues.

Acknowledgement

This work was supported in part by USPHS Grant GM-16598 from the National Institutes of Health.

References

BAMFORD, C. H., BARB, W. G., JENKINS, A. D., ONYON, P. F.: The Kinetics of Vinyl Polymerization by Radical Mechanisms. New York: Academic Press 1958.

BAWN, C. E. H.: The Chemistry of High Polymers. London: Butterworth 1948.

BERNHARD, W.: A new staining precedure for electron microscopical cytology. J. Ultrastruct. Res. 27, 250—265 (1969).

BEVINGTON, J. C.: Radical Polymerization. New York: Academic Press 1961.

BIRBECK, M. S. C., MERCER, E. H.: Applications of an epoxide embedding medium to electron microscopy. J. roy. micr. Soc. 76, 159—161 (1956).

BJORKSTEN, J.: Polyesters and Their Applications. New York: Reinhold Publishing Corp. 1956.

BORYSKO, E.: Recent developments in methacrylate embedding. I. A study of the polymerization damage phenomenon by phase contrast microscopy. J. biophys. biochem. Cytol. 2 Suppl., 3—14 (1956).

BORYSKO. E., SAPRANAUSKAS, P.: A new technique for comparative phase-contrast and electron microscope studies of cells grown in tissue culture, with an evaluation of the technique by means of time-lapse cinemicrographs. Bull. Johns Hopk. Hosp. 95, 68—79 (1954).

BREWSTER, J. H.: The effect of structure on the fiber properties of linear polymers. I. The orientation of side chains. J. Amer. chem. Soc. 73, 366—370 (1951).

BULL, H. B.: An Introduction to Physical Biochemistry, 2nd ed. Philadelphia: F. A. Davis, Co. 1971.

CASLEY-SMITH, J. R.: The preservation of lipids for electron microscopy; urea- formaldehyde as an embedding medium. Med. Res. 1, 59 (1963).

CASLEY-SMITH, J. R.: Some observations on the electron microscopy of lipids. J. roy. micr. Soc. **87**, 463—473 (1967).

COPE, G. H., WILLIAMS, M. A.: Quantitative studies on neutral lipid preservation in electron microscopy. J. roy. micr. Soc. **88**, 259—277 (1968).

COSSLETT, A.: Some applications of the ultraviolet and interference microscopes in electron microscopy. J. roy. micr. Soc. **79**, 263—271 (1960).

FARRANT, J. L., McLEAN, J. D.: Albumins as embedding media for electron microscopy. In Proc. 27th Ann. Mtg. Electron Micr. Soc. Amer. Ed.: C. ARCENEAUX, 422—423. Baton Rouge, La.: Claitor's Pub. Div. 1969.

FERNÁNDEZ-MORÁN, H., FINEAN, J. B.: Electron microscope and low-angle X-ray diffraction studies of the nerve myelin sheath. J. biophys. biochem. Cytol. **3**, 725—748 (1957).

FINCK, H.: Epoxy resins in electron microscopy. J. biophys. biochem. Cytol. **7**, 27—30 (1960).

FINEAN, J. B.: X-ray diffraction studies of the myelin sheath in peripheral and central nerve fibres. Exp. Cell Res. **5** Suppl., 18—32 (1958).

GIBBONS, I. R.: An embedding resin miscible with water for electron microscopy. Nature (Lond.) **184**, 375—376 (1959).

GILËV, V. P.: The use of gelatin for embedding biological specimens in preparation of ultrathin sections for electron microscopy. J. Ultrastruct. Res. **1**, 349—358 (1958).

GLAUERT, A. M., GLAUERT, R. H.: Araldite as an embedding medium for electron microscopy. J. biophys. biochem. Cytol. **4**, 191—194 (1958).

GLAUERT, A. M., ROGERS, G. E., GLAUERT, R. H.: A new embedding medium for electron microscopy. Nature (Lond.) **178**, 803 (1956).

HAYAT, M. A.: Principles and Techniques of Electron Microscopy: Biological Applications, Vol. 1. New York: Van Nostrand, Reinhold Co. 1970.

HENDRY, J. A., HOMER, R. F., ROSE, F. L., WALPOLE, A. L.: Cytotoxic agents: II, bis-epoxides and related compounds. Brit. J. Pharmacol. **6**, 235—255 (1951).

JOY, R. T., FINEAN, J. B.: A comparison of the effects of freezing and of treatment with hypertonic solutions on the structure of nerve myelin. J. Ultrastruct. Res. **8**, 264—282 (1963).

KELLENBERGER, E., SCHWAB, W., RYTER, A.: L'utilization d'un copolymère du groupe des polyesters comme matériel d'inclusion en ultramicrotomie. Experientia (Basel) **12**, 421—422 (1956).

KELLY, A.: Strong Solids. Oxford: Clarendon Press 1966.

KELLY, A.: The nature of composite materials. Sci. Amer. **217**, no. 3, 160—176 (1967).

KURTZ, S. M.: A new method for embedding tissues in Vestopal W. J. Ultrastruct. Res. **5**, 468—469 (1961).

KUSHIDA, H.: On an epoxy resin embedding method for ultra-thin sectioning. J. Electronmicroscopy **8**, 72—75 (1959).

KUSHIDA, H.: A new polyester embedding method for ultrathin sectioning. J. Electronmicroscopy **9**, 113—116 (1960).

KUSHIDA, H.: A study of cellular swelling and shrinkage during fixation, dehydration and embedding in various standard media. In Proc. 5th Int. Congr. Electron Micr. **2**, P-10. Ed.: S. BREESE, Jr.. New York: Academic Press 1962.

KUSHIDA, H.: A new method for embedding with epoxy resin at room temperature. J. Electronmicroscopy **14**, 275—283 (1965).

LANE, B. P., EUROPA, D. L.: Differential staining of ultrathin sections of Epon-embedded tissues for light microscopy. J. Histochem. Cytochem. **13**, 579—582 (1965).

LEDUC, E. H., BERNHARD, W.: Ultrastructural cytochemistry. Enzyme and acid hydrolysis of nucleic acids and proteins. J. biophys. biochem. Cytol. **10**, 437—455 (1961).

LEDUC, E. H., BERNHARD, W. B.: Water-soluble embedding media for ultrastructural cytochemistry. Digestion with nucleases and proteinases. In The Interpretation of Ultrastructure. Ed.: R. HARRIS (Symp. Int. Soc. Cell Biol. Vol. 1), 21—45. New York: Academic Press. 1962.

LEDUC, E. H., BERNHARD, W.: Recent modifications of the glycol methacrylate embedding procedure. J. Ultrastruct. Res. **19**, 196—199 (1967).

LEDUC, E. H., HOLT, S. J.: Hydroxypropyl methacrylate, a new water-miscible embedding medium for electron microscopy. J. Cell Biol. **26**, 137—155 (1965).

LEE, B.: Microtomist's Vade-Mecum. 9th Ed. Ed.: J. GATENBY and E. COWDRY. Philadelphia: Blakiston 1928.

LEE, H., NEVILLE, K.: Handbook of Epoxy Resins. New York: McGraw-Hill 1967.

LENARD, J., SINGER, S. J.: Alteration of the conformation of proteins in red blood cell membranes and in solution by fixatives used in electron microscopy. J. Cell Biol. **37**, 117—121 (1968).

LITTLE, K.: The action of electrons on high polymers. In Proc. Int. Conf. Electron Micr. London, 1954, 165—171. London: Roy Micr. Soc. 1956.

LOW, F. N., CLEVENGER, M. R.: Polyester-methacrylate embedments for electron microscopy. J. Cell Biol. **12**, 615—621 (1962).

LUFT, J. H.: Improvements in epoxy resin embedding methods. J. biophys. biochem. Cytol. **9**, 409—414 (1961).

MAALØE, O., BIRCH-ANDERSEN, A.: On the organization of the 'nuclear material' in *Salmonella typhimurium*, In Bacterial Anatomy, 6th Symp. Soc. Gen. Microbiol., 261 — 278. Ed.: E. SPOONER and B. STOCKER, Cambridge: Cambridge Univ. Press 1956.

McCLINTOCK, F. A., ARGON, A. S.: Mechanical Behavior of Materials. Reading, Mass.: Addison-Wesley 1966.

McLEAN, J. D., SINGER, S. J.: Crosslinked polyampholytes. New water-soluble embedding media for electron microscopy. J. Cell Biol. **20**, 518—521 (1964).

NEWMAN, S. B., BORYSKO, E., SWERDLOW, M.: New sectioning techniques for light and electron microscopy. Science **110**, 66—68 (1949a).

NEWMAN, S. B., BORYSKO, E., SWERDLOW, M.: Ultra-microtomy by a new method. J. Res. Nat. Bur. Standards **43**, 183—199 (1949b).

Nielsen, L. E.: Mechanical Properties of Polymers. New York: Reinhold Pub. Corp. 1962.

PEASE, D. C.: Histological Techniques for Electron Microscopy, 2nd ed. New York: Academic Press 1964.

PEASE, D. C., BAKER, R. F.: Sectioning techniques for electron microscopy using a conventional microtome. Proc. Soc. exp. Biol. (N.Y.) **67**, 470—474 (1948).

PETERSON, R. G., PEASE, D. C.: Features of the fine structure of myelin embedded in water-containing aldehyde resins. In Proc. 7th Int. Congr. Electron Microscopie **1**, 409—410. Ed.: P. Favard, Paris: Soc. Française de Microscopie Électronique 1970a.

PETERSON, R. G., PEASE, D. C.: Polymerizable glutaraldehyde-urea mixtures as watersoluble embedding media. In Proc. 28th Ann. Mtg. Electron Micr. Soc. Amer. Ed.: C. ARCENEAUX. 334—335, Baton Rouge, La.: Claitor's Pub. Div. 1970b.

PETERSON, R. G., PEASE, D. C.: Polymerizable glutaraldehyde-urea mixtures as watercontaining embedding media for electron microscopy. Anat. Rec. **169**, 401 (1971).

RABINOWICZ, E.: Friction and Wear of Materials. New York: John Wiley & Sons, Inc. 1965.

REIMER, L.: Quantitative Untersuchungen zur Massenabnahme von Einbettungsmitteln (Methacrylat, Vestopal und Araldit) unter Elektronenbeschuß. Z. Naturforsch. **14B**, 566—575 (1959).

REIMER, L.: Irradiation changes in organic and inorganic objects. Lab. Invest. **14**, 1082 — 1096 (1965).

Riddle, E. H.: Monomeric Acrylic Esters. New York: Reinhold Pub. Corp. 1954.

Robertson, J. G., Parsons, D. F.: A resorcinol-formaldehyde resin as an embedding material for electron microscopy of membranes. In Proc. 27th Ann. Mtg. Electron Micr. Soc. Amer. Ed.: C. Arceneaux, 328—329. Baton Rouge, La.: Claitor's Pub. Div. 1969.

Robertson, J. G., Parsons, D. F.: Myelin structure and retention of cholesterol in frog sciatic nerve embedded in a resorcinol-formaldehyde resin. Biochim. biophys. Acta (Amst.) 219, 379—387 (1970).

Rose, F. L., Hendry, J. A., Walpole, A. L.: New cytotoxic agents with tumour-inhibitory activity. Nature (Lond.) 165, 993—996 (1950).

Rosenberg, M., Bartl, P., Leško, J.: Water-soluble methacrylate as an embedding medium for the preparation of ultrathin sections. J. Ultrastruct. Res. 4, 298—303 (1960).

Ross, W. C. J.: Biological Alkylating Agents. London: Butterworth 1962.

Rust, J. B.: Copolymerization of maleic polyesters. Ind. Eng. Chem. 32, 64—67 (1940).

Ryter, A., Kellenberger, E.: L'inclusion au polyester pour l'ultramicrotomie. J. Ultrastruct. Res. 2, 200—214 (1958).

Saunders, J. H., Frisch, K. C.: Polyurethanes. Chemistry and Technology. Part I. Chemistry. New York: Interscience Publishers 1962.

Spurlock, B. O., Kattine, V. C., Freeman, J. A.: Technical modifications in Maraglas embedding. J. Cell Biol. 17, 203—207 (1963).

Spurr, A. R.: A low-viscosity epoxy resin embedding medium for electron microscopy. J. Ultrastruct. Res. 26, 31—43 (1969).

Sterzing, P. R., Scaletti, J. V., Napolitano, L. M.: Tissue cholesterol preservation: solubility of cholesterol digitonide in ethanol. Anat. Rec. 168, 569—572 (1970).

Strangeways, T. S. P., Conti, R. G.: The living cell in vitro as shown by dark-ground illumination and the changes induced in such cells by fixing reagents. Quart. J. micr. Sci. 71, 1—14 (1927).

Sunshine, I.: Handbook of Analytical Toxicology. Cleveland: Chem. Rubber Co. 1969.

Szubinska, B.: Swelling of *Amoeba proteus* during fixation for electron microscopy. Anat. Rec. 148, 343—344 (1964a).

Szubinska, B.: Electron microscopy of the interaction of ruthenium violet with the cell membrane complex of *Amoeba proteus*. J. Cell Biol. 23, 92A (1964b).

Szubinska, B.: "New membrane" formation in *Amoeba proteus* upon injury of individual cells. J. Cell Biol. 49, 747—772 (1971).

Szubinska, B., Luft, J. H.: Ruthenium red and violet. III. Fine structure of the plasma membrane and extraneous coats in Amoebae (*A. proteus* and *Chaos chaos*). Anat. Rec. 171, 417—441 (1971).

Szubinska, B., Luft, J. H.: Osmotic damage to cells from viscous embedding media. J. Ultrastruct. Res. Submitted for publication (1973).

Tooze, J.: Measurements of some cellular changes during the fixation of amphibian erythrocytes with osmium tetroxide solutions. J. Cell Biol. 22, 551—563 (1964).

Wachtel, A. W., Gettner, M. E., Ornstein, L.: Microtomy. In Physical Techniques in Biological Research, Vol. III A. Ed.: A. W. Pollister, 173—250. New York: Academic Press 1966.

Watson, M. L.: Explosionfree methacrylate embedding. J. appl. Physics. 34, 2507 (1963).

Watson, W. A. F.: Studies on a recombination-deficient mutant of *Drosophila*. II. Response to X-rays and alkylating agents. Mutation Res. 14, 299—307 (1972).

Watson, W. F.: Mechanochemistry. New Scientist 9, 548—550 (1961).

Zelander, T., Ekholm, R.: Determination of the thickness of electron microscopy sections. J. Ultrastruct. Res. 4, 413—419 (1960).

Substitution Techniques[1]

Daniel C. Pease

A. Introduction

A practical, working histologist or cytologist in a sense requires understanding of what he is doing at two different levels. For convenience and general productivity he knows very well he has to "fix" tissues and cells as a first preparative step. However, he is quite aware that this is basically a crime against nature, and no matter how sophisticated the fixation procedure may be, there surely are always lingering doubts as to how closely the final image resembles the living object. One has to define terms. I suppose originally fixation carried no more of a connotation than that biological products were rendered insoluble. Later it would have been appropriate to think of fixatives as being mainly protein coagulants. Now we talk of fixatives, at least good ones, as being cross-linking agents which do their job quite subtly, even at macromolecular levels of resolution. However, even in the last decade, we have radically changed our ideas about what constitutes a good fixative, as we relegated osmium tetroxide to a secondary role and advanced aldehyde fixation to a position of primacy. As objective scientists it at least behooves us to find and explore other ways of preserving tissue fine structure so that at the very least we have independent criteria for the evaluation of fixation procedures. If it is accepted that fixation by definition implies molecular alteration by chemical means, then we have no choice but to look to physical methods of tissue preparation for our solution.

Most (but not quite all) physical methods start with living tissue that is first frozen. This opens up several possibilities: 1) cleaving the frozen tissue, and perhaps etching the new surface, and then shadowing and replicating the exposed surface, 2) drying the tissue by sublimation at deep subzero temperature in a vacuum before embedding and sectioning it, 3) dissolving the ice by substitution with a suitable medium, also at low subzero temperature, and 4) directly sectioning frozen tissue without any embedment. It is only the third of these techniques that we are going to deal with in this chapter. However, when one talks of substituting some

---

[1] Supported in part by grants of the U.S. National Institutes of Health.

strange medium for tissue water, one is fundamentally talking about exchange, and as we proceed I will indicate that one can achieve a large measure of success with this at room temperature, without ever freezing the tissue. Thus, what is commonly called "freeze-substitution", is a technique allied to what I have called "inert dehydration" (PEASE, 1966a), and these methods, along with "freeze-drying", all have as their objective the intention of stealing water away from tissues and cells with minimal damage to native macromolecular organization. In a sense, any dehydration procedure does this, but when macromolecular structure has been stabilized by suitable fixation there seem to be few problems with dehydration, and it really does not matter very much whether you throw tissue into alcohol, acetone or any other water-miscible solvent. It, of course, does matter very much when the tissue has not been "fixed".

In a way, the "inert dehydration" of unfixed tissue at room temperature is the simplest of the physical techniques to be discussed in this chapter, and the results are to provide comparisons and controls for tissues that are "frozen-substituted". It therefore seems expedient and appropriate to consider this method first, although historically it is a late-comer to our armamentarium.

B. Inert-dehydration

The small glycols, glycerol, the small sugars and perhaps a few other molecules that bristle with hydroxyl groups constitute a rather remarkable family of molecules. By and large, they are not notably damaging to living cells, except perhaps at high concentrations. Thus, they can be included in tissue culture media, and the food industry can use them freely. They are "cryoprotective" agents *par excellence* which has bearing on the problem we are going to discuss, for these compounds are thought not to be "structure-breaking" insofar as water lattices are concerned (DOEBBLER, 1966; FRANK, 1965; KAROW and WEBB, 1965; KLOTZ, 1965; LING 1965). That is to say, these compounds can fit into the organized water shells associated with biological molecules, particularly proteins. Such "structured" water is now regarded as being an important consequence of the "hydrophobic bonding" that stabilizes native proteins. The presence of small glycols and glycerol does not seriously disturb this so that proteins in general are not denatured by their presence, even in substantial amount. It was this writer who first took advantage of these properties of this small group of molecules to fully dehydrate fresh, unfixed tissue at room temperature (PEASE, 1966a). Following that, it was of course easily feasible to transfer the dehydrated tissue into any embedding medium by choosing appropriate intermediate solvents. Fig. 1 illustrates "inertly dehydrated" tissue.

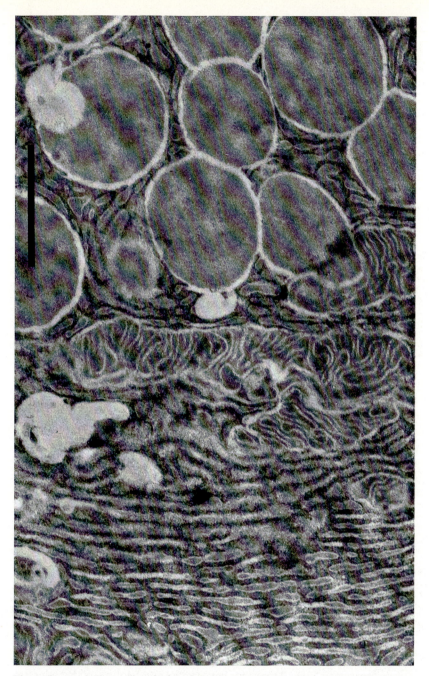

Fig. 1. Pancreatic tissue, "inertly dehydrated" at room temperature with ethylene glycol, illustrating certain of the artifacts which are commonplace. These presumably reflect differential shrinkage. Note the irregularity of the mitochondrial shapes, and the angularity of their cristae. The zymogen granules have pulled away somewhat from their surround. Discrete globular cavities are evident, sometimes associated with membranes. When nuclei are seen, these invariably have scalloped profiles comparable to those evident in Figs. 2, 3, 4 and 6. All of the tissue illustrated was embedded in hydroxypropyl methacrylate. Micron marks are included with each figure

I. Method

Empirical experiments soon demonstrated that the key to successful room temperature, glycolic dehydration was the rate at which glycol (or glycerol) was exchanged for tissue water. There appeared to be two opposing deleterious forces at work. On the one hand, too rapid an exchange led to severe osmotic shock. On the other hand, too slow an exchange presumably damaged the plasma membrane too soon by solvent action, before the cellular contents had been sufficiently stabilized by the removal of water. A high degree of stability is, in fact, achieved when cells and tissue develop a glassy appearance. This colloid change becomes evident as glycolic concentrations approach 60%, and seems nearly complete at the 70% level. Empirically, the best results are obtained when such a concentration of glycol is achieved in about 3 minutes after the exchange starts. In practice, we usually cheat a little and generally collect the specimens before starting the dehydration in a balanced salt solution (usually Hanks') to which 10% glycol has already been added.

Efforts to move tissue stepwise from one concentration of glycol to another never turned out very satisfactorily although, as a prelude to freezing or fixation, other investigators have successfully employed a stepwise schedule to build up 50 or 60% concentrations of glycol (BULLIVANT, 1965; COPE, 1967). In my experience far better results are achieved by increasing the concentration of glycol continuously while the specimen chamber is being vigorously agitated. For small quantities of tissue, Syracuse watch glasses or small Stender dishes prove most convenient. The tissue samples to be dehydrated are placed in such a container in 5 ml of Hanks' solution with the added 10% glycol. In a graduate or other container nearby, there is 7 ml of pure glycol to be pipetted into the tissue container over a 3 minute interval while the latter is being rotated to swirl its contents. When both solutions are combined the actual glycol content of the mixture is 66%. The rest, of course, is by then a much diluted Hanks' solution (experiments have been performed in which the ionic strength was maintained simply by adding an appropriate amount of sodium chloride directly to the otherwise pure glycol. It appears that this has a decidedly deleterious effect).

It can be assumed that tissue is well stabilized when it is in 66% glycol, and it is not apt to be damaged by subsequent treatments if these are at all reasonable. Such tissue can be stored reasonably well, and at least some histochemical reactions can be performed readily in this medium. Some kinds of fixation and block staining can be performed at this point for special purposes. Several times we have quite successfully used 5% glutaraldehyde dissolved in 66% glycol for this purpose. This is also the basis of Cope's (1967) procedure although he has also advocated dropping the temperature to $-20\,°C$ before fixation, and completing infiltration and embedding at subzero temperatures. This has been in an effort to preserve lipids better than otherwise would have been the case.

Tissue can be osmicated in 66% glycol if trimethylene glycol is used as the substituting glycol rather than ethylene glycol for osmium tetroxide is effectively bound and removed from solution by substances with hydroxy groups in a 1-, 2-position, as in either ethylene or propylene glycol, or in glycerol. It is not, however, so bound by hydroxyl groups in a 1-, 3-position (MILAS et al., 1959).

Most generally, however, one does not want to stop the dehydration at the 66% glycol level, but wants to complete it and move on to an embedment for ultrathin sectioning. This is best done by successively decanting the substituting medium, and adding increasing amounts of glycol until the tissue is in pure glycol. Incidentally, tissue can be stored indefinitely in pure glycol in a refrigerator without it undergoing any recognizable morphological deterioration.

Glycol (or glycerol) is directly miscible with either hydroxyethyl or hydroxypropyl methacrylate (glycomethacrylates), and so there is no problem with embedments in these media. These solvents are not miscible, though, with epoxy resin mixtures or with Vestopal. Then an intermediate solvent has to be used. It was originally suggested that cellosolve (the monomethyl ether of ethylene glycol) would serve (PEASE, 1966a). However, we have since felt that this solvent can have deleterious effects, and now when we wish to embed in an epoxy resin we pass the tissue through propylene oxide. For Vestopal we go first to propylene oxide and then through toluene.

It should be noted that an inertly dehydrated specimen (or for that matter any other sort of a specimen) need never again be exposed to water during the staining and sectioning processes. Any one of the common embedments can be sectioned with ethylene glycol as the trough fluid (PEASE, 1966b). Sections are picked up as is usual by pressing the grids (coated or not) down upon the sections from above. There are stains, particularly phosphotungstic acid and uranyl acetate, which can be dissolved in ethylene glycol (PEASE, 1966b). Finally, after washing the sections by dipping the grids in pure glycol, this can be removed by first blotting them, and then evaporating the glycol with a simple mechanical vacuum pump. A high vacuum is not required.

For most purposes as suggested above, ethylene glycol is the choice of solvents to use for "inert dehydration". It is the smallest possible glycol, and thus can be expected to exchange most readily with water. However, we have also used propylene and trimethylene glycol quite effectively, as well as glycerol. Even the small sugars can be used in the form of a concentrated syrup which permits adequate stabilization of the tissue before it becomes necessary to switch to glycol or glycerol to complete the dehydration (Pease, 1966a). Attempts to use dimethyl sulfoxide have been partially successful in the sense that small areas of cytoplasm may be reasonably preserved. However, most of the tissue has generally been destroyed, perhaps because the cytomembranes are damaged too soon by the larger solvent action of this substance. Minor quantities of DMSO in mixtures are tolerable. Tissue substituted with strong urea solutions is not totally destroyed, but there are bizarre changes in the cytomembrane systems, particularly of the mitochondria. Urea, of course, is a strong protein denaturant, and is regarded as a "structure-breaking" substance, possibly because its triangular shape prevents it from participating in the tetrahedral bonding of structured water (Frank, 1965). It thus appears that the desirable agents for inert dehydration are represented by a small group of cryoprotective substances which have in common multiple exposed hydroxyl groups which can serve as appropriately oriented hydrogen bonding sites, and thus have a minimal effect upon structured water (Doebbler, 1966).

II. Artifacts

The artifacts characteristic of inertly dehydrated tissues seem to be of three types, although perhaps all relate fundamentally to inequities in the rate of exchange between tissue water and glycol. In any given tissue block there is a zone of optimal preservation that can be expected to be about 0.5 mm wide. Deeper tissue is destroyed. Also, it is invariably found that the surface layer, about 0.1 mm thick, is destroyed too. This has been true even when a natural surface has been examined, so the damage cannot be attributed to simple mechanical injury. Whatever exactly occurs that destroys the surface of a tissue block being inertly dehydrated, these same forces apparently work similarly on exposed cells in suspension and at the surface of tissue culture explants. For the present, therefore, these types of preparations cannot be effectively preserved by inert dehydration although conceivably a protective envelope or shell might be devised.

A second type of artifact expresses itself simply as shrinkage, even in those areas optimally preserved (Fig. 1). This is not generally very noticeable except in nuclear profiles. These become scalloped rather than remaining circular. Scalloping can also occur around inclusions such as secretory granules, and irregularities in the profiles of some large mitochondria such as occur in the cortex of the kidney may be evident. What appear to be angular displacements of endoplasmic reticulum also occur. These are really relatively minor disturbances, and it seems likely that they are strains reflecting differences in the degree of hydration of neighboring areas, that some compartments are able to shrink more than others.

The smallest order of obvious artifact in inertly dehydrated tissue probably also results from internal tensions set up within cells as a result of varying forces, including the degree of hydration. This results in the appearance of small cavities (Fig. 1). Quite commonly these are seen as a splitting of cytomembranes to produce a more or less lens-shaped cavity. It, of course, has been the experience of people working with "freeze-fracturing" techniques that the bi-layers of lipid in cytomembranes are easily cleaved. The forces which maintain these sorts of molecular arrangements are notoriously weak, and if mechanical stresses are set up within cells by differential shrinkage of neighboring areas, it should be no wonder that they might be relieved by this kind of tearing. Globular cavities are also seen in cytoplasm that cannot always be identified with a membrane source (Fig. 1). However, this does not preclude the latter as a point of origin.

Insofar as inertly dehydrated tissue is unfixed, it is not surprising that lipids seem to be fairly completely extracted by the time sections are obtained. Cytomembranes are always seen as white lines, in negative contrast, in ultrathin sections irrespective of the stains employed. However, membrane lipids can at least to some extent be preserved by osmicating material that has been substituted with trimethylene glycol to about the 60—70% level.

III. Fine Structure

A great deal of tissue fine structure is retained by tissue infiltrated without fixation by inert dehydration. This is evident in Fig. 1, and previously published figures (PEASE, 1966a, 1966b, 1968). This includes protein, nucleoprotein, and glycoprotein systems. Tissues so prepared in general look remarkably like tissue well preserved with an aldehyde, but without postfixation in osmium tetroxide. It accepts stains in a similar way. There are some subtle differences, however (PEASE, 1966a, 1968). Thus, ribosomes attached to endoplasmic reticulum are seen as flattened structures, although free ribosomes remain spherical. This corresponds to some findings from frozen dried tissue and suggests that in life these structures may be reasonably plastic. Polysaccharide moieties, including chondroitin sulfate and glycogen, are preserved in unparalleled quantities, and at times with superior precision. Thus, the glycocalyx in some situations is seen as a very well-defined layer of uniform thickness. Some cytoplasmic proteins are preserved in quantities not ordinarily seen. This has been particularly obvious in the I-disk of striated muscle where the great amount of stainable material, otherwise commonly more or less lost, hides filamentous detail. In such respects this is, of course, a disadvantage.

In retrospect it is curious that glycerol and the small glycols had not been used at an earlier date than 1966 (PEASE, 1966a) for the complete dehydration of fresh tissue at room temperature. The cryoprotective

properties of these substances had been appreciated since the work of
Polge, Smith and Parks (1949). Occasionally investigators had used high
concentrations of these compounds as a prelude to the rapid freezing of
specimens. Thus, Fernández-Morán (1960, 1961a, 1961b) worked
tissue into buffered salt baths containing as much as 30—60% glycerol.
Later, Bullivant (1965) also studied frozen tissues which were pretreated
with as high as 60% glycerol solutions. Rebhun (1965) was interested in
the freezing process in relation to freeze-substitution, and in the course of
his studies examined the effects of plunging living tissues into full strength
glycerol for brief periods (15 sec to 1 min) before quenching them. In a
different context, it has been commonplace since Szent-Györgyi's (1951)
studies to store fresh striated muscle in 50% glycerol in a refrigerator for
months at a time before using it for physiolgical or biochemical experiments.
It was this sort of a preparation that Huxley (1957) employed in his
classical studies of skeletal muscle which led him to the sliding filament
theory of contraction. This kind of preparation, of course, beautifully
preserved the filamentous structures, although other cytoplasmic proteins
had been very slowly, largely extracted.

C. Freeze-substitution

Freeze-substitution was invented and introduced almost incidentally
by Simpson (1941) in a paper entitled "An Experimental Analysis of the
Altmann Technique of Freeze-Drying". As the title suggests, Simpson was
really interested in exploring the various parameters involved in freezing
and drying tissues. In the course of this, he looked at quite a variety of
tissues that had been frozen in various ways, and subsequently handled
with many different technical variations. At one place or other along the
way, most of his tissues ended up being fixed, and he introduced the term
"freeze-fixation" for these. As far as low temperature substituting media
were concerned, he explored methyl cellosolve, absolute ethanol, anhydrous
diethyl ether, and anhydrous chloroform at temperatures ranging from
—40 °C to —78 °C. It is hard to understand how he thought of the latter
two compounds as being reasonable low temperature solvents for ice.
His paper was a short one, illustrated by a single, unimpressive plate. It is
perhaps no wonder that the article did not have much original impact.

Feder and Sidman (1958) published a definitive paper which certainly
advanced the art as far as light microscopy went, and they also reviewed
the literature on freeze-substitution up until that time. They were able
to cite only 21 applications of the technique, along with 12 additional
modifications. They, themselves, were interested only in the best possible
preservation of morphology, and emphasized using fixatives in the substituting media. Thus, osmium tetroxide was dissolved in acetone, and

mercuric chloride and picric acid in ethanol. They did, indeed, succeed quite well in preserving tissue for conventional microscopy. At about the same time, papers by PATTEN and BROWN (1958) and HANCOX (1957) also included critical reviews of the state of the art.

It was about the same time that FERNÁNDEZ-MORÁN (1957, 1959a, 1959b) published his first accounts of efforts to apply this technique to electron microscopy. His original biological interest seems to have been a fairly esoteric attempt to localize rhodopsin in retinal rods. There quickly followed more general papers recounting a complex of experiences with a variety of tissues (FERNÁNDEZ-MORÁN, 1960, 1961a, 1961b). Almost from the beginning, FERNÁNDEZ-MORÁN had conceived the idea of quenching tissue in liquid helium II at —272 °C, followed by the substitution of the tissue with alcohol at low temperature, and then also infiltrating and embedding the tissue with methacrylate mixtures at low temperature by polymerizing with ultraviolet radiation.

During a part of this developmental period STANLEY BULLIVANT worked in close association with FERNÁNDEZ-MORÁN, although he was established in an independent laboratory. In 1960 he published experiments using essentially the same technique as FERNÁNDEZ-MORÁN. This latter work was an uncomplicated study of pancreatic tissue, and did indeed demonstrate remarkably good preservation. Although FERNÁNDEZ-MORÁN seems to have turned in other directions after 1961 (his later papers added no new material), BULLIVANT did maintain some further interests in freeze-substitution. Two subsequent papers by BULLIVANT (1965, 1970) include important reviews. It remains a curious fact, though, that freeze-substitution has never received anything like the attention and effort that has been given to freeze-drying. One can argue that the latter technique has attracted investigators for many reasons other than just the preservation of morphology, yet this perhaps is a specious argument, for biologically interesting macromolecules may survive as well in "inert" media as in a dessicated state. So far we have very little information one way or the other concerning this possibility.

In more recent years there have been three more groups particularly interested in exploring freeze-substitution techniques. Chronologically, the first of these was REBHUN (1961, 1965; REBHUN and SANDER, 1971) whose primary interests have been in preserving marine eggs; then followed VAN HARREVELD and his coworkers, CROWELL and MALHOTRA (VAN HARREVELD and CROWELL, 1964; VAN HARREVELD, CROWELL and MAL-HOTRA, 1965 and VAN HARREVELD and KHATTAB, 1968; MALHOTRA, 1968a, 1968b; MALHOTRA and VAN HARREVELD, 1965a, 1965b, 1966) whose main goal was to preserve the extracellular space in neuropil with its original geometry (and who also had some interest in mitochondrial morphology), and finally, the present author (PEASE, 1967a, 1967b), who had been exploring the potential uses of the small glycols as substituting media. The different objectives of these various investigators have been so diverse that no single methodology has evolved. A part of the confusion lies in the fact that there are at least three fundamentally distinct, but inextricably intertwined, problems involved in the total technique. There is a question of how best to freeze tissue in the first place, which is being discussed in detail elsewhere in this volume. Then, there is the question of how substitution should be performed, including the best medium of choice and the temperature to work at, which is the principal problem to be considered here. And, finally, one has the possibility of adding fixatives to the substituting media, and for certain objectives this is clearly desirable, but it is not so for others.

In what follows I want to speak of my own, fairly recent work first, including some unpublished material, mainly because the technique and the results parallel in large measure those of "inert dehydration". The latter technique serves as an almost ideal control for comparison with frozen-substituted tissue, for in essence it differs only in that in the one case the entire processing is done at room temperature, and in the other the tissue is first deeply frozen. It will then be possible more properly to place in perspective the efforts of the earlier investigative teams which also will be reviewed.

I. Experiments of this Author

The success achieved by the present author dehydrating fresh tissue at room temperature with the small glycols (PEASE, 1966a, 1968) suggested that these solvents might be close to ideal as substituting media for deeply frozen tissue. In particular, it seemed likely that one could design experiments which would distinguish between such damage as might be done during the initial freezing process and subsequent damage that conceivably might be related to the substitution phase.

Propylene glycol, trimethylene glycol and glycerol are all substances that are very difficult to freeze in a crystalline form. Instead, they become glass-like at extremely low temperatures. Indeed these substances become so viscous below about $-40\,°C$ that it hardly seems reasonable they could be used successfully as substituting media at lower temperatures. Ethylene glycol behaves differently, and is easily frozen as a crystalline material at a temperature not much below $-13\,°C$. However, ethylene glycol added to water has remarkable properties as an antifreeze. A 70% solution will not freeze above $-50\,°C$, and it can be supercooled at dry-ice temperatures $(-77\,°C)$ for hours, even in the presence of ice crystal seeds. Although technically it may not be quite correct to speak of a 70% solution of ethylene glycol as a eutectic, it is effectively that for our purposes, and will in fact readily dissolve crystalline ice at —50 °C. The results of our experiments with inert dehydration had shown us that a high degree of tissue stability was achieved in 70% glycol. Thus, it became obvious that this might serve as a useful substituting medium for frozen tissues at temperatures as low as —50 °C. One has the obvious option of using supercooled 70% glycol at —77 °C for quenching. The dehydration finally can be completed with one of the other small glycols at —40 °C, or with glycerol at a temperature still substantially below freezing. Time and again the results have proven to be essentially indistinguishable from those of inert dehydration.

The basic equipment that we have used is a low temperature freezer with a motorized device inside to roll containers, and thus stir their contents. We have learned that we can very reliably freeze small bits of tissue by plunging them into 70% glycol, supercooled with dry ice to —77 °C, in the same container to be used for substitution. Thus, transfer problems are eliminated. Results of this simplest of all procedures are illustrated in Fig. 2.

In a series of experiments comparing this procedure with much more drastic quenching processes, no differences could be detected (PEASE, 1967b). The more severe procedures involved either quenching the small bits of tissue in frozen Freon 22 in equilibrium with its own melt, at —160 °C (Fig. 3), or by making essentially instantaneous contact of a tissue surface with a dry polished brass block, precooled to —196 °C with liquid nitrogen.

Small tissue samples stirred in 70% ethylene glycol at —50 °C generally become markedly transparent within a few hours, indicating that substitution is at least well underway. Indeed, gross ice shavings and chips will dissolve in such a bath in a similar length of time. However, in general, we have allowed an overnight period to achieve equilibrium. On occasion frozen tissue has been exposed to these conditions for as much as a week without it being possible subsequently to detect any improvement. I, therefore, categorically reject a criticism of BULLIVANT (1970, p. 121) that the tissue may have been "allowed to warm before substitution was complete, any ice melted, and reconstitution took place." The general method has proven to be a remarkably simple and consistent procedure.

The use of a 70% ethylene glycol solution as the basic substituting medium provides the potential for additional routes for modification and control. Thus, it is perfectly feasible to incorporate an aqueous fixative in the medium, such as glutaraldehyde or formaldehyde. That the former, at least, is active at low temperatures is evident from the fact that tissue is decidedly yellowed by it even at — 50°C (PEASE, 1967 b). (Osmium tetroxide cannot be added to ethylene glycol without combining with it (MILAS et al., 1959). It can, however, be added to any 1-, 3-glycol such as

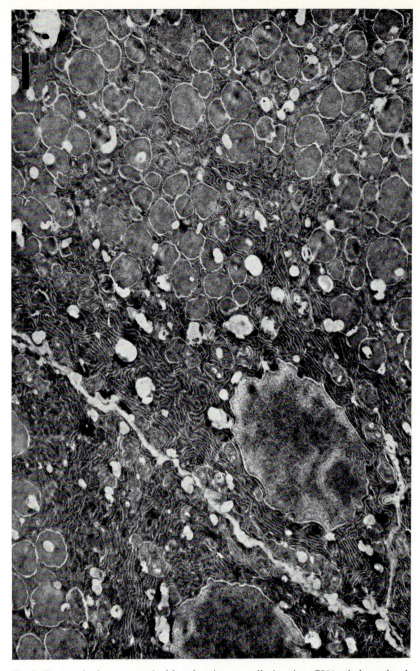

Fig. 2. Pancreatic tissue, quenched by plunging a small piece into 70% ethylene glycol, supercooled to —77 °C, and then substituted in this bath at —50 °C. The character of the artifacts resemble those of inert dehydration, and there is no evidence of massive ice crystal damage. (The extraction of unfixed chromatin evident here is a common, but not universal, artifact of the alkalinity of lead citrate staining.)

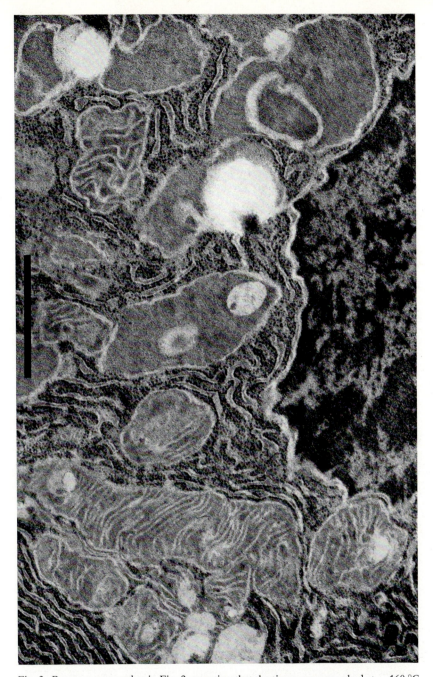

Fig. 3. Pancreas prepared as in Fig. 2 excepting that the tissue was quenched at −160 °C in Freon 22, before transferring it to 70% ethylene glycol for substitution at −50 °C. The angularity of the membrane systems, including the profile of the nucleus, indicates differential shrinkage. Basically, though, the evident artifacts are identical with those of "inertly dehydrated" material that was never frozen, as illustrated in Fig. 1

trimethylene glycol.) Block staining can be attempted, and we have done some work along this line adding uranyl salts or phosphotungstic acid to the medium.

Allusion has already been made to the fact that frozen tissue substituted with glycol (Figs. 2—6) looks almost exactly like the "inertly dehydrated" tissue (Fig. 1) described in the previous section, which never experienced freezing. Although originally I thought that some of the cavities that were observable in frozen-substituted tissue might possibly represent ice crystal damage (PEASE, 1967b), I am now inclined to think that most are the same sort of cavities that seem to be unavoidable in inertly dehydrated tissue. In any case, this is not a serious factor, for quantitative measurements of the total cavity volume in frozen-substituted pancreatic cell cytoplasm at most amounted to no more than 6% of the total volume (PEASE, 1967b).

In general, assuming that the initial quenching process was reasonably performed, the artifacts of glycolic freeze-substitution are uncannily like those of inert dehydration. At the semi-gross level in both cases the surface layer (even natural surfaces) invariably is destroyed. The deeper parts of an overly thick block also are destroyed. There is a zone about a half a millimeter thick, a few cell layers below the surface, that constitutes the well-preserved tissue.

On the finer scale, both sorts of preparations show the same order of magnitude of shrinkage, including nuclei that have scalloped edges (Figs. 2—6). Splits in cytomembranes are commonplace. Scattered small cavities, sometimes related to split membranes, are common. In frozen tissue, there is no evidence of any particular disruption or disturbance of extracellular stromal spaces or contents (Figs. 2 and 4). Thus, there is no reason to believe that at any time ice crystals preferentially developed in those areas.

The reliability of glycolic freeze-substitution is in marked contrast to what other investigators have reported using different substituting media, particularly the lower alcohols and acetone. Furthermore, the results are not immediately in harmony with the considerable body of conventional physical knowledge about the freezing process that generally indicates that ice crystal formation is an inevitable concomitant of deep freezing. Originally, I interpreted my own findings on the basis that little or no crystalline ice formed. Instead, tissue water must be largely vitrified under the circumstances of the experiments. Further experiments which have so far not been published have tended to confirm this. The experiments in question have compared the effects of different substituting media on tissue samples quenched identically and simultaneously.

A variety of different experiments have been performed including such relatively drastic ones as quenching fresh tissue in Freon 22 at —160 °C, and then moving the tissue into liquid nitrogen, storing it there for hours, and then finally moving the tissue into either acetone, alcohol, or 70% glycol at —50 °C. In all of these cases several days were allowed for substitution to be sure that equilibration was achieved before further processing the tissue. The tissue substituted with glycol was entirely typically preserved with no indication of massive ice crystal damage (Fig. 6). On the other hand, when the other substituting media were used, preservation was absolutely hopeless with angular cavities everywhere comparable to those seen in Figs. 7 and 8.

In what was regarded as a particularly drastic experiment, fresh tissue was quenched in Freon 22 at —160 °C, and then this bath was warmed up to —50 °C (well above the "transition temperature"). The tissue was left in the Freon for 5.5 hours before finally dividing it into the three different substituting baths being tested. Once more the tissue dehydrated with 70% glycol was perfectly well preserved (Figs. 4 and 5), whereas after alcohol or acetone substitution ice crystal artifacts totally dominated the tissue.

Another series of experiments involved quenching tissue in the three different media at —77 °C, and then allowing time for substitution at —50 °C. Again, the glycolic substitution was entirely typical (Fig. 2), while the other samples were hopelessly destroyed by massive ice crystal damage (Figs. 7 and 8). Another related experiment involved quenching all of the tissue in 70% glycol at —77 °C, but then transferring some of the tissue to either prechilled acetone or alcohol. Once again the material substituted in glycol was well preserved whereas the material that had been transferred was for all practical purposes totally destroyed.

Final experiments essentially repeated an earlier effort on the part of AFZELIUS (1962) to use a pure fixative as a substituting medium, which therefore might be expected to fix simultaneously with its penetration. Acrolein, with a freezing point of —87.7 °C, makes this feasible. I have frozen pieces of pancreas either directly in acrolein at —77 °C, or transferred them into acrolein after freezing in Freon 22 at —160 °C. In either case ice crystal cavities completely dominated the final picture. This is in accord with AFZELIUS' (1962) unhappy experience.

Finally, it should be noted that careful precautions have been taken to prevent tissue dehydration before quenching. As will be discussed later, REBHUN (1965) and REBHUN and SANDER (1971) have clearly demonstrated that if tissue is allowed to dry somewhat before freezing it, a degree of protection is afforded. Anticipating this, I was careful in my original experimental procedure (PEASE, 1967a, 1967b), and in the more recent unpublished work referred to above, there is no possibility that significant evaporation influenced the results.

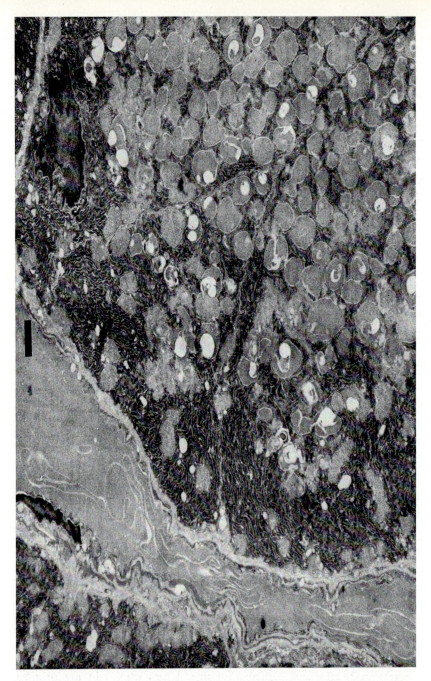

Fig. 4. This pancreatic tissue was first frozen at —160 °C in Freon 22, which was then warmed to —50 °C and held at that temperature for 5.5 hrs before proceeding with the —50 °C substitution in 70% glycol (with glutaraldehyde added). There is no evidence of secondary ice crystal formation, even in the plasma and red cells of the included blood vessel

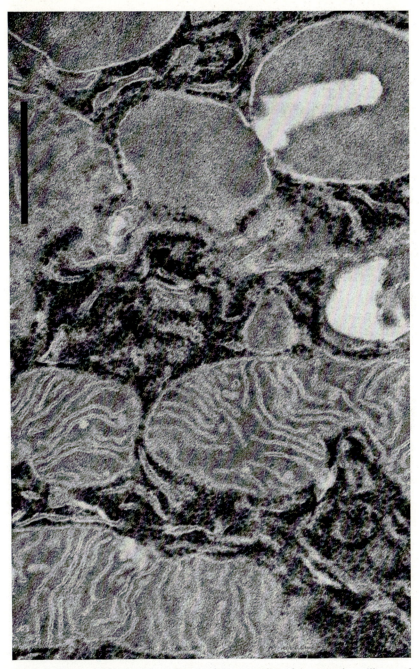

Fig. 5. A high magnification view of some of the organelles of the same tissue illustrated in Fig. 4. A comparison with the inertly dehydrated tissue of Fig. 1 indicates no essential difference in spite of the temporal opportunity given this tissue for "secondary" ice crystal formation

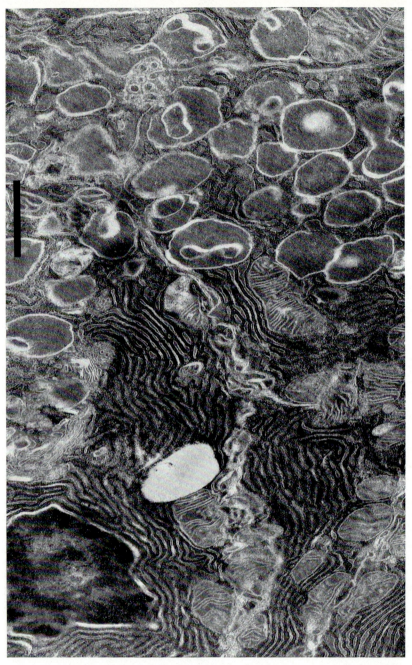

Fig. 6. Tissue quenched at −160 °C in Freon 22, and then transferred to liquid nitrogen at −196 °C for 4.5 hrs, before finally substituting at −50 °C in 70% ethylene glycol with glutaraldehyde added. There is no evidence of "secondary" ice formation

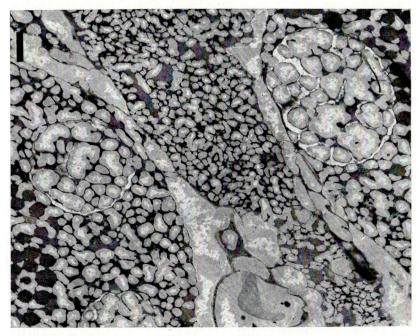

Fig. 7. Tissue prepared simultaneously with that illustrated in Fig. 2, but quenched directly in absolute alcohol precooled to —77 °C, instead of in 70% ethylene glycol. Extreme ice crystal damage is evident everywhere. Tissue quenched in Freon 22 at —160 °C, and then transferred to absolute alcohol for substitution at —50 °C demonstrated identical artifactual destruction

In the light of these and other experiments I am convinced that a most important factor that can be responsible for the observable ice crystal damage commonly seen in frozen-substituted tissue is the substituting medium itself, rather than the freezing process.

II. The Work of FERNÁNDEZ-MORÁN and BULLIVANT

FERNÁNDEZ-MORÁN's interest in the possibility of using freeze-substitution techniques for electron microscopy dates back to before(1957), and to 1959 when he published abstracts dealing with what was then current work (1959a, 1959b). In the next two years three important papers detailed his efforts (FERNÁNDEZ-MORÁN, 1960, 1961a, 1961b). The most comprehensive and important of these is (1961a). I am not aware of his adding any new material subsequently, although occasional reference to this original work reappears from time to time in his more recent publications.

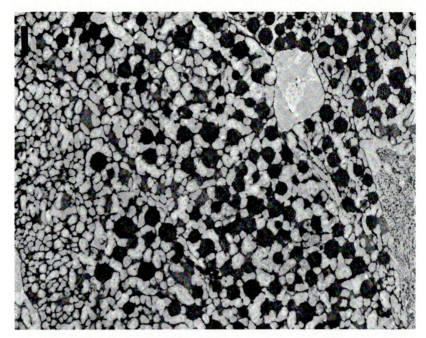

Fig. 8. Material prepared as in the case of Figs. 2 and 7, excepting that the quenching was directly into anhydrous acetone precooled to —77 °C. As with alcoholic quenching (unlike glycolic quenching) the tissue was utterly destroyed by ice crystal artifacts. Also the same destructive artifact was present in tissue quenched in Freon 22 at —160 °C, and subsequently transferred to acetone at —50 °C

FERNÁNDEZ-MORÁN quickly made life rather complicated for himself. For one thing, he had the notion of using liquid helium II at 1—2 °K as the quenching medium, although he also worked with Freon 22 at —160 °C. He experimented with partially glycerinated tissues to afford a measure of cryoprotection. He experimented with block staining of his tissues at —150 °C with halogens and organo-metallic compounds dissolved in isopentane. His basic substitution medium was an alcohol-acetone-ethyl chloride mixture, used at temperatures varying from —130 °C to —80 °C, into which he sometimes incorporated such potential staining substances as platinum chloride, osmium tetroxide, gold chloride, and "other heavy metal salts". He decided to infiltrate resin mixtures and polymerize embedments at as low temperatures as possible also, so that after the water presumably had been replaced by the original substituting medium, the specimens were then infiltrated at temperatures varying from —100 °C to —75 °C with mixtures of methylacrylate and methacrylate monomers, which was followed by ultraviolet polymerization using benzoin as a catalyst at $-80\,°C$ to $-20\,°C$. These various operations required successive changes in media, and obviously it was important that water not be introduced along the way. Thus, FERNÁNDEZ-MORÁN had to develop the substantial hardware to make this possible. Basically, he used filled and sealed vessels, and transfers were made by puncturing flexible membranes with hypodermic needles for the withdrawal or addition of fluids, operations being performed within a freezer. He was also much afraid of initiating ice crystal formation in his specimens if his various media were not strictly anhydrous, so that he went to rather extraordinary precautions to achieve this.

At the time when FERNÁNDEZ-MORÁN was doing this work, his scientific objective was almost exclusively to study retinal rod receptors. There is no question but that he obtained some very interesting micrographs of these, yet the highly specialized nature of this material precluded an assessment of how generally useful the procedures might be. This was left for BULLIVANT's (1960) original work in this field. Another major difficulty in assessing what might have been really essential for successful freeze-substitution in FERNÁNDEZ-MORÁN's work is that he has never published anything like enough detail, particularly of his failures, so that the outsider could know what may have been significant for success. Of particular pertinence in the present context is how much preglycerination he found necessary. He (FERNÁNDEZ-MORÁN, 1961a) indicates using a controlled glycerination of 30—60% glycerol in a Veronal solution. In another publication he spoke of infiltrating fresh specimens with buffered glycerol-Ringer solutions "gradually increasing concentration up to 50—60% at $-20\,°C$" (FERNÁNDEZ-MORÁN, 1961b). We now know, of course, that tissue worked into 60% glycerol has in effect been "inertly dehydrated". Freezing at that stage can be regarded as having been almost redundant, and it is doubtful that the subsequent manipulation of the tissue at very low temperatures may have done much to benefit its preservation. Unfortunately, I am not aware of any of FERNÁNDEZ-MORÁN's micrographs providing specific information on this aspect of their preservation.

BULLIVANT (1960) set about to apply the basic freeze-substitution procedure of FERNÁNDEZ-MORÁN to pancreatic tissue, and succeeded remarkably well. At the time BULLIVANT was privy to unpublished information from FERNÁNDEZ-MORÁN's laboratory, and he recognized his indebtedness to him by including his name in the title of the article which was published. Perhaps the most significant change that BULLIVANT introduced into the methodology was to employ methanol as the substituting media, and also he abandoned the notion of very low temperature polymerization of the methacrylate embedment that he employed. No mention was made in the original article of using preglycerination or any other sort of cryoprotective agent before quenching the tissue. Indeed, it is implied that small pieces of pancreas were immediately transferred to aluminum foil, and these then were plunged directly into helium II. No particular efforts were made to investigate the possibilities of block staining during the substitution process, but BULLIVANT did subsequently explore the use of osmium tetroxide, lead acetate and the Gomori lead method for acid phosphatase applied as stains to sections, and he analyzed the effects of the total procedure in terms of the different organelles present.

The principal artifact evident in the published micrographs was a tendency for cytomembranes to split and open up cavities. This surely can be regarded as a relatively minor defect. More significantly, as will be considered later in discussing REBHUN's work, at least some of his micrographs show evidence of a shrinkage artifact characterized by angularities in the mitochondria and the pattern of the endoplasmic reticulum. The most likely interpretation is that some drying of the fresh tissue occurred as it was removed from the animal, transferred to the aluminium supports, and then finally plunged into the quenching bath. Unfortunately, in this original work, BULLIVANT gave no indication of how often he was reasonably successful in preserving tissue this way. In a subsequent paper (BULLIVANT, 1965), commenting on his past experience, he observed that fresh tissue without cryoprotection "occasionally yielded specimens with acceptable preservation, results in general were not as satisfactory as could be attained by protecting tissue with glycerol and then freezing." Thus, inferentially, his original technique lacked reliability.

In this second BULLIVANT (1965) paper the author reported a schedule for, and results of, prequenching glycerination. Varying concentrations of glycerol were made up in isotonic Veronal-acetate buffer, and fresh tissue samples were passed through these successively at 4 °C until a 45% level of glycerol was achieved. Then the tissue was further glycerinated to the 60% level at −25 °C. This sounds very much like what FERNÁNDEZ-MORÁN (1961 a) may have used earlier. In any case, once again, it would now seem in retrospect that BULLIVANT had nearly achieved an "inert dehydration" and that subsequent low temperature freezing and dehydration may have been largely redundant.

In this latter work BULLIVANT also decided to abandon quenching with helium II on the basis of some model experiments he performed which indicated that actually liquid propane chilled to −175 °C demonstrated a more rapid heat transfer. BULLIVANT no doubt correctly attributed this to the insulating gas film that almost surely must form around the specimen if it is plunged directly into substances such as liquid helium or liquid nitrogen with very low boiling points. BULLIVANT also, for no very obvious reasons, switched to ethanol as the substituting medium in this later series of experiments.

The micrographs presented by BULLIVANT (1960, 1965) demonstrated excellent preservation, but not without an obvious shrinkage artifact producing such distortions as crenulated nuclei very similar indeed to my own results with "inert dehydration". Thus, I attribute this shrinkage to the glycerination procedures used before the tissue was quenched. The number of micrographs presented attests to the reliability of the final overall procedure.

Subsequently, BULLIVANT has turned his attention to other problems particular relating to freeze-etching and freeze-fracturing. However, he has written a recent review entitled "Present Status of Freezing Techniques" (BULLIVANT, 1970). This contains an excellent bibliography into 1969.

III. The Experiments of REBHUN and Associates

REBHUN's first extended attempt to apply techniques of freeze-substitution was reported in 1961 (REBHUN, 1961). He was interested in marine eggs, and for the quenching process, decided that Freon 12 at —150 °C was superior to using Freons 13 or 14, even though the latter can be subjected to lower temperatures. Like BULLIVANT as described above, he spread his eggs first on aluminum foil. His substituting fluid was specially dried acetone to which he sometimes added 1% osmium tetroxide. After the substitution was deemed complete, the containers were brought to room temperature before infiltrating and embedding was started. It was evident that occasionally he achieved good cytoplasmic preservation, however his own comments on the frequency of this are quite illuminating, "The frequency of successful embeddings which we have obtained in the total project is low. With Freon 12 it varies with the experiment but never exceeds more than 1 in 20 blocks of embedded material. Within the successful blocks, at most 1 or 2 or 3 eggs is devoid of obvious ice crystals." Perhaps it was this low yield that prompted REBHUN subsequently to devote considerable effort to studying the various parameters that might be involved in successful freeze-substitution. REBHUN (1965) concluded in a general way that he could "reproduce, in the main, almost all types of structures seen in published papers of other authors using freezing techniques by controlling intracellular water prior to freezing." Rat pancreas and liver were used along with marine eggs for this work.

To speak of REBHUN's (1965) biological experiments and conclusions first, water content was controlled in a variety of ways. In one series of experiments tissue was dipped into full strength glycerol for periods ranging from 15—60 sec before freezing. REBHUN was quite aware that this did not yield equilibrated samples, and thought of it simply as a way of removing some water. Another series of experiments involved soaking the tissue in 5—20% glycerol, ethylene glycol or dimethyl sulfoxide for

5—15 min. Marine eggs (*Spisula*) were more exactly osmotically dehydrated to known degrees before substitution. After quenching, the specimens were substituted with either acetone or ethanol-acetone at —70 °C to —85 °C. The results were most informative. In no case was fully hydrated tissue reasonably preserved. REBHUN's (1965) Fig. 1 illustrates the "best" preservation under those circumstances, and it is evident that ice crystals had destroyed the cytoplasm and nucleoplasm. Only zymogen granules and perhaps mitochondria survived. On the other hand, circumstances which removed more and more water prior to freezing yielded a better and better fine structure. The more extreme treatments, of course, evidenced substantial shrinkage artifact in the form of scalloped nuclei and angular patterns of endoplasmic reticulum and mitochondria. Thus, he established that the quality of morphological preservation clearly can be related to the amount of (free) water in the system. The exact meaning of this is a question I will return to in the concluding remarks.

REBHUN (1965) also investigated in model systems the cooling rates of a variety of media that might be useful for quenching biological material. He compared a number of hydrocarbons and halogenated hydrocarbons (Freon and Genetron). Although the fastest velocities were obtained with propane and propylene, Freon 22 and Genetron 23 were almost as fast, and are so much safer to use that they must be regarded as the media of choice.

Subsequently, REBHUN and SANDER (1971) made a detailed study of the protection partial dehydration can provide against freezing damage. They chose clam (*Spisula*) eggs whose cellular water content was varied by osmotic means simply by equilibrating them in sea water variously concentrated. A single layer of eggs was flattened on a Formvar film, and then a second film added on top so that they were sandwiched. This sheet was "shot" into Freon 12 or Freon 22 at about —150 °C. The principal substituting medium consisted of a 1 : 1 mixture of acetone and ethanol. Often osmium tetroxide was added to this. Some experiments involved other substituting media, including acetone, methanol, ethanol, tetrahydrofuran, and various mixtures of these. Substitution was generally at —80 °C. These experiments dramatically showed that when eggs were frozen in less than double strength seawater, ice crystallization ordinarily occurred throughout all cells. In eggs frozen in double to triple strength sea water, 10 to 75% of the eggs were devoid of ice crystals in the cytoplasm, and in triple strength sea water, also in the nucleus. These results seem unequivocal, and provide one essential clue to what is required for successful morphological preservation by freeze-substitution.

IV. Experiments of VAN HARREVELD, CROWELL and MALHOTRA

VAN HARREVELD for many years has been interested in the physiology of the extracellular compartment of the brain. He teamed up first with

CROWELL and then with MALHOTRA in an effort to determine its dimensions by employing substituting and drying techniques to brain tissue which had been very rapidly frozen. Since he was interested primarily in the geometry involved, he had no hesitation about fixing the tissue at as low a temperature as possible. VAN HARREVELD and CROWELL (1964) developed a freezing device which had some distinctive features, but basically involved slapping a piece of tissue against a dry, polished, silver block precooled to about $-207\,°C$ with liquid nitrogen at reduced pressure. Substitution was then generally performed in acetone at $-85\,°C$ to which 2% osmium tetroxide had been added. (However, osmium tetroxide does not blacken tissue below about $-20\,°C$, and thus fixation probably only occurred during the warm-up period). It was noted, though, that after substitution at $-50\,°C$ "reasonable preparations were still obtained. No passable preparations were obtained by substitution at $-25\,°C$. With methyl- and ethyl alcohol as solvents, the results were poor when carried out at $-50\,°C$; some passable blocks were obtained at $-85\,°C$". Good preservation was limited to a layer about 10 microns deep, and the authors originally suggested that this surface layer may have been vitreously frozen in contrast to deeper portions of the material. Subsequently, these investigators applied this basic procedure to various other neural tissues (MALHOTRA and VAN HARREVELD, 1965b, 1966; VAN HARREVELD, CROWELL and MALHOTRA, 1965; VAN HARREVELD and KHATTAB, 1968), and MALHOTRA in particular has also studied pancreatic tissue with special emphasis upon mitochondria (MALHOTRA, 1968a, 1968b; MALHOTRA and VAN HARREVELD, 1965a).

VAN HARREVELD and CROWELL (1964) explored an important question. As the temperature drops, what is the solubility of water in acetone? They have found that at $-50\,°C$ a 14% solution about represents saturation, and that at $-80\,°C$ the solubility drops to 2—3%. Thus, at quite low temperatures acetone becomes a poor substituting medium, and to be effective must originally be truly anhydrous.

The considerable success of VAN HARREVELD and his group, particularly using an acetone substituting medium with presumably fully hydrated tissues, is at considerable variance with other experience. They give the impression of attributing their success largely to the extremely fast freezing that they undoubtedly achieved. I think that neither REBHUN and SANDER (1971) nor I are prepared to accept that fully, although neither of us is in a position to provide a satisfactory alternate explanation.

V. Pertinent Findings of Other Investigators

It is evident from what has been said above that a number of different solvents have been explored as possible substituting media, and there has been no general agreement as to what solvent may be best. Indeed, one must consider the desired end result, for instance, whether or not "fixation" is desired or whether one is striving to leave intact the native proteins. A basic requirement of the solvent is that it be miscible with water, and also remain fluid at the temperature at which the exchange is to occur. Besides solvents already mentioned, such substances as propylene oxide, dioxane, dimethyl formamide, dimethyl sulfoxide, and isobutyric acid are possibilities which so far seem inadequately studied in terms of electron microscopic morphology. All of these compounds have freezing points below $-40\,°C$, with several below $-70\,°C$. BARTL (1962) made an interesting, but largely unsuccessful, attempt to use glycomethacrylate (hydroxyethyl methacrylate) as a substituting medium which subsequently then could be directly polymerized at low temperature.

Aside from my own use of small glycols and glycerol as substituting media, as described above, I am only aware of one previous effort to employ these compounds. This was the work of BLANK, McCARTHY, and DE LAMATER (1951) who, however, were interested only in the gross retention of radioisotopes and bacterial morphology. They worked only at $-20\,°C$, and paid no attention to tissue preservation.

ZALOKAR (1966) developed a simple model system for estimating the speed at which ice might be dissolved in substituting media. This consisted simply of freezing a piece of Millipore filter paper soaked with an aqueous solution of Methylene Blue, and dipping it into the substituting medium. As the ice disappears, the stain leaches out and leaves the paper colorless. At dry ice temperature he found that acetone was unable to dissolve ice for several days. Alcohols were better solvents, and methyl alcohol much better than ethyl, so that the ice disappeared in a few hours. A combination of methyl alcohol and acrolein was even more effective, and for his own work he used 3 parts of absolute methanol containing 20% uranyl acetate, and one part of acrolein. He emphasizes that the acrolein must be added after chilling to avoid the rapid formation of acetal. ZALOKAR obtained quite good preservation of *Drosophila* salivary glands and *Neurospora* with this mixture as a substituting medium, but the angular profiles of mitochondria, zymogen granules, and endoplasmic reticulum of the former indicate that substantial shrinkage had occurred. REBHUN and SANDER (1971), I think correctly, believed that ZALOKAR's technique of picking up his specimens on Millipore filter paper inevitably partially dehydrated them before they were initially frozen so that the character of the substituting fluid may not have been the critical factor in his success.

D. Conclusions

I believe we now have at hand the information to interpret in a general way most of the seemingly disparate facts, although this is not the place for a very sophisticated discussion. Let us begin with REBHUN's contention that if tissue is initially sufficiently dehydrated, either by drying or by the osmotic removal of water, ice crystal artifacts can be avoided without difficulty (REBHUN, 1961, 1965).

There is yet another way that *free* water can in fact be effectively removed. This is by incorporating appropriate cryoprotective agents. Best known and understood are the polyhydroxy compounds including glycerol, glycols, and sugars. Of course, in part these substances simply displace some water in an equilibrated system. But more importantly, there is now general agreement between the cryobiologists and physical chemists that these substances protect by hydrogen bonding with water (DOEBBLER, 1966; HECHTER, 1965; KAROW and WEBB, 1965; LEHMAN, 1965; MERYMAN, 1966; ROWE, 1966). Each hydroxyl group in essence is capable of binding at least one molecule of water which is then incapable of freezing. What is significant here is the conclusion that then none, or only a minor fraction of the total tissue water of equilibrated tissue may be *readily* available as a potential source of crystalline ice. The questions we need to ask are exactly *when* and *why* can tissue water indeed yield distinctive ice crystal artifacts under some circumstances. There are, of course, forces at work that make the initiation of crystallization a special sort of an event.

These questions relate to the fact that I would insist upon, that within reasonable limits fully hydrated tissue can be deeply frozen in a variety of ways and not be seriously damaged morphologically. We have seen that this is the case whenever the small glycols or glycerol have been used as substituting media in the subsequent processing. It is also the common experience of those who employ freeze-etching or freeze-fracture techniques not to encounter great difficulties with the freezing process itself (BULLI-VANT, 1970). If they generally encountered crystallization damage comparable to what is seen, for example, in Figs. 7 and 8, those techniques would be well-nigh useless. There is also another curious observation that has been made repeatedly, that tissue can be deeply frozen, and then be rewarmed without there being serious loss of structural detail (BAKER, 1962; STOWELL et al., 1965; TRUMP et al., 1965). In general, in the past, these latter experiments have been interpreted as due to cytoplasmic reorganization after an assumed destruction. However, it is that basic assumption that I would challenge, and suggest instead that the residual morphology, if well preserved, survived the freezing process essentially intact, that there never was need for reconstitution.

The conclusion seems inescapable that damaging ice crystals need not be and will not be produced in suitably frozen tissue if inappropriate ("structure-breaking") solvents are avoided. Thus, the fine structure of quenched tissue moved into an acetone or alcohol bath is apt to be promptly destroyed by "secondary" ice crystal formation, while it is admirably preserved if instead it is moved into a glycol or glycerol environment. It is my contention, then, that the damage commonly occurs not at the time of quenching, but rather as substitution begins.

There exists a wealth of experimentation with model systems and physical instrumentation that has consistently demonstrated how difficult it is to achieve even a modest amount of "supercooling" when aqueous systems are involved. Excellent accounts of this evidence are to be found in numerous reviews and symposia including these examples (LUYET, 1966; MAZUR, 1966, 1970; MERYMAN, 1966). In the final analysis, however, all this effort ultimately is not entirely convincing, for there are two factors which cannot be assessed adequately in models. The first questions the legitimacy of extrapolation of these results to the problem of crystal nucleation within tiny, well-isolated compartments such as cells or their organelles. The second raises the more fundamental question as to how water really exists in the environment of tissue, and within cytoplasm.

It is quite apparent that there is not much truly free water to be found in most cells and tissues. In fact, most of it is more or less "bound" or, to use a more modern term, "structured". This concept is concomitant to the theory of "hydrophobic bonding", and develops the theme that highly ordered lattices of hydrogen-bonded water surely form and exist around the apolar radicals of large molecules such as proteins and polysaccharides (DOEBBLER, 1966; FRANK, 1965; KAROW and WEBB, 1965; KLOTZ, 1965; LING, 1965; SINSNOGLU and ABDULNER, 1965). The structuring clearly becomes less with distance (KAROW and WEBB, 1965), but it is not easy to decide when it has become negligible. Indeed, JACOBSON (1955) has demonstrated that some membrane proteins, at least, are able to structure water over distances of several micra. Although in a sense the lattice arrays can be compared to ice, they lack the degree of order and permanence of conventional ice, and very importantly can incorporate small molecules and ions. They are based upon approximately tetrahedral frameworks rather than having the hexagonal arrangements of what we usually recognize as being ice. The point of this, though, is that "the water immediately adsorbed on the proteins is not readily freezable because its molecular arrangement is dictated by the unique structure of the protein at that location, and its transformation into another structure (i.e., ice) is energetically unfavorable" (LING, 1965). However, the structured water would constitute a large *potential* reservoir of free water, available if its lattice were to be destroyed. The final problem, then, is to examine the circumstances that might release this flood of free water.

The small glycols and glycerol undoubtedly can fit into the lattice structure of ordered water without seriously disturbing it. This is one reason living cells can tolerate such high concentrations of these substances, and surely it is the basis of successful inert dehydration. In relation to freeze-substitution, these substances do no harm. This is in contrast to solutes which lack strong and multiple hydrogen bond acceptor capacity, and which are therefore inevitably "structure breaking" (DOEBBLER, 1966; KAROW and WEBB, 1965). Concomitantly, the latter also tend to be protein denaturants because of the interdependence between hydrophobic bonding and the hydration shells (LING, 1965). These latter solutes and solvents are the common ones, including the monohydric alcohols and acetone. It is my present belief that free water for hexagonal ice formation is created when substitution starts with such a solvent.

In summary, the present interpretation implies that freezing tissue is relatively easily accomplished without the formation of morphologically damaging ice crystals. So long as the chilled tissue is isolated in an immiscible shell of liquid nitrogen, a hydrocarbon, or a fluorocarbon, the system is apparently stable at least for hours. If low temperature substitution is initiated with a medium that is not "structure-breaking" damaging ice crystals will not develop, but if a "structure-breaking" solvent (such as a monohydric alcohol or acetone) is used instead, belated "secondary" ice crystal formation is apt to wreck the specimen. It is postulated that this happens in the latter case because of the appearance then of much free water, available for rearrangement as definitive crystals. How much free water becomes released, of course, depends in part on the original state of tissue hydration.

In all cases, whether inert dehydration at room temperature, or freeze-substitution at subzero temperatures, successful morphological preservation implies stealing a cell's water away from it, while at the same time the hydrophobic bonding is reinforced. There is no fundamental difference in inert dehydration and freeze-substitution with glycol or glycerol.

References

AFZELIUS, B.: Chemical fixatives for electron microscopy. In "The Interpretation of Ultrastructure". Ed.: R. J. C. HARRIS, pp. 1—20, New York: Academic Press 1962.

BARTL, P.: Freeze-substitution method using a water-miscible embedding medium. In "Electron Microscopy", Proc. V Internat. Congr. EM, 1962. Ed.: S. S. BREESE, Jr., P-4. New York: Academic Press 1962.

BAKER, R. F.: Freeze-thawing as a preparatory technique for electron microscopy. J. Ultrastruct. Res. 7, 173—184 (1962).

BLANK, H., McCARTHY, P. L., DeLAMATER, E. D.: A non-vacuum freezing-dehydrating technique for histology, autoradiography and microbial cytology. Stain Tech. 26, 193—197 (1951).

BULLIVANT, S.: The staining of thin sections of mouse pancreas prepared by the Fernández-Morán helium II freeze substitution method. J. biophys. biochem. Cytol. 8, 639—647 (1960).

BULLIVANT, S.: Freeze substitution and supporting techniques. Lab. Invest. 14, 440/1178—457/1195 (1965).

Bullivant, S.: Present status of freezing techniques. In "Some Biological Techniques in Electron Microscopy". Ed.: D. F. Parsons, p. 101. New York: Academic Press 1970.

Cope, G. H.: Low-temperature embedding in water-miscible methacrylates after treatment with antifreezes. J. roy. micr. Soc. **88**, 235—254 (1967).

Doebbler, G. F.: Cryoprotective compounds. Review and discussion of structure and function. Cryobiol. **3**, 2—11 (1966).

Doebbler, G. F., Rinfret, A. P.: Rapid freezing of human blood. Cryobiol. **1**, 205—211 (1965).

Feder, N., Sidman, R. L.: Methods and principles of fixation by freeze-substitution. J. biophys. biochem. Cytol. **4**, 593—602 (1958).

Fernández-Morán, H.: Electron microscopy of nervous tissue. In "Metabolism of the Nervous System". Ed.: D. Richter, pp. 1—34. Oxford: Pergamon Press 1957.

Fernández-Morán, H.: Electron microscopy of retinal rods in relation to localization of rhodopsin. Science **129**, 1284 (1959a).

Fernández-Morán, H.: Cryofixation and supplementary low-temperature preparation techniques applied to the study of tissue ultrastructure. J. appl. Physiol. **30**, 2038 (1959b).

Fernández-Morán, H.: Low temperature preparation technique for electron microscopy of biological specimens based on rapid freezing with liquid helium II. Ann. N. Y. Acad. Sci. **85**, 689—713 (1960).

Fernández-Morán, H.: Lamellar systems in myelin and photoreceptors as revealed by high-resolution electron microscopy. In "Macromolecular Complexes". Ed.: M. V. Edds, Jr., pp. 113—159. New York: Ronald Press 1961a.

Fernández-Morán, H.: The fine structure of vertebrate and invertebrate photoreceptors as revealed by low-temperature electron microscopy. In "The Structure of the Eye". Ed.: G. K. Smelser, pp. 521—556. New York: Academic Press 1961b.

Frank, H. S.: The structure of water. Fed. Proc. **24** (Suppl. 15), S-1—S-11 (1965).

Hancox, N. M.: Experiments on the fundamental effects of freeze-substitution. Exp. Cell Res. **13**, 263—275 (1957).

Hechter, D.: Role of water structure in the molecular organization of cell membranes. Fed. Proc. **24** (Suppl. 15), S-91—S-102 (1965).

Huxley, H. E.: The double array of filaments in cross-striated muscle. J. biophys. biochem. Cytol. **3**, 631—647 (1957).

Jacobson, B.: On the interpretation of dielectric constants of aqueous macromolecular solutions. Hydration of macromolecules. J. Amer. chem. Soc. **77**, 2919—2926 (1955).

Karow, A. M., Webb, W. R.: Tissue freezing. A theory for injury and survival. Cryobiol. **2**, 99—108 (1965).

Klotz, I. M.: The role of water structure in macromolecules. Fed. Proc. **24** (Suppl. 15), S-24—S-33 (1965).

Lehmann, H.: Changes in enzymes at low temperatures; long-term preservation of blood. Fed. Proc. **24** (Suppl. 15), S-66—S-69 (1965).

Ling, G. N.: Physiology and anatomy of the cell membrane: Physical state of water in the living cell. Fed. Proc. **24** (Suppl. 15), S-103—S-112 (1965).

Luyet, B. J.: Anatomy of the freezing process in physical systems. In "Cryobiology". Ed.: H. T. Meryman, pp. 115—138. New York: Academic Press 1966.

Malhotra, S. K.: Freeze-substitution and freeze-drying in electron microscopy. In "Cell Structure and its Interpretation". Ed.: S. M. McGee-Russell and K. F. A. Ross, pp. 11—21. London: Edward Arnold Pub. 1968a.

Malhotra, S. K.: Initial post-mortem changes in mitochondria and endoplasmic reticulum. In "Cell Structure and its Interpretation". Ed.: S. M. McGee-Russell and K. F. A. Ross, pp. 381—391. London: Edward Arnold Pub. 1968b.

MALHOTRA, S. K., VAN HARREVELD, A.: Some structural features of mitochondria in tissues prepared by freeze-substitution. J. Ultrastruct. Res. **12**, 473—487 (1965 a).

MALHOTRA, S. K., VAN HARREVELD, A.: Dorsal roots of the rabbit investigated by freeze-substitution. Anat. Rec. **152**, 283—292 (1965 b).

MALHOTRA, S. K., VAN HARREVELD, A.: Distribution of extracellular material in the central white matter. J. Anat. (Lond.) **100**, 99—110 (1966).

MAZUR, P.: Physical and chemical basis of injury in single-celled micro-organisms subjected to freezing and thawing. In "Cryobiology". Ed.: H. T. MERYMAN, pp. 214 to 315. New York: Academic Press 1966.

MAZUR, P.: Cryobiology: The freezing of biological systems. Science **168**, 939—949 (1970).

MERYMAN, H. T.: Review of biological freezing. In "Cryobiology". Ed.: H. T. MERYMAN, pp. 1—114. New York: Academic Press 1966.

MILAS, N. A., TREPAGNIER, J. H., NOLAN, J. T., ILIOPOLUS, M. I.: A study of the hydroxylation of olefins and the reaction of osmium tetroxide with 1,2-glycols. J. Amer. chem. Soc. **81**, 4730—4733 (1959).

PATTEN, S. F., BROWN, K. A.: Freeze-solvent substitution technique. A review with an application to fluorescence microscopy. Lab. Invest. **7**, 209—223 (1958).

PEASE, D. C.: The preservation of unfixed cytological detail by dehytration with "inert" agents. J. Ultrastruct. Res. **14**, 356—378 (1966 a).

PEASE, D. C.: Anhydrous ultrathin sectioning and staining for electron microscopy. J. Ultrastruct. Res. **14**, 379—390 (1966 b).

PEASE, D. C.: Eutectic ethylene glycol and pure propylene glycol as substituting media for the dehydration of frozen tissue. J. Ultrastruct. Res. **21**, 75—97 (1967 a).

PEASE, D. C.: The preservation of tissue fine structure during rapid freezing. J. Ultrastruct. Res. **21**, 98—124 (1967 b).

PEASE, D. C.: Unfixed tissue prepared for electron microscopy by glycolic dehydration In "Electron Microscopy 1968". Proc. 4th Europ. Reg. Confr. **2**, pp. 11—16. Ed.: D. S. BOCCIARELLI. Rome: Tipographia Poliglotta Vaticana 1968.

POLGE, C., SMITH, A. U., PARKS, A. S.: Revival of spermatozoa after vitrification and dehydration at low temperature. Nature (Lond.) **164**, 666 (1949).

REBHUN, L. I.: Applications of freeze-substitution to electron microscope studies of invertebrate oocytes. J. biophys. biochem. Cytol. **9**, 785—797 (1961).

REBHUN, L. I.: Freeze-substitution: fine structure as a function of water concentration in cells. Fed. Proc. **24** (Suppl. 15), 217—232 (1965).

REBHUN, L. I., SANDER, G.: Electron microscope studies of frozen-substituted marine eggs. I. Conditions for avoidance of intra-cellular ice crystallization. Amer. J. Anat. **130**, 1—15 (1971).

ROWE, A. W.: Biochemical aspects of cryoprotective agents in freezing and thawing. Cryobiol. **3**, 12—18 (1966).

SIMPSON, W. L.: An experimental analysis of the Altmann technique of freezing-drying. Anat. Rec. **80**, 173—189 (1941).

SINANOGLU, O., ABDULNUR, S.: The effect of water and other solvents on the structure of biopolymers. Fed. Proc. **24** (Suppl. 15), 12—23 (1965).

STOWELL, R. E., YOUNG, D. E., ARNOLD, E. A., TRUMP, B. F.: Structural, chemical, physical and functional alterations in the mammalian nucleus following different conditions of freezing, storage, and thawing. Fed. Proc. **24** (Suppl. 15), 115—141 (1965).

SZENT-GYÖRGYI, H.: "Chemistry of Muscular Contraction", 1—150. New York: Academic Press 1951.

TRUMP, B. F., YOUNG, D. E., ARNOLD, E. A., STOWELL, R. E.: The effects of freezing and thawing on the structure, chemical constitution, and function of cytoplasmic structures. Fed. Proc. **24** (Suppl. 15), 144—168 (1965).

VAN HARREVELD, H., CROWELL, J.: Electron microscopy after rapid freezing on a metal surface and substitution fixation. Anat. Rec. **149**, 381—385 (1964).

VAN HARREVELD, H., CROWELL, J., MALHOTRA, S. K.: A study of extracellular space in central nervous tissue by freeze-substitution. J. Cell Biol. **25**, 117—137 (1965).

VAN HARREVELD, H., KHATTAB, F. I.: Electron microscopy of the mouse retina prepared by freeze-substitution. Anat. Rec. **161**, 125—139 (1968).

ZALOKAR, M.: A simple freeze-substitution method for electron microscopy. J. Ultrastruct. Res. **15**, 469—479 (1966).

Freeze-Etching and Freeze-Fracturing

Stanley Bullivant

A. Introduction

The freeze-etching (or fracturing) technique involves the making of a platinum-carbon replica of the fracture face through frozen cells. This replica is examined in the electron microscope. There are two advantages compared with earlier techniques. Firstly, fixation and dehydration are not necessary and the replica is thus of cells which differ from the living state only in that they are frozen. (This statement will be qualified later.) Secondly, and possibly more importantly, the fracture proceeds along a structurally weak path within frozen membranes and thus three dimensional images of the interiors of membranes and of cells are obtained. These differ from those obtained by observation of the essentially two dimensional slice of a thin section.

The aim of this chapter is a practical one; to enable an electron microscopist who wishes to do so, to start using the technique, and to use it with understanding. Accordingly, I shall first consider the physical basis of the freezing of biological systems. A detailed account of the technique of freeze-etching, including a discussion of the different instruments which have been proposed, will follow. Finally, I shall discuss problems of interpretation, particularly those of membrane structure and contamination. I intend to do this only insofar as it is relevant to basic interpretation, and will not attempt a review of the large body of literature on specific biological applications of freeze-etching.

The most comprehensive review of the subject published so far is that by Koehler (1968).

B. Freezing of Biological Systems

As the subject of the freezing of biological systems is central to this discussion, it will be considered first. The ideal would be extremely rapid freezing with every molecule in a cell instantaneously immobilised. This is unattainable because of the low heat transfer resulting from the low thermal conductivity of biological specimens. The effects of freezing at attainable rates will now be described.

It is convenient to consider the freezing of a suspension of cells in an isotonic buffer. If cooling is sufficiently slow (1 °C/min), the extracellular solution freezes first. The water freezes out as pure ice crystals. This increases the concentration of solutes in the still-unfrozen regions and leads to osmotic removal of water from the cells. The final result of this process may be that the cells contain no unbound water, and hence no ice crystals, but they have a shrunken and distorted appearance (Rapatz and Luyet, 1961). The extracellular solution is frozen into two phases; one consisting of pure ice crystals and the other of a surrounding network of frozen eutectic containing the solutes. The effect of more rapid freezing (1000 °C/sec) is to produce small ice crystals (1 μm) in both the cells (Fig. 1a) and the extracellular solution. This cooling rate is the one normally obtained when small specimens are quenched in liquid propane or Freon (Bullivant, 1965). At higher cooling rates, such as at the outer edge of the specimen, where heat transfer is rapid; or when special techniques are used (Moor, 1971), extremely small ice crystals can be produced. These are so small that they do not interfere with the image at the electron microscope level. It is not proven that such rapid freezing leads to the vitreous state, which may be defined as the absence of ice as crystals. The only good evidence for freezing into the vitreous state is where deposition is from the vapour onto a surface below —130 °C (Lonsdale, 1958), or where water in a very finely divided state is frozen (Pryde and Jones, 1952). However, for electron microscopy the important point is to have the ice crystals below about 2 nm in size.

Good preservation of cells at conventional freezing rates (Freon) can be obtained if glycerol or other agents such as dimethyl sulphoxide or sucrose are added to the cell suspension. The action of such cryoprotectants in giving good morphological preservation seems to be that they bind water and hence make less available for freezing. This results in smaller ice crystals at a given cooling rate (Fig. 1b).

After biological systems are frozen they do not necessarily remain static. Larger ice crystals may grow at the expense of small ones at temperatures

Fig. 1. a Freeze-fracture of unfixed non-glycerinated mouse pancreas. The micrograph shows an area of endoplasmic reticulum, considerably distorted by ice crystal formation. In this unetched preparation, the ice crystals (I) appear smooth and are surrounded by a eutectic reticulum (E) containing cell components. × 70,000. b Freeze-fracture of unfixed glycerinated mouse pancreas. No disturbance due to ice crystals is seen. The fracture goes along a face of the inner nuclear membrane revealing nuclear pores (NP) and then goes stepwise across layers of endoplasmic reticulum (arrows). × 70,000. c Freeze-etch of frozen glycerol solution. The regions occupied by ice crystals (I) are deeply etched, while the surrounding glycerol eutectic (G) is smooth and unetched. ×35,000 All micrographs of replicas are printed as positives, that is the deposited platinum-carbon appears black and the shadow cast by protruding objects is white. The shadowing direction is indicated by an encircled arrow

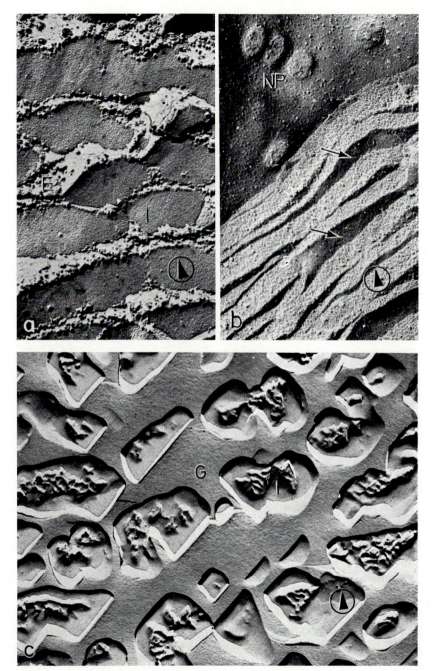

Fig. 1

as low as —130 °C (MERYMAN, 1957) and recrystallization from an amorphous phase may take place (MACKENZIE and RASMUSSEN, 1969). These facts should be borne in mind, for in the freeze-etching technique the specimen temperature is raised to —100 °C. However, with the technique at its present level of sophistication, this is of more academic than practical importance.

The freezing of biological systems has been discussed so far only to the extent that it is relevant to freeze etching. Freezing of cells and tissues to preserve them in a viable state is also of great importance. It is worth mentioning that cells frozen so as to be suitable for ultrastructural studies may not retain viability, and vice-versa. For example, cells which have been frozen slowly in the presence of glycerol will be shrunken and distorted, but may show high survival on thawing. Survival and good morphological preservation may go hand in hand for some cell types (MOOR, 1964). The reader interested in more details on all aspects of freezing of biological systems should consult the review by MAZUR (1970) and the book edited by MERYMAN (1966). The review of freezing methods used in electron microscopy by BULLIVANT (1970) also contains more detailed information on the freezing process than is given here.

C. Methods and Instrumentation

I. Historical Development

HALL (1950) made an apparatus in which the surface of a frozen aqueous suspension could be etched and replicated in a vacuum. Independently, MERYMAN (1950) carried this a step further by fracturing the frozen specimen prior to replication. Later, MERYMAN and KAFIG (1955) using a refined version of this apparatus with provision for fracturing in the vacuum, made attempts to look at freeze-fractured and etched biological material. STEERE (1957) produced freeze-etch replicas of plant virus crystals, and was the first worker to obtain good, recognisable replicas of biological material using the method. HAGGIS (1961) fractured frozen erythrocytes in the vacuum and the replicas showed cross-fractures through the cytoplasm. The pioneering work of these people provided the basis for the technique as we now use it.

The modern phase of exploitation of the technique began with the production of an advanced design of freeze-etching machine by MOOR et al. (1961). Essentially the machine embodies a cryo-ultramicrotome mounted in a vacuum evaporator. Using the cooled knife (—196 °C), thin sections are cut off the frozen specimen (—100 °C). The surface remaining is allowed to etch and then replicated with platinum-carbon. A simplified freeze-fracturing device was described by BULLIVANT and AMES (1966). This differs from the MOOR machine in that the frozen tissue is broken in a less precise fashion, and under the surface of liquid nitrogen, before being transferred to the vacuum evaporator for replication. All the differing freeze-etch and freeze-fracture machines can be considered to be variants on either the MOOR or the BULLIVANT and AMES approach.

II. Physical Basis of Technique

I am going to first of all consider the physical events which occur when cells are freeze-etched. These events are the same no matter what method of freeze-etching is employed. Details of instrumentation will be given later.

Pretreatment and freezing have already been referred to, and further technical details will be given later. For the purpose of this description we shall assume that we have cells pretreated and frozen in such a way that intracellular ice crystals are so small that they have caused little disturbance to cell ultrastructure. It is also convenient to assume that the extracellular ice crystals are larger, as indeed they often are. Fracturing, etching, replicating and cleaning of the replica will now be considered.

1. Fracturing

According to WACHTEL, GETTNER and ORNSTEIN (1966), when substances are sectioned, the cutting involves the rupturing of molecular bonds which lie in the path of the advancing knife edge. With substances of higher viscosity, a different process, brittle fracturing, occurs. The yield is not necessarily in line with, nor near the knife edge. Observations indicate that freeze-fracturing can be regarded as brittle fracturing. The fracturing produces a series of roughly hemispherical breaks in front of the microtome knife edge (MOOR, 1969). I have observed that fracturing with a razor blade has a similar effect, although the width of the breaks may be greater. Areas that are actually scratched by the knife are of poor quality, while areas within the hemispherical break are good. The fracture also shows a marked tendency to go along interior planes of membranes (BRANTON, 1966, see section D.I). This is because there is a plane of weakness and the energy required to break unit area of new membrane face is less than that required to produce unit area in ice or frozen protoplasm. Hence the fracture results from both brittle fracture and fracturing along membrane planes. A fracture, passing from the extracellular space and through a cell is shown diagrammatically in Fig. 2a. There are also excursions of the fracture plane around macromolecules and there is some plastic deformation.

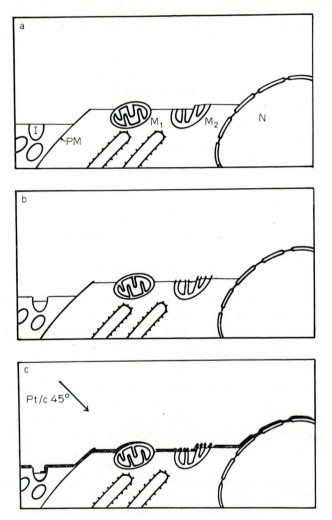

Fig. 2. Diagrammatic representation of the effects of fracturing, etching and replicating
Details, such as membrane particles and membrane splitting are not shown. a Fracturing
The material remaining after fracturing is shown. The fracture has proceeded through the
extracellular space exposing ice crystals (I), along a face of the plasma membrane (PM)
and then through the cytoplasm. One mitochondrion (M 1) is fractured along its outer
membrane and another (M 2) is cross-fractured. The fracture finally goes along a face of
the outer membrane of the nucleus (N). b Etching. In the extracellular space the large
ice crystals etch deeply. In the cytoplasm the etching of the smaller crystals given a rough
surface and leaves cross fractured membranes, such as those of the mitochondrion M 2,
standing slightly above the surface. c Shadowing. The Pt-C coat varies in thickness
according to the angle of the shadowed surface

2. Etching

In freeze-etching the specimen, after fracturing, is kept at $-100\,°C$ in the vicinity of a cold trap for a short while before replication. The vapour pressure of ice at $-100\,°C$ is 1×10^{-5} mm Hg and if the partial pressure of water vapour surrounding the specimen is very low (as would be expected in high vacuum in the vicinity of the cold trap), then a pure ice surface would sublime to give a recession rate of about 2 nm per second (KOEHLER, 1968). The eutectic, consisting of glycerol, salts, proteins etc. will not sublime and the effect of etching is to lower the surface of ice crystals. This will be readily apparent with large extracellular ice crystals (Fig. 1c). The effect on the cytoplasm, where the ice crystals are very small, will be to leave a eutectic network, thus imparting a granular appearance to the smooth fracture face. The effect of etching can be seen in Figs. 6 and 7. The cytoplasmic cross fracture face will be lowered slightly by etching, so that membrane edges and other non-etchable features stand above it. Membrane fracture faces are not affected by etching. The changes brought about by etching of extracellular fluid and a cell are shown diagrammatically in Fig. 2b.

A comment on terminology is in order at this point. The whole technique is commonly referred to as "freeze-etching", although the etching step is not essential. Since this is common useage, I will continue to use the term "freeze-etching" for the technique, except in instances where I am referring particularly to the fracturing step, or to the whole process carried through without an etching step.

3. Replicating

It is usual to shadow-cast the cold specimen surface with a mixture of platinum and carbon (BRADLEY, 1958), at an angle of 45°. Shadowing produces a film 2—5 nm thick on surfaces normal to the shadow. The thickness of platinum-carbon at other points is directly related to the slope of the fracture surface at that point, and hence the topology of the surface is converted into variations in thickness of shadow and finally into variations of density in the electron microscope negative (CHALCROFT, 1971). The eye can qualitatively interpret these in the final micrograph. The platinum-carbon shadow has a grain size of approximately 2 nm, and this is thus the limit of specimen detail which can be resolved in the replica. Following shadowing, a carbon backing is evaporated normal to the surface, to strengthen the replica. The process of replication is shown diagrammatically in Fig. 2c.

4. Cleaning

Cleaning simply involves chemical removal of the tissue and washing of the replica before picking it up and examining it in the electron microscope. The techniques for doing this will be described in Section III.5.

III. A Simple Freeze-Fracture Device

I will describe first the use of the simple freeze-fracture device introduced by Bullivant and Ames (1966). I intend to describe its use more fully because I am more familiar with it. Most of the steps to be described are applicable to any freeze-etching technique. Other devices, and their operation, will be considered later.

The device we use at present employs a central tier with tunnels to protect the specimen. It was described as a footnote to the original paper (Bullivant and Ames, 1966) and has been described in more detail subsequently (Bullivant, Weinstein and Someda, 1968; Bullivant, 1969, 1970). It is generally known as a type II device to distinguish it from the earlier version without tunnels.

1. Pre-Treatment

Small pieces of tissue (3×1 mm) are soaked for at least 20 minutes in 25% glycerol in isotonic phosphate buffer at $4\,°C$. This is the standard treatment, but as an alternative tissue is often fixed in glutaraldehyde prior to glycerination. As will be described later, this does not cause extensive changes. Fresh tissue suffers considerable change if left in the glycerol solution for a long period. More rarely, all pretreatment is omitted and tissue is frozen directly from the animal. Suspensions of cells can be treated in any of the ways described for whole tissue.

2. Freezing

The tissue is inserted in the hole in the 2.3 mm diameter cylindrical specimen holder so that it protrudes above the holder (Fig. 3, step 1). When using suspensions of cells, they are centrifuged out of the glycerol solution and the pellet is resuspended in the same solution to give a thick suspension. Some of this is put in the hole in the holder and a piece of polythene tubing inserted. The suspension rises in the tubing to a level above the top surface of the holder. When frozen, a fracture can be made through both tubing and suspension by drawing a razor blade across. Freezing is done by dropping the specimen holder into liquid Freon 12 at its melting point of $-155\,°C$ (Fig. 3, step 2). The liquid Freon is dispensed from a can into a plastic test tube equipped with a wire-loop handle. The Freon is frozen by immersion in liquid nitrogen. It is removed from the nitrogen and once it begins to melt the specimen is dropped in. Meanwhile, a pair of forceps have been pre-cooled in liquid nitrogen and these are used to pick the frozen specimen out of the Freon. The specimen is then blotted very briefly, so as to remove Freon but not cause it to warm up appreciably, and is then dropped on to the specimen block which is in liquid nitrogen (Fig. 3, step 3).

3. Fracturing

The apparatus in which fracturing, vacuum pumpdown and replication are performed will now be described. The main parts consist of three separate tiers machined out of 7.5 cm cylindrical brass stock. The lower tier has a small hole central in its upper surface, into which the specimen holder will fit snugly. This tier fits in a metal container which in turn rests on a lucite base. Screws go through the base and the container into the tier

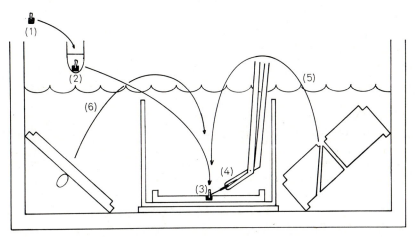

Figs. 3 and 4 show the steps in the simple freeze-fracturing technique (see section C.III). — Fig. 3. Freezing and fracturing. Step 1: specimen in holder. Step 2: freezing in liquid Freon. Step 3: insertion in cold block under liquid nitrogen. Step 4: fracturing. Steps 5 and 6: placing central tier and lid in position

and secure all three together. The central tier has two 2.3 mm diameter tunnels drilled through it; one central and coaxial, and the other at 45° to the axis. These tunnels intersect at a point in the centre of the lower face of the tier. When this tier is assembled on top of the lower tier, it locks into position on a pin and the tunnels then intersect at the top of the specimen holder hole. The central tier fits quite closely into the lower tier. The upper tier is a lid with a metal wire loop attached to its upper surface. It fits on the central tier.

In use, the three parts are cooled unassembled in liquid nitrogen in a styrofoam container (Fig. 3). After freezing in Freon, the specimen in its holder is fitted in the hole in the lower tier (Fig. 3, step 3). A razor blade is held in forceps and cooled under the nitrogen. After liquid nitrogen temperature is reached, the blade is pushed across the surface of the tier to fracture off the protruding part of the specimen (Fig. 3, step 4). The central tier and the lid are then put on (Fig. 3, steps 5 and 6). The whole apparatus is the lifted out of the styrofoam container, some of the nitrogen is poured off, and it is placed in the evaporator (Fig. 4, step 7). I use a Kinney KSE 2 evaporator, but any good quality evaporator equipped with a nitrogen trap and capable of 10^{-6} mm Hg should be satisfactory.

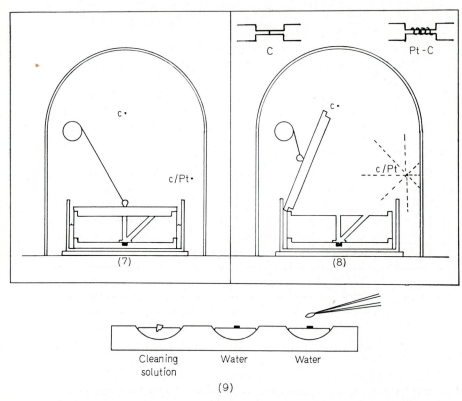

Fig. 4. Replication and cleaning of replica. Step 7: assembly in position on evaporator base-plate at beginning of pump-down. Step 8: lid lifted and platinum-carbon evaporation taking place. The inserts at the top show the carbon (C) and platinum-carbon (Pt-C) electrodes (not drawn to scale). Step 9: cleaning of replica. After the cleaning solution the replica is floated on distilled water and then picked up on a grid

4. Replication

The assembly fits on the baseplate of the electrode set (Fig. 4, step 7). Its lucite base is made so that it will fit in only one position. The electrodes can thus be pre-aligned so that the platinum-carbon one is in line with the 45° tunnel and the carbon one in line with the vertical tunnel. After the assembly is in position, the loop on the lid is hooked to a small electrically operated crane in the evaporator. The bell-jar is placed in position and rough pumping commenced with the rotary pump. During roughing the liquid nitrogen trap is filled. The nitrogen in the apparatus boils off and is pumped away and after about five minutes the pressure in the bell-jar falls to about 1×10^{-3} mm Hg. The main valve is then opened, thus bringing the diffusion pump into communication with the bell-jar. During the remaining 15 minutes of pumping, any frost previously formed on the apparatus sublimes; and the pressure usually falls to better than 1×10^{-5} mm Hg. The lid is lifted at this point and platinum-carbon shadowing done through the 45° tunnel, followed by carbon backing through the 90° tunnel (Fig. 4, step 8). Measurements show that the specimen temperature is below $-140\,°C$ at the time of replication, and hence no etching takes place. Even with the lid lifted the fractured specimen surface is protected from contamination because it is within the cold blocks, and at the bottom of the narrow tunnels.

The shadowing method is quite similar to that of MOOR (1959), which is a modification of BRADLEY's (1958) method. Both the carbon-platinum and carbon electrodes are made by machining 3.1 mm (1/8 inch) diameter spectrographic carbon rods (Union Carbide) down to 1 mm diameter for a distance of about 4 mm at one end. A dual electrode set (Ladd Research) is used. It has the advantage that the positions of all electrode holders can be adjusted so that it is easy to pre-align with respect to the tunnels. The carbon source consists of two of the machined carbon rods aligned so that they butt on each other (Fig. 4, step 8, insert). They are held in this position by light pressure of the leaf spring. The carbon-platinum source is the same, with the addition of a helix of platinum wire symmetrically over the carbon points (Fig. 4, step 8, insert). The helix is made by winding 5 cm of 0.1 mm diameter pure platinum wire on a 1 mm diameter metal former. It is removed and put on the electrodes, and should fit fairly tightly. Both the platinum-carbon and carbon sources are arranged so that they are approximately 15 cm from the position of the specimen. It is usual to pre-pump the bell-jar, prior to opening it to put the freeze-fracture apparatus in. During pre-pumping the platinum carbon electrodes are carefully heated until the platinum melts. This can conveniently be observed through a dense glass filter. Usually the platinum melts into a few globules at the junction between the carbon points. Occasionally all the melted platinum is at the far side from the future position of the specimen. If this happens, the electrodes are rotated until the platinum is on the correct side.

To carry out the platinum-carbon shadowing, the variable transformer controlling the primary is set at 70%. This is equivalent to a secondary voltage of approximately 10 V across the electrodes. The power is switched on and the transformer turned up to 80% over a period of about 2 secs.

The power is switched off as soon as a medium purple colour is seen on the polished top of the central tier. This is a good indication that the amount of shadow is correct. Carbon backing is done in a similar way, with the voltage on the electrodes set just short of the sparking point, until that part of the top of the tier not shadowed with platinum-carbon, is covered with a light brown coat. Once replication is complete, air is admitted to the bell-jar and the specimen holder is taken out and allowed to warm up.

5. Cleaning of Replica

For a suspension of cells fractured in a piece of plastic tubing, the specimen holder is immersed in the cleaning solution until the replica floats.

For a piece of tissue, the tissue with adhering replica is removed from the holder and placed on a wax sheet. Under a dissecting microscope, a thin slice including the replica is cut off with a razor blade. This slice is floated, replica side up, in the cleaning solution contained in a depression in a porcelain spotting plate (Fig. 4, step 9). It is an advantage to dip the replica end of the tissue in 1% collodion in amyl acetate before slicing off the end. This leaves a thin coat of collodion on the replica, helping to hold it together through subsequent steps.

The choice of cleaning solution depends on the tissue and should be decided by experiment. For animal tissues I routinely use a strong household hypochlorite bleach (sodium hydroxide and sodium hypochlorite) for a period of 30 minutes. Chromic acid (40%) is sometimes more effective, but its dark colour is a disadvantage as it is difficult to see the replica on it. For micro-organisms SLEYTR (1970b) has found 10% sodium hypochlorite followed by 70% sulphuric acid to be effective. KOEHLER (1966) used enzyme treatment. BRANTON and MOOR (1964) used household bleach followed by 70% sulphuric acid, and similar treatments have been used by others for plant tissues. For tissues containing a high proportion of lipid, the replica should be washed with a suitable solvent, although it is often best to leave this until the replica is on the grid. It may be helpful to take the replica through increasing concentrations of the cleaning solution to avoid it breaking up. However, if the replica is protected by a collodion coat, as described above, this is not necessary.

Once the replica is judged clean, it is transferred in a platinum loop to an adjacent depression containing distilled water (Fig. 4, step 9). This wash is repeated once. The replica should be kept floating at all times. If it sinks it is very difficult to recover it intact. The replica is picked up from the surface of the distilled water either on an uncoated grid (200 or 400 mesh), or on a formvar coated large hole grid. This is done under the dissecting microscope, by bringing the grid, held in forceps, down on to the floating replica. The replica sticks to the grid or film and the surplus water is blotted off using lens tissue. If collodion protection has been used, the grid is immersed in amyl acetate for one minute to remove it. The film on large hole grids will stand up to this treatment better if the grid is put into the microscope and the film irradiated with electrons before it is used to pick up the replica.

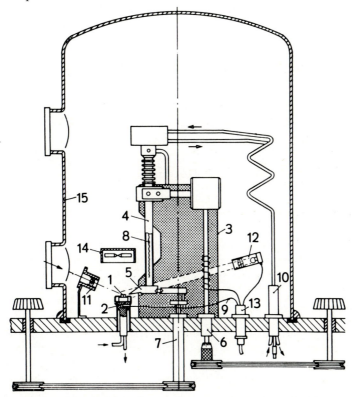

Fig. 5. Diagram of MOOR-BALZERS freeze-etch unit. 1: specimen, 2: specimen stage, 3: microtome stand, 4: microtome arm, 5: knife, 6: advance mechanism, 7: microtome drive, 8: liquid nitrogen in arm, 9: thermo-couple lead, 10: liquid nitrogen supply, 11: lens, 12: lamp, 13: electrical feed-through, 14: carbon electrodes (platinum-carbon not shown), 15: metal bell jar. The diagram is taken from MOOR (1965) and is by courtesy of Dr. H. MOOR and BALZERS AG

IV. A Microtome Freeze-etch Device

The first really successful freeze-etch device was that described by MOOR et al. (1961). It contains a liquid nitrogen cooled ultra-microtome, a specimen stage whose temperature can be controlled, and shadowing and backing electrodes; all mounted in a high vacuum system. A machine based on this design is available commercially from Balzers AG, Liechtenstein as model BA 360 M. This machine and its operation is described in a Balzers publication by Moor (1965).

The way in which its operation differs from that of the simple freeze-fracture device will now be described. A schematic diagram of the machine is shown in Fig. 5.

1. Freezing

The specimen is put on a thin copper disc and dropped into liquid Freon 12 at its melting point. After freezing, the specimen disc is clamped on to the specimen stage which has already been pre-cooled to —100 °C. The stage contains both a heater and a liquid nitrogen cooling coil and the temperature is kept constant by operation of these under the control of a temperature measuring and feedback system.

2. Fracturing

, The sharpened steel cutting blade is cooled to approximately liquid nitrogen temperature, as the microtome arm is filled with liquid nitrogen. The cutting stroke is in a horizontal plane. Provision is made for both mechanical and thermal advance, although the latter is not normally used by most operators. Cutting can be carried out with the specimen at any particular controlled temperature down to —180 °C.

3. Etching

After the cut is made, and with the specimen at —100 °C, the rear under-surface of the knife is positioned over the specimen, for a period of 30 secs to 1 min. Etching time is measured from the end of the last cutting stroke to the start of shadowing. If no etching is required, the specimen temperature is held at —120 °C, and replication follows immediately on fracturing.

4. Replication

The knife is moved out of the way to permit replication, and hence the specimen is not protected from contaminants in the atmosphere of the evaporator at this point. The carbon-platinum shadowing and carbon backing procedures are quite similar to those used with the simple freeze-fracture device, except that the carbons are sharpened to a cone rather than a thin rod (Moor, 1959). In the original apparatus there was no method of monitoring the amount of shadow. The shadow thickness is controlled by setting a stop on the electrode movement so that a certain amount of platinum-carbon evaporates. A quartz crystal thin film measurement device is now available from Balzers.

5. Cleaning of Replica

It is possible to transfer a thin piece of tissue on one of the specimen discs straight to the cleaning solution, without any intermediate steps. Microtoming the specimen gives a flatter surface than fracturing, and quite often the replica floats well with the tissue adhering to its underside. Floating is done quite simply by inserting the disc at a shallow angle through the surface of the solution.

V. Other Simple Devices

These are all similar in principle to the BULLIVANT and AMES (1966) device. The first two to be described were developed completely independently. The other two are modifications of the BULLIVANT and AMES apparatus to permit etching. The various devices will be described with the names of the workers responsible as headings.

1. GEYMEYER

GEYMEYER (1966, 1967) described an early version of his apparatus and the present version and its operation are described by GEYMEYER (1971) and SLEYTR (1970a, 1970b). The specimen is cut with a simple microtome under the surface of liquid nitrogen. It is then covered with a metal lid and transferred, still under liquid nitrogen, into the evaporator. The specimen holder is clamped on a stage in the evaporator. Only after high vacuum is obtained is the lid removed and the specimen exposed for etching. The specimen stage is heated to $-100\,^\circ$C with an electric heater and after etching is over it can be re-cooled. Platinum is evaporated at 45° and a carbon backing film is deposited while the specimen revolves and swivels. The apparatus is marketed by Leybold-Heraeus, Cologne, as model EPA 100. The freeze-etching unit fits in a modular high vacuum evaporator equipped with a cold trap within the chamber.

2. WINKELMANN

In this device (WINKELMANN and MEYER, 1968; WINKELMANN and WAMMETSBERGER, 1968) suspensions of cells are frozen between two pieces of aluminium foil. The specimen is clamped in a two-part brass block pre-cooled in liquid nitrogen. The block is carried over to the evaporator. During pump-down the block is heated electrically and its temperature measured with a thermocouple. When etching temperature ($-90\,^\circ$C to $-100\,^\circ$C) is reached, the upper part of the block is lifted, thereby tearing away the upper foil and fracturing the specimen. After the etching period, the specimen is shadowed with platinum-carbon and backed with carbon. There is no cold trap over the specimen during etching.

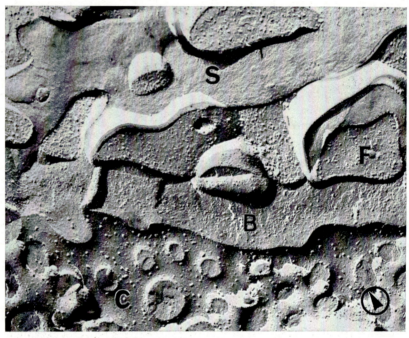

Fig. 6. Freeze-fracture replica of mouse kidney glomerulus The basement membrane (B) is revealed as a band of fine granularity, which can be compared with the smooth fracture face within the urinary space (S). Podocyte foot process (F); glomerular fenestrated capillary (C). The tissue was fixed in glutaraldehyde and glycerinated and the replica made in a type II device. ×64,000

3. WEINSTEIN

WEINSTEIN and SOMEDA (1967) modified the BULLIVANT and AMES type I device (without middle tunneled block) by putting a heater and thermocouple in the specimen block. A similar modification was made by BULLIVANT (1967). The reason behind such a modification is to produce etching. The heater is turned on and is turned off again when the temperature of the block is just below — 100 °C. The temperature of the block settles on a plateau near — 100 °C, while the lid above it is considerably cooler. Under these conditions etching takes place. The temperature is not as precisely controlled as in the MOOR freeze-etch machine. More recently

Weinstein and McNutt (1970) have modified the type II device (with middle block and tunnels) by the addition of heater and thermocouple to the specimen block, and have got good results (McNutt and Weinstein, 1970). An example of etching produced by their modification is shown in Fig. 7. Ladd Research (Burlington, Vt., U.S.A.) have designed an

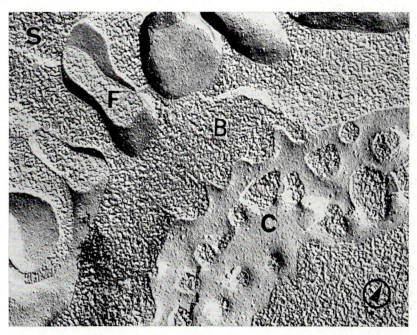

Fig. 7. Replica of heat-etched guinea pig kidney glomerulus. Various aspects of podocyte foot processes (F) can be seen. Etching masks ultrastructural features of the basement membrane (B) and the texture in this region is indistinguishable from that in the urinary space (S). Glomerular fenestrated capillary (C). The tissue was fixed in paraformaldehyde-glutaraldehyde and soaked in a balanced salt solution containing 1 M sucrose. Heat etching was carried out in a type II device according to the curve published in Weinstein and McNutt (1970). × 64,000. This micrograph is by courtesy of Dr. R. S. Weinstein

apparatus with heater and thermocouple and a fuse-operated upper lid to cover the tunnels (Ladd and Ladd, 1970). This is based on the Bullivant and Ames design. The Ladd device produces good freeze-fracture replicas but has not been extensively tested yet in connection with etching. Dempsey (1971) has used a type II device fitted with a thermocouple, but without a heater, to give etching. He finds that the specimen block warms up faster than the block with tunnels and if shadowing is done shortly after the specimen block reaches —100 °C, then good etching is produced. All modifications of the simple freeze-fracture block to give etching suffer from the same drawback, namely the lack of precise temperature control. However, good results may be obtained if some care is taken.

4. McAlear

McAlear and Kreutziger (1967) made a modification of the Bullivant and Ames type II device. The part of the block containing the tunnels can be rotated and a razor blade fastened to its lower surface fractures the specimen. The apparatus is pre-cooled in liquid nitrogen, but the fracturing is done in the vacuum. Etching is done either by heating the specimen (Kreutziger and McAlear, 1967) or by radiant heat (McAlear and Kreutziger, 1967). A difficulty with the device is that it is not within a container and is thus not covered with liquid nitrogen during transfer to the evaporator. A layer of frost is picked up during transfer and this may lead to contamination. Also, as pointed out later (section D.III), fracturing within the enclosed space between the blocks may give condensation of water vapour released by fracturing.

VI. Other Microtome Devices

The devices to be described are all similar to the Moor machine in that the cold specimen is cut within the evaporator and there are facilities for etching.

1. Koehler

Koehler (1966, 1968) developed an instrument similar to the Moor prototype (Moor et al., 1961) but accommodated in a Mikros rather than a Balzers evaporator. The specimen temperature is adjusted by admitting liquid nitrogen manually rather than by automatic valving.

2. Steere

Steere (1966, 1969) has described a freeze-etching module which can be mounted on any evaporator. The frozen specimen is clamped on a holder within the module. The specimen temperature can be controlled and at about $-100\,°C$ the specimen is fractured within the vacuum by a cooled blade manipulated from the exterior via a feed-through. During etching the specimen is surrounded by a cold trap at liquid nitrogen temperature. The module is marketed by Denton Vacuum Inc., (Cherry Hill, N.J., U.S.A.) in conjunction with their vacuum evaporator.

3. Preston

This instrument was developed by Preston and Barnett (Barnett, 1969). It is similar in most respects to the Moor-Balzers machine, but as with the KOEHLER design, is based on a different evaporator. An interesting innovation is that the microtome knife is driven backwards and forwards by two solenoids. The instrument (model FE 600) is available from NGN Ltd. (Accrington, Lancs., Great Britain). Recently it has been modified so that there are two knives on the microtome, thus enabling the final cut before etching to be done with a new sharp edge (ROBARDS, AUSTIN and PARISH, 1970).

4. EDWARDS

Edwards High Vacuum, Ltd., (Crawley, Sussex, Great Britain) are producing a freeze-etch machine quite similar to the MOOR-BALZERS one. To my knowledge this has not been described in the scientific literature, but only in a brochure (Edwards High Vacuum, 1971).

VII. Complementary Replicas

As will be evident later in the section concerned with interpretation (section D), it is advantageous to look at replicas from both the faces produced by a single freeze-fracture, and to be able to match specific complementary regions. The significance of the results of such experiments will be discussed later. At the moment only the technical means of achieving them will be described. All methods are essentially similar in that the specimen is fractured and both fracture faces are replicated. For reasons of familiarity I will describe the method used in our laboratory first. This does not imply that it is superior. Because of the nature of the problem, all the experimental solutions must be nearly the same.

1. CHALCROFT

CHALCROFT and BULLIVANT (1970) observed details of complementary fracture faces in liver. They used a simple specimen holder, which is shown, together with the mode of operation, in Fig. 8. The specimen holder (Fig. 8a) consists of two standard cylindrical holders filed lengthwise until semicylindrical halves, with the specimen holes exposed, are left. Semi-circular copper wings are added to the holders to facilitate handling and orientation. Very small strips ($0.1 \times 0.1 \times 2$ mm) of fixed, glycerinated liver are laid lengthwise in the specimen holder, which is held in its end-to-end orientation (Fig. 8a, b) with locking forceps. The holder and specimens are frozen in this orientation by plunging into melting Freon 12 at $-155\,°C$. After freezing, the holder is quickly blotted dry of Freon and dropped into the liquid nitrogen bath containing the pre-cooled blocks of the freeze-fracture device. The tissue is fractured by gripping the holder with two pairs of cooled forceps underneath the liquid nitrogen and bending and pulling (Fig. 8c). The two halves are then placed fracture face uppermost in the specimen hole in the lower brass block of the freeze-fracture device so that the dividing line between the halves is at right angles

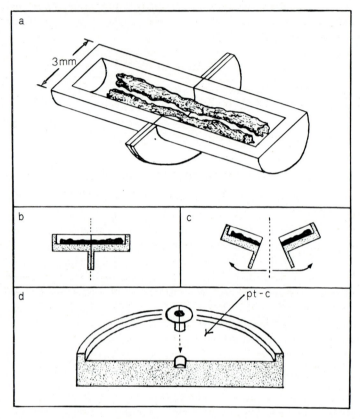

Fig. 8. Diagram illustrating method of obtaining complementary replicas. a View of specimen holder in end-to-end position with two strips of tissue in place. b Longitudinal section of holder before fracturing. c Longitudinal section of holder after fracturing. d Cutaway view of lower block showing insertion of specimen holders. The diagram is modified from one in CHALCROFT and BULLIVANT (1970) by courtesy of the Journal of Cell Biology and the Rockefeller University Press

to the direction of platinum-carbon shadowing (Fig. 8d). This orientation ensures that complementary faces are shadowed at the same angle. From this point, the previously described operating procedure (section C.III) is followed until the replicated specimens are removed from the evaporator. The specimens are allowed to thaw and matching pairs are selected. It is helpful to coat the replica surface with a film of collodion as previously described. The specimens are floated, replica side up on the cleaning solution to remove the tissue, and then on several changes of distilled water. Considerable care is needed to keep the replicas intact and they should be kept afloat throughout. Replicas are picked up on separate formvar-coated 2×1 mm single hole grids.

Carrying through the technique to this stage is quite a delicate operation. Finding areas that match can be an additional test of patience. It is best to find some easily recognisable feature on one replica, a surface-fractured nucleus for example, and then search for it on the paired replica. It helps to have one grid upside down in the microscope specimen holder. This ensures that the two halves appear as straight rather than mirror images of each other, and makes the location of matching features easier. Searching for such matching features certainly convinces one that it is necessary to keep the work within bounds by starting off with very tiny pieces of tissue.

WEINSTEIN, CLOWES and McNUTT (1970) used a device quite similar to that of CHALCROFT and BULLIVANT to make complementary replicas of a red cell suspension.

2. STEERE

STEERE and MOSELEY (1969) were the first to obtain complementary replicas. Their device consists of a hinged holder in which the specimen can be broken in the evaporator. It has been used in studies of both suspensions (DEMSEY and STEERE, 1971) and tissue (STEERE, 1971). The device is marketed by Denton Vacuum.

NANNINGA (1971b) has made complementary replicas of bacteria using a similar hinged device.

3. WEHRLI

This technique was described by WEHRLI, MÜHLETHALER and MOOR (1970). A fine tube scored to form a fracture line is filled with the cell suspension. After freezing, it is fitted in a hinged specimen holder and broken in high vacuum in the MOOR-BALZERS freeze-etch machine. After replication the replicas are reinforced by coating with evaporated naphthalene, which can be removed subsequently by sublimation at 70 °C. A later, and very useful innovation is the mounting of the specimen so that it penetrates through the holes in two gold London finder grids clamped in alignment (MÜHLETHALER, WEHRLI and MOOR, 1970; MOOR, 1971). The specimen fractures between the two finder grids. To find complementary areas all one has to do is locate corresponding grid holes. The complete complementary replica device is marketed by BALZERS.

4. Sleytr

In this method (Sleytr, 1970a, 1970b) the fracturing is done under liquid nitrogen, and the two halves, covered with the nitrogen are carried over into the Leybold-Heraeus freeze-etching unit. The two halves of the specimen holder have flats on them so that they can be aligned with the folding line of the break at right angles to the shadowing direction. This device is marketed by Leybold-Heraeus.

5. Winkelmann

In this method (Winkelmann and Wammetsberger, 1971) the suspension is frozen in a pre-scored tube, inserted in a hinged device and fractured in the Winkelmann freeze-etching machine.

VIII. Technical Variations

In this section, technical variations, either not previously described or very recent, will be considered.

1. Pretreatment

The standard, and by far the most frequently used pretreatment is soaking in glycerol in buffer. Dimethyl sulphoxide has been used as an alternative (Weinstein and Someda, 1967). The use of 1 M sucrose leads to the formation of smaller ice crystals and more uniform etching (Weinstein, personal communication, 1971).

Plant tissue is often difficult to impregnate with glycerol. Two ways to overcome the problem have been tried. Some plants can be grown in glycerol (Branton and Moor, 1964; Northcote and Lewis, 1968; Fineran, 1970). For example, Branton and Moor allowed onion roots to grow in a glycerol solution for four days. Fineran (1970) has carried out extensive experiments on methods of glycerination for plant tissues. Alternatively, glutaraldehyde pre-fixation may be used, and this makes the cells permeable to glycerol (Moor, 1966; Matile and Moor, 1968; Fineran, 1970).

2. Freezing

As emphasised by Moor (1971) no matter how good a coolant is used, that is no matter how high the heat conductivity is at the interface, there is a limit to the rate at which heat can be extracted from a specimen. This limit is imposed by the low thermal conductivity of watery biological systems. He says that a region only a few μm in thickness is "vitrified" in specimens frozen with no cryoprotectant. This is in line with the freeze-substitution experiments of van Harreveld and his colleagues (van Harreveld and Crowell, 1964; van Harreveld, Crowell and Malhotra, 1965) who obtained a 10 μm thick layer free of ice crystal damage; using the ultimate in cooling methods which entailed bringing fresh tissue into sudden contact with a piece of polished silver at liquid nitrogen temperature.

As described earlier, the routine solution to the ice crystal problem is a cryoprotectant. Cryoprotectants may have damaging effects (Moor, 1964, 1971) and also prevent deep etching to reveal membrane surfaces and other structures. A way of increasing the thickness of the useable layer in un-protected tissue is to freeze the specimen while it is under high pressure. The specimen in suspension form is put in a thin walled metal tube and held under a pressure of 2000 atmospheres (Moore and Riehle, 1968; Moor, 1971). The specimen is frozen by blowing liquid nitrogen at high speed at the outside of the tube, thereby eliminating any vapour layer which would reduce the cooling rate. This arrangement allows the use of liquid nitrogen at atmospheric pressure while the specimen is under high pressure. The rationale behind the technique is that the freezing point of pure water is lowered to $-22\,^{\circ}C$ and the rate of nucleation and ice crystal growth are reduced. Good results have been obtained with glycerol/water test specimens (Moor and Riehle, 1968) and with cells (Moor and Hoechli, 1970; Moor, 1971). The lethal effect of high pressure on cells is a disadvantage from Moor's point of view, as he has always been especially interested in freeze-etching as a method to show the ultrastructure of potentially viable cells. It may be possible to limit the lethal effect by using very short exposure times to the high pressure (Moor, 1971).

3. Replication

The standard shadowing method uses simultaneous evaporation of platinum and carbon. The platinum wire is wound onto carbon points and a current is passed (Moor, 1959). I have found this method superior to the use of platinum-carbon pellets. Using the pre-melting technique described in section C.III, about 75% of the replicas are well shadowed.

Experiments have been carried out using electron beam evaporation sources (Moor, 1971). A beam of electrons is accelerated through 2 to 2.5 KV and hits a small area on the anode source. Ta-W, Pt-C and Ir-C combination sources have proved useful. Although Ta-W produces a very nearly grainless film and a Pt-C film is more grainy, the latter contains a greater amount of structural information as judged by optical diffraction (Moor, 1971). This is because specimen detail is "decorated".

Moor (1971) gives three main reasons for using an electron beam source. They are a) to obtain finer grain size and increase resolution, b) to limit heat damage, and c) to give reproducibility. He has shown that the Pt-C film retains more information and I believe that Pt-C evaporated in the standard way is equally as good in this respect. ZINGSHEIM, ABERMANN and BACHMANN (1970) have shown that the ratio of heat falling on the specimen to amount of shadow can be reduced by electron beam evaporation employing a small higher temperature source. However, it may be that optimum quality of shadow (ratio of platinum to carbon) is obtained with a particular source temperature. If this is so, the electron beam method has no advantage. MOOR's (1971, Fig. 18) micrograph of destruction allegedly due to heat from standard shadowing could also be interpreted as showing ice crystal contamination. As already mentioned, reproducibility of 75% can be obtained with the standard Pt-C method. Unless the electron beam source can be shown to have more practical advantages than at present demonstrated, then it does not seem worth the considerable extra expense.

D. Interpretation

Two problems concerning the interpretation of freeze-etch images of membranes which have vexed workers for several years will be considered first. These are the locus of the fracture with respect to membranes, and the nature of the particles associated with membranes. The first problem appears solved, and some interesting information about the particles is currently being obtained. A good review on the interpretation of freeze-etched membranes has been published recently by BRANTON (1971). The section will conclude with some consideration of the nature and sources of contamination seen in the replicas.

I. The Membrane Fracture Face

We now know that membranes split along a unique interior plane. I will describe the notation I wish to use to denote the faces produced by membrane fracture. This can best be done with reference to Fig. 9, which shows the fracture faces of the plasma membranes of two adjoining liver cells. I will use the term "face" to refer to a fracture face within the membrane and the term "surface" to refer to either of the two natural surfaces of the membrane. Fracturing of a membrane produces two faces. One is the face of the membrane portion left frozen to the cytoplasm, and this will be called the A face. The other is the face of the membrane portion left frozen to the extracellular space, and this will be called the B face. Although this is not apparent in Fig. 9, the A face generally appears convex as if viewed from outside the cell, and the B face appears concave as if viewed from within the cytoplasm. Usually the A face is covered with more of the 8.5 nm particles than the B face, which is smoother but bears some particles.

Another notation has been used in the literature; namely (+) for A, and (−) for B (MEYER and WINKELMANN, 1969; CHALCROFT and BULLI-

Fig. 9. Freeze-fracture replica showing fractures faces of the plasma membranes of two adjoining cells from mouse liver. The fracture reveals the A face of the general plasma membrane (A) of the underlying cell. This face is covered with the characteristic 8.5 nm. particles. The A face of a gap junction (gA) is also present on this face. The fracture steps up across the extracellular space onto the B face (B) of the general plasma membrane of the upper cell. This faces shows fewer particles than the A face and has some small pits (circled). The B face of a gap junction (gB) is present on this face, and at the arrow the fracture steps down onto the A face of the other membrane within the same gap junction. The tissue was fixed in glutaraldehyde and glycerinated. × 97,000. This figure, with the labelling altered, is taken from CHALCROFT and BULLIVANT (1970) by courtesy of the Journal of Cell Biology and the Rockefeller University Press

VANT, 1970). I now prefer the A, B notation of McNUTT and WEINSTEIN (1970) as it has the provision of C and D to refer to the intracellular and extracellular membrane surfaces respectively, which can be revealed by etching. In addition the use of (+) (—) has the connotation of more or less particles and is not so appropriate because there are some situations where the A face has less particles than the B face (FLOWER, 1972; GILULA, 1972). PINTO DA SILVA and BRANTON (1970) use "convex" and "concave" to describe the faces seen in red cell membranes, but this is not readily applicable in all situations.

Originally it was thought that fracturing revealed two membrane surfaces, rather than each membrane fracturing to leave either an A or a B face to be replicated. Since this view was the predominant one in the literature for several years, it will be discussed. The idea was put forward in several forms by MOOR and MÜHLETHALER (1963); MÜHLETHALER, MOOR and SZARKOWSKI (1965) and MOOR (1969). On this interpretation face A in Fig. 9 would be the outer (extracellular) surface of the plasma membrane of the underlying cell and face B the inner (cytoplasmic) surface of the plasma membrane of the upper cell. The apposing extracellular and cytoplasmic ice surfaces were implicity assumed to have been lost along with the chips fractured off.

A different view was proposed by BRANTON (1966). He suggested that the fracture splits the membrane along an inner hydrophobic region. He had good evidence to back up his proposal, notably the non-etchability of the membrane fracture faces and the appearance of a small ridge, representing half a membrane, at the base of fractures running along membrane planes. DEAMER and BRANTON (1967), using radioactive labelling, provided evidence that artificial membranes are split by the fracture. BULLIVANT (1969, 1970) and MEYER and WINKELMANN (1969) argued on logical grounds in favour of BRANTON's idea. On BRANTON's interpretation, the A and B faces are as I have defined them. The complementary B and A faces are assumed to have been lost along with the chips fractured off.

Overwhelming evidence in favour of BRANTON's proposal has come recently from three types of experiment.

Fig. 10. These two micrographs, a and b, are complementary replicas of the fracture through mouse liver plasma membranes near the bile canaliculus. They are mirror images about the horizontal dividing line. The A face (A, Fig. 10b) of the general plasma membrane is the complement of the B face (B, Fig. 10a). The A and B faces of the gap junction (gA, gB) are complementary, as are the ridges (tA) and the furrows (tB) of the tight junction, which runs along the edge of the bile canaliculus. A group of ice crystals (X) has fallen on the face after fracturing (Fig. 10b). The tissue was fixed in glutaraldehyde and glycerinated. × 97,000. This figure, with the labelling altered, is taken from CHALCROFT and BULLIVANT (1970) by courtesy of the Journal of Cell Biology and the Rockefeller University Press

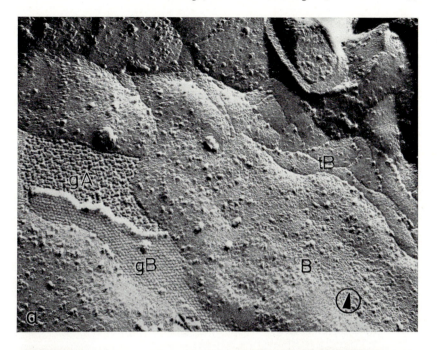

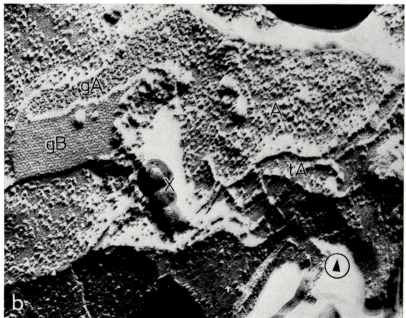

Fig. 10

1. Complementary Replicas

On MÜHLETHALER, MOOR and SZARKOWSKI's (1965) and MOOR's (1969) interpretation the fracture could be along either surface of the membrane. Complementary replica techniques have shown that there is a unique fracture plane and the two characteristic faces are complementary; hence fracturing along two different surfaces is ruled out. These experiments have been performed for a wide variety of membranes and have all given the same result, namely that membranes fracture along a unique plane. The experiments were done by STEERE and MOSELEY (1969) on nerve myelin; WEHRLI, MÜHLETHALER and MOOR (1970) on chloroplasts; CHALCROFT and BULLIVANT (1970) on liver; WEINSTEIN, CLOWES and McNUTT (1970) and WINKELMANN and WAMMETSBERGER (1971) on red cells; SLEYTR (1970a, 1970b) on yeast and bacteria; NANNINGA (1971a) on bacteria; MOOR (1971) on yeast; and STEERE (1971) on heart muscle.

I shall use a complementary replica experiment done in this laboratory as an illustration (CHALCROFT and BULLIVANT, 1970). The matching halves from a liver cell plasma membrane fracture are shown in Figs. 10a and 10b. The A face of the membrane (with many particles, Fig. 10b) is the complement of the B face (with fewer particles, Fig. 10a). Junctional regions also show complementarity.

2. Surface Labelling

By themselves the complementary replica experiments show only that there is a unique fracture plane. They do not show whether it is within the membrane or at its surface. An experiment which will decide this involves labelling the true surface with a recognisable marker and showing that the fracture face and this surface (revealed by deep etching) are at different levels and are separated by a small ridge. The experiment has been carried out on red cell membranes by PINTO DA SILVA and BRANTON (1970) using covalently bound ferritin as a marker, and TILLACK and MARCHESI (1970) using fibrous actin. PINTO DA SILVA and BRANTON's experiment had the additional refinement that both outer and inner surfaces of the red cell ghost membranes were labelled. The fracture face was below the labelled surface whichever aspect of the membrane was seen, and hence must be between the two surfaces and within the membrane. A micrograph from PINTO DA SILVA and BRANTON's paper is shown in Fig. 11. The fracture face shows the familiar 8.5 nm particles, while the outer surface is covered with ferritin marker particles. In TILLACK and MARCHESI's micrographs, the periodicity of the actin on the etch face could be clearly seen. This is good evidence that the layer interpreted as the etch face is not a layer of material deposited during the etching process, but is the true surface.

Without using a label, FLOWER (1971) has shown that the A face particles of the ciliary necklace cause protruberances on the surface revealed by deep etching.

Fig. 11. Surface labelling experiment. The micrograph shows a fractured and deep etched red cell ghost membrane. The convex fracture face (A), with many particles, is separated from the outer surface (D), revealed by etching, by a small ridge. This etch surface bears the ferritin particles which are used as a marker. ×52,000. This figure, with the labelling altered, is taken from PINTO DA SILVA and BRANTON (1970) by courtesy of Dr. P. PINTO DA SILVA, the Journal of Cell Biology and the Rockefeller University Press

3. Thin Sectioning

A third approach is to freeze-fracture tissue; thaw it, embed it and cut sections at right angles to the plane of fracture. "Half membranes" should be visible if splitting has occurred. The tissue has to be pre-fixed in glutaral-dehyde (otherwise the fractured cells break up on thawing) and then post-fracture fixed with osmium tetroxide. Early experiments (BULLIVANT and WEINSTEIN, 1969; BULLIVANT, 1969; WEINSTEIN, 1969; and LEAK, 1969) all showed "unit membranes" at the fracture face and hence it was assumed that the fracture had gone along the surface. However there may be the possibility that during thawing a lipid half bi-layer may form on top of the half bi-layer left by fracturing. To prevent this happening I evaporated a thin protective carbon layer on the fracture face immediately after fracturing. Thin sections show a "half membrane" at this fracture face (Fig. 12) (Bullivant, unpublished, 1970). Nanninga (1971b) has shown evidence of a remaining "half membrane" even when a protective coat was not applied.

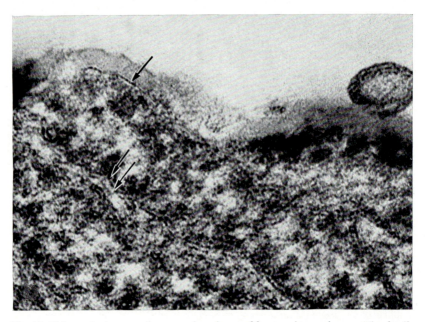

Fig. 12. Thin sectioning experiment to show locus of fracture in membranes. The details of the preparation of this specimen are given in Section D.I.3. Membranes within the tissue show the typical trilaminar appearance (double arrow). At the fracture face the membrane only retains a single dense line (single arrow). The fracture face is covered with an amorphous layer of evaporated carbon. × 280,000

The above experiments bear out Branton's (1966) original statement that frozen membranes split along an inner hydrophobic region. The reason for the split is that hydrophobic bonds are relatively unimportant compared with bonds in ice (Branton and Park, 1967). Freeze-etching has provided important information on membrane structure. There is always the possibility that freezing may alter the arrangement of membrane lipids, but Deamer et al. (1970) have shown that hexagonal and lamellar phases as demonstrated by X-ray diffraction, retain their structures after freeze-etching. This reinforces the idea that freeze-etching is a satisfactory method for investigating membrane structure.

Experimental perturbation of membrane structure by chemical fixation has also given information on the fracture process. BRANTON and PARK (1967) used the observation that acetone extraction of lipids prevented membrane splitting to support the idea that the fracture normally proceeded along the inner lipid region. WEINSTEIN and BULLIVANT (1967) showed that glutaraldehyde pre-fixation made it less likely that the fracture went along membrane planes. However, this effect has not been quantitated and for most membranes there is little difference between glutaraldehyde fixed and unfixed material. STAEHELIN, MUKHERJEE and WYNN-WILLIAMS (1969) and WEINSTEIN et al. (1970) showed that glutaraldehyde fixation caused the intramembraneous fibrils of the tight junction to remain with the A face on fracturing, whereas without fixation they were broken into segments and the parts distributed between the faces. For some specific membranes, the bonding of the membrane associated particles to the faces is affected by glutaraldehyde fixation (DEMPSEY, 1971). The above observations relate to particles or fibrils within membranes and it appears that in general the predisposition to fracture along the interior of the lipid bi-layer is not affected by glutaraldehyde fixation.

Osmium tetroxide fixation has a more pronounced effect. JAMES and BRANTON (1971) list a number of published and unpublished observations showing that OsO_4 fixed chloroplast, mitochondrial and yeast membranes no longer fracture along membrane faces but only yield cross-fractures. MEYER and WINKELMANN (1970) showed that both OsO_4 and $KMnO_4$ had this same effect on yeast plasma membranes. Conversely, NANNINGA (1968, 1969, 1971a) showed that OsO_4 fixation had no effect on the fracturing properties of *Bacillus subtilis* plasma and mesosomal membranes. This contradiction has been resolved by JAMES and BRANTON (1971) who showed that inability to fracture along membrane planes after OsO_4 fixation was correlated with the proportion of unsaturated fatty acids in the membrane lipids. In *B. subtilis* there are almost no unsaturated fatty acids, whereas in chloroplasts almost 95% are unsaturated.

II. Particles in Membranes

The membrane faces shown in Fig. 9 again serve as a good example. It is seen that there are more of the 8.5 nm particles on the A face than on the B face, which bears a few particles (and occasionally a few depressions) but is relatively smooth. This asymmetry of particle distribution between the two faces has been commented on by several investigators (WEINSTEIN and SOMEDA, 1967; BRANTON, 1969; BULLIVANT, 1969; MEYER and WINKELMANN, 1969). It is now generally accepted that the particles represent a real structure and are not due to contaminants condensed in the vacuum (see section D.III). This being so, there are two problems connected with the particles. These are, firstly the lack of pits on the B face corresponding to particles on the A face, and secondly the nature of the particles and their relation to membrane structure.

1. Lack of B Face Pits

If the fracture proceeds without distortion and there are no subsequent alterations prior to or during replication, then there should be small pits on the B face corresponding to particles on the torn-off A face. Certainly if pits were readily seen it would support the reality of the particles. There are two main explanations which can be advanced to account for the lack of pits.

a) Some process fills in pits and enhances particles. The shadowing could do this, the for first increment of shadowing material decreases the size of a pit and increases the size of a particle. Contamination could also be responsible. KREUTZIGER (1968) showed that the B face pits in the gap junction could be obscured by contamination. However, in replicas which were relatively free from contamination, and in which the gap junction pits were readily seen, it was still only possible to see a few pits in the B face of the general plasma membrane (CHALCROFT and BULLIVANT, 1970; see also Fig. 9). Others (STAEHELIN, 1970; STEERE, 1971) have claimed that they see sufficient pits on the B face to match the particles on the A face. McNUTT and WEINSTEIN (1971) have shown that in thin, well shadowed, contamination-free replicas, the gap junction B face pits are easily seen. In addition, scattered pits in yeast and red blood cell plasma membrane B faces and hexagonally arrayed pits in the yeast B face all appear as ring-shaped depressions. Particles on the A face of the plasma membrane of one yeast cell are shown in Fig. 13a. The B face ring-shaped depressions of another yeast cell are shown in Fig. 13b.

The process which fills in pits, selectively obscures those in the general membrane compared to those in the gap junction. It must be a specific process. Alternatively the gap junction pits must originally be larger. It is not entirely possible to differentiate between shadowing and contamination as the mechanism for filling up the pits. An experiment has been done which shows that in some instances pits must have originally been present in the

---

Fig. 13. The two micrographs are of the plasma membrane fracture faces of yeast, *Saccharomyces cerevisiae*. The A face (Fig. 13a) bears both scattered particles and hexagonal arrays of particles (arrows). Apart from a few particles, the B face (Fig. 13b) shows hexagonal arrays of ring shaped depressions (arrows) and also scattered ring-shaped depressions (circled). There is some doubt that the arrays are members of complementary sets, for the centre-to-centre spacing is 18 nm. for the A face particles and 11 nm. for the B face depressions. The two micrographs are not a complementary pair. Fig. 13a. 136,000. Fig. 13b. × 141,000. Both micrographs are by courtesy of Dr. N. S. McNUTT

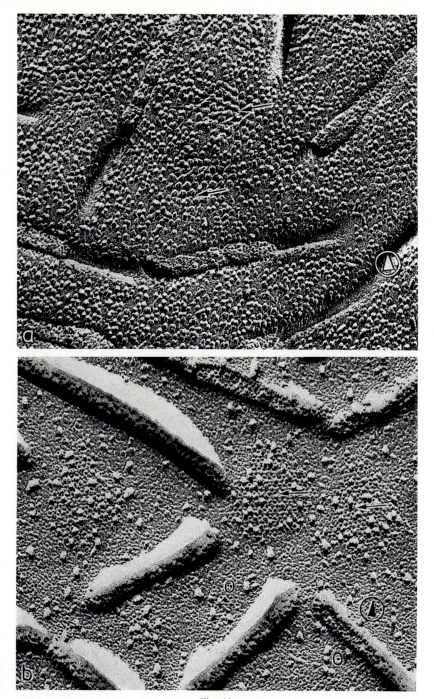

Fig. 13

B face. I carried out a two stage replica test experiment (Bullivant, unpublished 1971). A carbon film was evaporated vertically on to a freeze-fracture face, without prior shadowing. The tissue was digested off, and the replica picked up, turned over, and shadowed from what had originally been the tissue side. Particles on the replica of the B face were somewhat more numerous than on the replica of the A face. These B face particles must have resulted from pits in that face which had been filled with carbon.

b) The particles or the pits plastically deform during fracture. This explanation would not require B face pits. There is evidence that deformation of polystyrene spheres occurs during freeze-fracturing (Clark and Branton, 1968). A model involving particle deformation during membrane fracture has been advanced by Meyer and Winkelmann (1969). Deformation of myosin filaments has been used as an explanation for the appearance of oblique fractures through muscle (Bertaud, Rayns and Simpson, 1968). We have confirmed that this can happen by looking at complementary replicas of freeze-fractured muscle (Bullivant et al., 1972). An alternative may be that pits are formed, but that the tails of the exposed lipid molecules on the surrounding B face collapse prior to or during shadowing (Chalcroft, 1971).

Although there is some evidence for the deformation of other structures during fracturing, it is difficult to decide whether the membrane particles themselves are deformed. If particles were seen opposite particles in a complementary pair of replicas, this would be good evidence for deformation. This has not been seen for the scattered non-junctional membrane particles. We originally interpreted the B face gap junction structures as pits (Chalcroft and Bullivant, 1970) but in some cases the shadows could equally well represent particles (Figs. 10a, 10b). However, they are usually quite clearly pits (Fig. 9). Where they appear as particles it could be an effect of the shadowing angle (McNutt and Weinstein, 1970). The mechanism by which the two faces in gap junctions are generated is not really understood at present. It seems likely that there may be different classes of membrane particles. Some may easily deform, while others may leave pits.

2. The Nature of the Particles

Originally Bullivant and Ames (1966) suggested that the particles could be due to condensation of contaminants. The use of the type II apparatus with the tunnels reduced contamination and particles (Bullivant, 1970 and see section D.III). However, in the best replicas some particles remained; more on the A face than the B face, as already described. The best argument for the reality of the particles is the correlation of their number in a particular membrane with the functional activity of that membrane (BRANTON, 1969).

In the specific case of the arrayed particles of the gap junction, it is very likely that they are involved in cell-to-cell communication (GOODENOUGH and REVEL, 1970; CHALCROFT and BULLIVANT, 1970; McNUTT and WEIN-STEIN, 1970). GOODENOUGH and REVEL (1970) showed that the gap junction face A particles were not seen in lipid-extracted membranes. This cannot be taken as evidence that the particles are lipid, because the fracture may no longer be deviated by the membrane to reveal the particles (McNUTT and WEINSTEIN, 1970). Further experiments by GOODENOUGH and REVEL (1971) which show that isolated junctions are sensitive to phospholipase also may not indicate that the particles are lipid. By analogy with the scattered particles (see next paragraph) it seems more likely that they are protein.

BRANTON (1971) reports experiments by ENGSTROM (1970) who digested red cell ghosts with pronase and looked at the membranes by freeze-etching. Removal of up to 45% of the original membrane protein caused little decrease in particle numbers, but did result in extensive aggregation of the particles. Removal of 70% of the protein resulted in almost complete loss of particles, but did not destroy the ability of the membrane to fracture along a smooth plane. This can be taken to indicate that the particles are protein, but BRANTON (1971) is at pains to emphasise the preliminary nature of the experiments. TILLACK, SCOTT and MARCHESI (1970a) did similar experiments and noted that 70—80% of the membrane protein could be extracted without significantly reducing the number of particles. They also showed (TILLACK, SCOTT and MARCHESI, 1970b) that treatment with an agent that appears to dissociate lipid-protein complexes, removed the particles. They suggest that the particles may be formed by a complex between hydrophobic portions of glycoproteins and membrane lipids. TILLACK, CARTER and RAZIN (1970) compared native and reformed *Mycoplasma* membranes. The native membranes showed particles on freeze-etching, while the reformed ones were smooth, although they contained all the lipid and more than 85% of the protein of the native membranes. This would suggest that it is not the mere presence of protein in membranes that leads to particles on freeze-etching, but the way in which the protein is complexed with lipid. This all fits in well with BRANTON's (1966) original idea that fracturing occurs along regions of hydrophobic bonding.

PINTO DA SILVA, DOUGLAS and BRANTON (1971) labelled A antigen sites on red cell ghosts with ferritin conjugated antibody. They showed that regions bearing particles on the fracture face were confluent with regions on the etch surface labelled with ferritin. They concluded that the particles contain glycoprotein or glycolipid and that they must protrude to the outer surface where they carry the heterosaccharide ABO determinants. They speculated that as the particles apparently traverse the membrane, they must have other functions, possibly including transport.

Although it is far from complete, the evidence does point to biological membranes consisting of lipid bi-layers containing scattered or arrayed protein particles.

III. Contamination

An understanding of the nature of contamination, the mechanism of its production, and its elimination are essential for producing good freeze-etch replicas and evaluating them.

Contamination can be divided into two categories. The first consists of material other than condensate from the vacuum. The second consists of water and hydrocarbons condensed on the cold fracture face in the vacuum.

In the first category are ice crystals which fell on the surface under the liquid nitrogen, undigested specimen sticking to the replica, and crystals and dirt from the cleaning solution. Ice crystals are easily recognised (Fig. 10b). They often have an angular shape. Undigested specimen appears dense and amorphous and tends to "blow up" under the electron beam. It may preferentially remain on certain structures, for example on the replicas of broken-off myosin filaments. Crystals and dirt are easily recognised. The contamination in this category is not really a problem, for it can usually be prevented.

The second category of contamination is more subtle. It is only recently that experiments have been done to trace its sources. There are two types of contamination within this category. The first consists of a build-up of small particles (rather similar to the membrane particles) on membrane faces and elsewhere. The second consists of flat plaques exclusively on membrane fracture faces.

1. Particulate Contamination

Bullivant (1969) noted that the use of the type II device (with the cryo-trap tunnels) reduced the number of particles on membranes compared with the situation when the original device (without such extensive shielding) was used. McNutt and Weinstein (1970) also commented on the effectiveness of the type II device in reducing contamination. Very little contamination is evident in Figs. 9, 13a, and 13b prepared with this device. BULLIVANT and FIELDS (unpublished, 1971) modified the type II device to permit the specimen to be fractured within the blocks and below the tunnels while in vacuum. Some replicas showed many particles on membrane fracture faces (Fig. 14c), and it was thought that these were due to water vapour condensation. It was presumed that water vapour was released due to the dissipation of the energy of fracturing, and this vapour, confined within the small chamber in the blocks, readily condensed on the available cold face of the specimen.

DEAMER et al. (1970) investigated the source of similar particulate contaminants in the BALZERS machine, using frozen lipid-water preparations as test specimens. They were alerted to the problem by the observation that one freeze-etch machine gave particulate contamination, whilst another did not. A thorough analysis showed that if the knife was allowed to warm up during the final cutting, then lipid faces were contaminated. The contamination was described by the authors as an "orange peel effect" and was quite similar to that shown in Fig. 14c. If the knife was not positioned directly above the sample after the final fracture, contamination of both lipid and surrounding ice was found. When the knife was allowed to warm up it was assumed that condensable vapours trapped on its lead-in coils may have been the source of contamination. With the knife not above the specimen, contamination presumably resulted from vapour in the vacuum chamber.

STAEHELIN and BERTAUD (1971) did similar experiments using frozen glycerol/water preparations as test specimens. Their experiments showed that the condensation patterns were highly substrate specific and temperature dependent. Condensation could occur on the ice or on the glycerol phases or on both (Figs. 14a, 14b). The position of the knife during etching was critical. If it was left behind the specimen, then contamination resulted. If it was left in some other forward position, where the specimen could not "see" the knife edge, there was no contamination. They concluded that small specimen chips located on the edge of the liquid nitrogen cooled microtome knife were the major source of water vapour contamination in the BALZERS freeze-etch apparatus operated under normal conditions.

It thus appears that the particulate contamination is due to water vapour condensing in a very specific fashion. At present, this is just an observation and the basis of this specificity has not been investigated. Eventually it may be possible to use specific contamination as a label for particular chemically defined regions on the fracture face.

To avoid contamination in the BALZERS machine, the time between fracture and replication should be a minimum. For observation of membrane fracture faces it is advisable to omit the etching step. Apart from vapour present in the evaporator, etching releases more, which may condense on

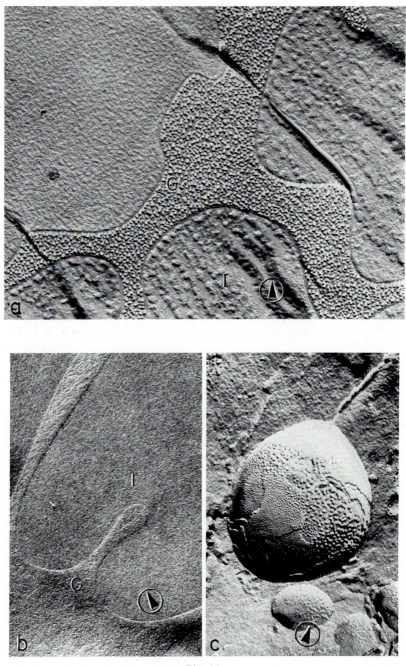

Fig. 14

susceptible parts of the specimen. We have not found water vapour contamination to be a great problem with the type II freeze-fracture apparatus, with protecting cold tunnels, used in the standard way without etching as described in section III. Some replicas do show slight contamination and we have not been able to trace the source of this.

2. Plaque Contamination

STAEHELIN (1968) showed plaques on membrane fracture faces of both artificial and natural membranes. He interpreted them as resulting from the fracture plane jumping backwards and forwards between the interior face and a surface, leaving half-membrane plateaus standing above the internal fracture face. A difficulty with this interpretation is that other faces should be seen, in the same replica, on which there are depressions complementary to the plaques (SLEYTR, 1970a). These have not been observed. No one has yet made a complementary pair of replicas where the structures corresponding to the plaques could be seen, and the question resolved. I have never seen plaques using the type II freeze-fracture device. It is becoming generally recognised that the plaques are contamination artifacts. STAEHELIN (personal communication, 1971) informs me that he has never observed them in replicas produced by the BALZERS machine he is using at the University of Colorado. He believes that they have no bearing on the structure of membranes and may be artifacts resulting either from contamination or the relaxation of strains in the lipid after the passage of the knife. STOECKENIUS (personal communication, 1971) has shown that he can only obtain plaques when the water-cooled baffle above the oil diffusion pump of the BALZERS machine is allowed to run warm. This observation, taken

---

Fig. 14. The effects of contamination are shown. a Freeze-fracture of 20% glycerol-water mixture made in a BALZERS machine with the aim of producing contamination. The specimen was fractured, exposed for 15 seconds — with the knife edge carrying specimen chips positioned approximately 3 mm away — and replicated at — 120 °C. The glycerol eutectic (G) shows many small contaminant particles. The ice (I) shows some contamination. ×35,000. The micrograph is by courtesy of Dr. L. A. STAEHELIN. For further details of the experiment, the paper by STAEHELIN and BERTAUD (1971) should be consulted. b Freeze-fracture of 20% glycerol-water mixture produced in Type II apparatus. Both glycerol (G) and ice (I) are free from contaminants. It should be emphasised that similar contamination-free replicas can be made with the BALZERS machine. The replica in Fig. 14a was produced under experimental conditions designed to produce contamination (again, see STAEHELIN and BERTAUD, 1971). ×35,000. c Small particle contamination due to ice condensation on lipid droplet in freeze-fractured liver cell. The replica was produced in a Type II apparatus modified to cut the specimen in the vacuum. For details refer to Section D.III.1. ×35,000

together with the appearance of the plaques as very low plateaus on the fracture faces, makes one assume that the plaques represent hydrocarbon condensate. Apart from their specificity for lipid fracture faces, Fineran (1970) has shown that they are present or absent on tonoplast membrane fracture faces depending on whether or not glutaraldehyde pre-fixation is employed. In the same replicas, the plasma membrane fracture faces never show plaques (Fineran, personal communication, 1971). Again, as in the case of particles, this specific contamination may prove a useful tool.

E. Conclusions

I. Choice of Equipment

As so rightly pointed out by Koehler (1968) extensive arguments as to which freeze-etching system is "best" are not productive. However, electron microscopists new to the field will certainly want help in making a decision on equipment. The choice is between the sophisticated and expensive machines which cut the specimen in the vacuum, and the simple and inexpensive ones where the cutting is done under liquid nitrogen. The former is exemplified by the Moor-Balzers machine, and the latter by the Bullivant and Ames type II device.

At the present time the Moor machine gives more reproducible etching and larger replicas than the Bullivant and Ames machine. It has the disadvantages of expense and of greater contamination (unless used very carefully). More effective shielding should reduce contamination.

The Bullivant and Ames machine has the advantages of simplicity, cheapness and low contamination. It has the disadvantages that etching is not easily controlled, and the replicas fragment more than with the Moor technique. As mentioned in this chapter, both these disadvantages are some way to being overcome.

For a laboratory wanting a ready-made machine, needing to do etching to reveal cytoplasmic components, and having the money; then the Moor-Balzers type of machine is indicated. For a laboratory interested in the internal structure of membranes and having facilities for making equipment, then the Bullivant and Ames type of machine is more appropriate. There is one word of warning. It has been our experience that freeze-etching is a technique where novices do not have early success. Considerable acquired skill is needed. It is hoped that the outline of the technique and the discussion in this chapter will help people who are new to the field, but such people would also be well advised to spend some time in a laboratory where freeze-etching is well established before starting out on their own.

II. Future

Freeze-etching has been developed to the point where it can now be considered to be a routine technique. Many of the problems of interpretation have been resolved. The technique is particularly applicable to membrane structure investigations and should show great potential when such studies are correlated with results obtained by other physical and chemical methods as discussed by BRANTON (1971).

Technical advances can be expected in rapid freezing methods, increased resolution shadowing, and in the understanding of the nature and sources of contamination and its cure.

Acknowledgements

During the preparation of this Chapter I have had discussions either personally or by letter with many people active in freeze-etching. Although they are too numerous to list here, I would like to thank everyone concerned. Without their ideas, information, private communications, preprints and micrographs I would not have been able to write a chapter that stood any chance of being up-to-date. In my own laboratory, Dr. J. P. CHALCROFT and Mr. G. P. DEMPSEY have gladly given their ideas and results. I would like to thank Dr. R. E. F. MATTHEWS for reading and commenting on the manuscript. Mr. J. J. FIELDS and Mr. G. F. GRAYSTON provided excellent technical and photographic assistance.

References

BARNETT, J. R.: Physical studies of cellulose biosynthesis. Ph. D. Thesis. University of Leeds (1969).

BERTAUD, W. S., RAYNS, D. G., SIMPSON, F. O.: Myofilaments in frozen-etched muscle. Nature (Lond.) **220**, 381—382 (1968).

BRADLEY, D. E.: Simultaneous evaporation of platinum and carbon for possible use in high-resolution shadow-casting for the electron microscope. Nature (Lond.) **181**, 875—877 (1958).

BRANTON, D.: Fracture faces of frozen membranes. Proc. nat. Acad. Sci. (Wash.) **55**, 1048—1056 (1966).

BRANTON, D.: Membrane structure. Ann Rev. Plant Physiol. **20**, 209—238 (1969).

BRANTON, D.: Freeze-etching studies of membrane structure. Phil. Trans. B **261**, 133—138 (1971).

BRANTON, D., MOOR, H.: Fine structure in freeze-etched *Allium cepa L.* root tips. J. Ultrastruct. Res. **11**, 401—411 (1964).

BRANTON, D., PARK, R. B.: Subunits in chloroplast lamellae. J. Ultrastruct. Res. **19**, 283—303 (1967).

BULLIVANT, S.: Freeze-substitution and supporting techniques. Lab. Invest. **14**, 1178—1195 (1965).

BULLIVANT, S.: Freeze-fracturing and freeze-etching. N. Z. med. J. **66**, 387—388 (1967).

Bullivant, S.: Freeze fracturing of biological materials. Micron **1**, 46—51 (1969).

Bullivant, S.: Present status of freezing techniques. In "Some biological techniques in electron microscopy", 101—146. Ed.: D. F.Parsons. New York-London: Academic Press (1970).

Bullivant, S., Ames, A.: A simple freeze-fracture replication method for electron microscopy. J. Cell Biol. **29**, 435—447 (1966).

Bullivant, S., Rayns, D. G., Bertaud, W. S., Chalcroft, J. P., Grayston, G. F.: Freeze-fractured myosin filaments. J. Cell Biol. **55**. 520—524 (1972).

Bullivant, S., Weinstein, R. S.: A thin section study of the path of fracture planes along frozen membranes. Anat. Rec **163**, 296 (1969).

Bullivant, S., Weinstein, R. S., Someda, K.: The type II simple freeze-cleave device. J. Cell Biol. **39**, 19a (1968).

Chalcroft, J. P.: Studies on cell membrane ultrastructure, and the use of the freeze-fracturing technique in electron microscopy. Ph. D. Thesis. University of Auckland (1971).

Chalcroft, J. P., Bullivant, S.: An interpretation of liver cell membrane and junction structure based on observation of freeze-fracture replicas of both sides of the fracture. J. Cell Biol. **47**, 49—60 (1970).

Clark, A. W., Branton, D.: Fracture faces in frozen outer segments from the guinea pig retina. Z. Zellforsch. **91**, 586—603 (1968).

Deamer, D. W., Branton, D.: Fracture planes in an ice-bilayer model membrane system. Science **158**, 655—657 (1967).

Deamer, D. W., Leonard, R., Tardieu, A., Branton, D.: Lamellar and hexagonal lipid faces visualized by freeze-etching. Biochim. biophys. Acta (Amst.) **219**, 47—60 (1970).

Dempsey, G. P.: A freeze-fracture study of membranes involved in transport. M. Sc. Thesis. University of Auckland (1971).

Demsey, A. E., Steere, R. L.: Complementary replicas of bacteria. In: Proceedings Electron Microscopy Society of America, 440—441, Baton Rouge: Claitor's Publishing Division 1971.

Edwards High Vacuum: Edwards freeze-etching device — advance publication 11884· Edwards High Vacuum, Crawley, England (1971).

Engstrom, L. H.: Structure in the erythrocyte membrane. Ph. D. Thesis. University of California, Berkeley (1970).

Fineran, B. A.: The effects of various pre-treatments on the freeze-etching of root tips. J. Microscopy **92**, 85—97 (1970).

Flower, N. E.: Particles within membranes: a freeze-etch view. J. Cell Sci. **9**, 435—441 (1971).

Flower, N. E.: A new junctional structure in the epithelia of insects of the order *Dictyoptera*. J. Cell Sci. **10**, 683—691 (1972).

Geymeyer, W.: Die gleichzeitige elektronenmikroskopische Erfassung von Oberfläche und Querschnitt gefriergeschockter kolloidaler Systeme. Proc. 6th Int. Conf. Electron Micr.**I**, 577—578 (1966).

Geymeyer, W.: Die elektronenmikroskopische Untersuchung temperaturempfindlicher Kolloide. Staub-Reinhaltung d. Luft **27**, 237—240 (1967).

Geymeyer, W.: Cutting, fracturing and drying of specimens in the frozen state. Publication of Leybold-Heraeus, Cologne (1971).

Gilula, N. B.: Cell junctions of the crayfish hepatopancreas. J. Ultrastruct. Res. **38**, 215 (1972).

Goodenough, D. A., Revel, J.-P.: A fine structural analysis of intercellular junctions in the mouse liver. J. Cell Biol. **45**, 272—290 (1970).

GOODENOUGH, D. A., REVEL, J.-P.: The permeability of isolated and *in situ* mouse hepatic gap junctions studied with enzymatic tracers. J. Cell Biol. **50**, 81—91 (1971).

HAGGIS, G. H.: Electron microscope replicas from the surface of a fracture through frozen cells. J. biophys. biochem. Cytol. **9**, 841—852 (1961).

HALL, C. E.: A low temperature replica method for electron microscopy. J. appl. Physics. **21**, 61—62 (1950).

JAMES, R., BRANTON, D.: The correlation between the saturation of membrane fatty acids and the presence of membrane fracture faces after osmium fixation. Biochim. biophys. Acta (Amst.) **233**, 504—512 (1971).

KOEHLER, J. K.: Fine structure observations in frozen-etched bovine spermatozoa. J. Ultrastruct. Res. **16**, 359—375 (1966).

KOEHLER, J. K.: The technique and application of freeze-etching in ultrastructure research. Advanc. biol. med. Phys. **12**, 1—84 (1968).

KREUTZIGER, G. O.: Specimen surface contamination and the loss of structural detail in freeze-fracture and freeze-etch preparations. In: Proceedings Electron Microscopy Society of America, 138—139. Baton Rouge: Claitor's Publishing Division 1968.

KREUTZIGER, G. O., McALEAR, J. H.: Three dimensional images of cardio-vascular elements with freeze-etching. In: Proceedings Electron Microscopy Society of America, 118—119. Baton Rouge: Claitor's Publishing Division 1967.

LADD, W. A., LADD, M. W.: New equipment for the electron microscopist. Norelco Rep. **17** (2), 12—16 (1970).

LEAK, L. V.: Path of fracture planes along membrane surfaces in frozen-etched tissue. In: Proceedings Electron Microscopy Society of America, 334—335. Baton Rouge: Claitor's Publishing Division 1969.

LONSDALE, K.: The structure of ice. Proc. roy. Soc. **A 247**, 424—434 (1958).

MACKENZIE, A. P., RASMUSSEN, D. H.: Low temperature studies with reference to conditions commonly used in "freeze-etching". Biophys. J. **9**, A 193 (1969).

MATILE, P., MOOR, H.: Vacuolation: origin and development of the lysosomal apparatus in root-tip cells. Planta (Berl.) **80**, 159—175 (1968).

MAZUR, P.: Cryobiology: the freezing of biological systems. Science **168**, 939—949 (1970).

McALEAR, J. H., KREUTZIGER, G. O.: Freeze-etching with radiant energy in a simple cold block device. In: Proceedings Electron Microscopy Society of America. Baton Rouge: Claitor's Publishing Division. 116—117 (1967).

McNUTT, N. S., WEINSTEIN, R. S.: The ultrastructure of the nexus. A correlated thin-section and freeze-cleave study. J. Cell Biol. **47**, 666—688 (1970).

McNUTT, N. S., WEINSTEIN, R. S.: Useful resolution standards for freeze-cleave and etch replication techniques. In: Proceedings Electron Microscopy Society of America, 444—445. Baton Rouge: Claitor's Publishing Division 1971.

MERYMAN, H. T.: Replication of frozen liquids by vacuum evaporation. J. appl. Physics. **21**, 68 (1950).

MERYMAN, H. T.: Physical limitations of the rapid freezing methods. Proc. roy. Soc. **B 147**, 452—459 (1957).

MERYMAN, H. T.: Ed. "Cryobiology". New York-London: Academic Press 1966.

MERYMAN, H. T., KAFIG, E.: The study of frozen specimens, ice crystals and ice crystal growth by electron microscopy. Naval Med. Res. Inst. Rept. NM 000 018.01.09 **13**, 529—544 (1955).

MEYER, H. W., WINKELMANN, H.: Die Gefrierätzung und die Struktur biologischer Membranen. Protoplasma (Wien) **68**, 253—270 (1969).

MEYER, H. W., WINKELMANN, H.: Die Darstellung von Lipiden bei der Gefrierätz-präparation und ihre Beziehung zur Strukturanalyse biologischer Membranen. Exp. Path. **4**, 47—59 (1970).

Moor, H.: Platin-Kohle-Abdruck-Technik angewandt auf den Feinbau der Milchröhren. J. Ultrastruct. Res. **2**, 393—422 (1959).

Moor, H.: Die Gefrier-Fixation lebender Zellen und ihre Anwendung in der Elektronen-mikroskopie. Z. Zellforsch. **62**, 546—580 (1964).

Moor, H.: Freeze-etching. Balzers High Vacuum Rept. **2**, 1—23 (1965).

Moor, H.: Use of freeze-etching in the study of biological ultrastructure. Int. Rev. exp. Path. **5**, 179—216 (1966).

Moor, H.: Freeze-etching. Int. Rev. Cytol. **25**, 391—412 (1969).

Moor, H.: Recent progress in the freeze-etching technique. Phil. Trans. B **261**, 121—131 (1971).

Moor, H., Hoechli, M.: The influence of high-pressure freezing on living cells. Proc. 7th Int. Conf. Electron Micr. **1**, 445—446 (1970).

Moor, H., Mühlethaler, K.: Fine structure in frozen-etched yeast cells. J. Cell Biol. **17**, 609—628 (1963).

Moor, H., Mühlethaler, K., Waldner, H., Frey-Wyssling, A.: A new freezing ultra-microtome. J. biophys. biochem. Cytol. **10**, 1—13 (1961).

Moor, H., Riehle, U.: Snap-freezing under high pressure: A new fixation technique for freeze-etching. Proc. 4th Europ. Reg. Conf. Electron Micr. **2**, 33—34 (1968).

Mühlethaler, K., Moor, H., Szarkowski, J.: The ultrastructure of the chloroplast lamellae. Planta (Berl.) **67**, 305—323 (1965).

Mühlethaler, K., Wehrli, E., Moor, H.: Double fracturing methods for freeze-etching. Proc. 7th Int. Conf. Electron Micr. **1**, 449—450 (1970).

Nanninga, N.: Structural features of mesosomes (chondrioids) of *Bacillus subtilis* after freeze-etching. J. Cell Biol. **39**, 251—263 (1968).

Nanninga, N.: Preservation of the ultrastructure of *Bacillus subtilis* by chemical fixation as verified by freeze-etching. J. Cell Biol. **42**, 733—744 (1969).

Nanninga, N.: The mesosome of *Bacillus subtilis* as affected by chemical and physical fixation. J. Cell Biol. **48**, 219—224 (1971a).

Nanninga, N.: Uniqueness and location of the fracture plane in the plasma membrane of *Bacillus subtilis*. J. Cell Biol. **49**, 564—570 (1971b).

Northcote, D. H., Lewis, D. R.: Freeze-etched surfaces of membranes and organelles in the cells of pea root tips. J. Cell Sci. **3**, 199—206 (1968).

Pinto da Silva, P., Branton, D.: Membrane splitting in freeze-etching. J. Cell Biol. **45**, 598—605 (1970).

Pinto da Silva, P., Douglas, S. D., Branton, D.: Localization of A antigen sites on human erythrocyte ghosts. Nature (Lond.) **232**, 194—196 (1971).

Pryde, J. A., Jones G. O.: The properties of vitreous water. Nature (Lond.) **170**, 685—688 (1952).

Rapatz, G., Luyet, B.: Electron microscope study of erythrocytes in rapidly frozen frog's blood. Biodynamica **8**, 295—314 (1961).

Robards, A. W., Austin, W. R., Parish, G. R.: The role of the microtome assembly in the freeze-etching technique. Proc. 7th Int. Conf. Electron Micr. **1**, 447—448 (1970).

Sleytr, U.: Die Gefrierätzung korrespondierender Bruchhälften: ein neuer Weg zur Aufklärung von Membranenstrukturen. Protoplasma (Wien) **70**, 101—117 (1970a).

Sleytr, U.: Fracture faces in intact cells and protoplasts of *Bacillus stearothermophilus*. A study by conventional freeze-etching and freeze-etching of corresponding fracture moieties. Protoplasma (Wien) **71**, 295—312 (1970b).

Staehelin, L. A.: The interpretation of freeze-etched artificial and biological membranes. J. Ultrastruct. Res. **22**, 326—347 (1968).

Staehelin, L. A.: Stereo electron microscopy applied to high resolution freeze-etch replicas. In: Proceedings Electron Microscopy Society of America, 306—307. Baton Rouge: Claitor's Publishing Division 1970.

STAEHELIN, L. A., BERTAUD, W. S.: Temperature and contamination dependent freeze-etch images of frozen water and glycerol solutions. J. Ultrastruct. Res. **37**, 146—168 (1971).

STAEHELIN, L. A., MUKHERJEE, T. M., WYNN WILLIAMS, A.: Freeze-etch appearance of the tight junctions of small and large intestine of mice. Protoplasma (Wien) **67**, 165—184 (1969).

STEERE, R. L.: Electron microscopy of structural detail in frozen biological specimens. J. biophys. biochem. Cytol. **3**, 45—60 (1957).

STEERE, R. L.: Development and operation of a simplified freeze-etching unit. J. appl. Physics **37**, 3939 (1966).

STEERE, R. L.: Freeze-etching simplified. Cryobiology **5**, 306—323 (1969).

STEERE, R. L.: Three dimensional complementarity of freeze-fracture and freeze-etch specimens as revealed by stereo-pairs obtained with a goniometer stage. JEOL News **9**, 1—10 (1971).

STEERE, R. L., MOSELEY, M.: New dimensions in freeze-etching. In: Proceedings Electron Microscopy Society of America, 202—203. Baton Rouge: Claitor's Publishing Division 1969.

TILLACK, T. W., CARTER, R., RAZIN, S.: Native and reformed *Mycoplasma laidlawii* membranes compared by freeze-etching. Biochim. biophys. Acta (Amst.) **219**, 123—130 (1970).

TILLACK, T. W., MARCHESI, V. T.: Demonstration of the outer surface of freeze-etched red blood cell membranes. J. Cell Biol. **45**, 649—653 (1970).

TILLACK, T. W., SCOTT, R. E., MARCHESI, V. T.: Studies on the chemistry and function of the intramembraneous particles observed by freeze-etching of red cell membranes. J. Cell Biol. **47**, 213a (1970a).

TILLACK, T. W., SCOTT, R. E., MARCHESI, V. T.: Cell membrane ultrastructure studied by freeze-etching. Fed. Proc. **29**, 489a (1970b).

VAN HARREVELD, A., CROWELL, J.: Electron microscopy after rapid freezing on a metal surface and substitution fixation. Anat. Rec. **149**, 381—385 (1964).

VAN HARREVELD, A., CROWELL, J., MALHOTRA, S. K.: A study of extracellular space in central nervous tissue by freeze substitution. J. Cell Biol. **25**, 117—137 (1965).

WACHTEL, A. W., GETTNER, M. E., ORNSTEIN, L.: Microtomy. In: Physical techniques in biological research. Ed.: A. W. POLLISTER. New York-London: Academic Press. 2nd Ed. **3 A**, 173—250 (1966).

WEHRLI, E., MÜHLETHALER, K., MOOR, H.: Membrane structure as seen with a double replica method for freeze-fracturing. Exp. Cell Res. **59**, 336—339 (1970).

WEINSTEIN, R. S.: Electron microscopy of surfaces of red cell membranes. In: The structure and function of red cell membranes, 36—76. Eds.: G. A. JAMIESON and T. J. GREENWALT. Philadelphia: Lippincott Co. 1969.

WEINSTEIN, R. S., BULLIVANT, S.: The application of freeze-cleaving technics to studies on red blood cell fine structure. Blood **29**, 780—789 (1967).

WEINSTEIN, R. S., CLOWES, A. W., McNUTT, N. S.: Unique cleavage planes in frozen red cell membranes. Proc. Soc. exp. Biol. (N.Y.) **134**, 1195—1198 (1970).

WEINSTEIN, R. S., McNUTT, N. S.: Heat etching with a Bullivant type II simple freeze-cleave device. In: Proceedings Electron Microscopy Society of America, 106—107. Baton Rouge: Claitor's Publishing Division 1970.

WEINSTEIN, R. S., McNUTT, NIELSEN, S. I., PINN, V. W.: Intramembraneous fibrils at tight junctions. In: Proceedings Electron Microscopy Society of America. Baton Rouge: Claitor's Publishing Division. 108—109 (1970).

WEINSTEIN, R. S., SOMEDA, K.: The freeze-cleave approach to the ultrastructure of frozen tissues. Cryobiology **4**, 116—129 (1967).

Winkelmann, H., Meyer, H. W.: A routine freeze-etching technique of high effectivity by simple technical means. Part I. The principle. Exp. Path. **2**, 277—280 (1968).

Winkelmann, H., Wammetsberger, S.: Eine mit einfachen Mitteln durchführbare Routinegefrierätztechnik hoher Effektivität. Teil II: Die technische Anordnung. Exp. Path. **3**, 113—116 (1968).

Winkelmann, H., Wammetsberger, S.: Die Darstellung komplementärer Bruchflächen bei der Gefrierätzung. Lecture to working session on electron microscopy. Berlin. January 18th (1971).

Zingsheim, H. P., Abermann, R., Bachmann, L.: Shadow casting and heat damage. Proc. 7th Int. Conf. Electron Micr. **1**, 411—412 (1970).

Electron Microscope Autoradiography

Analyses of Autoradiograms[1]

Miriam M. Salpeter and Frances A. McHenry

A. Introduction

Much has been written regarding standard procedures for preparing EM autoradiograms. These aspects of the technique will therefore not be dealt with in any depth in this chapter. (A few examples of recent reviews and assessments are: Caro, 1966, 1969; Salpeter, 1966; Reimer, 1966; Jacob, 1971; Budd, 1971; Bachmann and Salpeter, 1972; Salpeter and Bachmann, 1972.) Basically, one can put a section, prepared as for electron microscopy, either onto a metal microscope grid, or onto a flat substrate (we use a collodion coated slide). The section may then be stained and carbon coated. Emulsion can be applied in liquid or in pregelled form. For the former, the most common methods are either to dip the specimen into diluted emulsion or to flood it with emulsion using a medicine dropper. In either case it is then drained vertically. Partially pregelled emulsion layers are most commonly formed in wire loops and then applied to the specimen. If a flat substrate is used, the specimen is not transferred to a metal microscope grid until after photographic processing.

A schematic representation of these standard procedures is given in Fig. 1. Our preference is for mounting sections on the flat substrate, primarily because it allows the use of an external criterion (i.e. emulsion interference color) for the assessment of the quality and thickness of the emulsion layer over each section. This criterion is valid independent of the emulsion coating procedure employed and works equally well for dipping, flooding or looping. Secondly the flat substrate allows each section to be measured individually with an incident light interferometer before emulsion coating without the necessity of additional steps. For a brief description of how such measurements are made, see Salpeter and Bachmann (1972). Knowledge of both emulsion and section thicknesses is an essential requirement for optimum reliability when making quantitative assessments.

1 Supported in part by grant from the United States Public Health Service Research Grant GM 10422 from the Division of General Medical Sciences and a Career Development Award NB-K3-3738 from the Division of Neurological Diseases and Stroke.

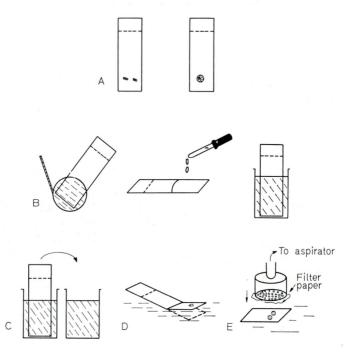

Fig. 1. Schematic of possible procedures in specimen preparation for EM autoradiography. A. Thin sections can be placed either on a flat substrate such as the collodion
coated slide illustrated here or directly onto an electron microscope grid which is attached
to a microscope slide. The sections can then be stained, measured for thickness using an
interferometer (if on a flat substrate) and vacuum coated with a thin layer of carbon.
B. Emulsion layers are applied either in pregelled form by a metal loop or liquid form by
flooding the slide from a medicine dropper or dipping the slide into the emulsion. In
all cases, the slide should be drained and dried vertically. Emulsion dilution should be
determined by using interference colors as the criterion of uniformity and thickness.
C. After exposure, slides are developed one or two at a time in small beakers. Chemicals
should be clean and changed frequently. D. and E. With the flat substrate procedure, the
collodion film supporting the sections is stripped after processing and a grid placed over
each ribbon. The specimen is then picked up by suction onto moistened filter paper. (For
greater detail on the flat substrate procedures see SALPETER and BACHMANN, 1972.) For
enhanced contrast we are now using a post-staining procedure which we have found
does not effect the developed grain distribution. This involves removal of the collodion and subsequent restaining of the sections. When thoroughly dry, the grid is
first immersed in amyl acetate for about 4 min and then transferred quickly to absolute alcohol for a few seconds. It is then floated, section side down, (for a few seconds each) on graded alcohols to distilled water. All transfers are made quickly, the
edge of the grid being touched to filter paper to drain off excess solution but not
allowing complete drying. In our experience staining directly in lead citrate tends to
disrupt the specimen. We therefore pre-stain in uranyl acetate.

On the other hand, the procedures which emulsion-coat sections already on microscope grids have simplicity in their favor, and are certainly adequate for qualitative judgments.

The problems of resolution in the EM autoradiographic technique and the factors which influence them have also been discussed in numerous publications (e.g. CARO, 1962; PELC, 1963; SALPETER, BACHMANN and SALPETER, 1969; SALPETER and SALPETER, 1971, among others) and will again not be reviewed here.

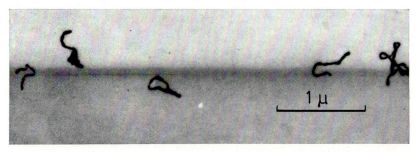

Fig. 2. Developed grains scattered around a radioactive line source used for testing resolution. The source was made by sandwiching a thin film of H³-polystyrene between plastic blocks and sectioning it at right angles. In this autoradiogram, a 500 Å section was coated with Ilford L4 and developed with Microdol X. 30,000 ×

In this chapter we want to deal primarily with an area of EM autoradiography which has not yet been extensively assessed in the literature, i.e. the analysis of autoradiograms. The discussion is not intended for the beginner but for the reader who is already well-versed in the literature and in the technical aspects of EM autoradiography.

The text will consist of three major parts. First, a consideration of what we know about the expected distribution of developed grains in relation to radioactive sources in the tissue. Second, a consideration of how one can use this information to interpret autoradiograms, including a brief discussion of some standard methods of analysis. And finally a consideration of the extent to which, and with what accuracy, information on grain distributions can be converted to specific activity in the specimen.

B. Distribution of Developed Grains Around Radioactive Sources

Let us become more specific by taking a general look at some autoradiograms (Figs. 10—14, 17, 20, 22). In EM autoradiography both the developed grains and finestructure are in focus at the same time. This results in a false sense of seeing the source of the radioactivity, especially if fine grained developers are used. However, it must be emphasized that the distribution of developed grains is related to, but is not coincident with, the distribution of radioactivity in the tissue. The extent of the discrepancy is of course a function of the resolution of the technique. This is illustrated in Fig. 2. Developed grains are seen scattered around a well-defined line source. (The source is a thin film of radioactive polystyrene laid between two smooth surfaces of plastic and sectioned at right angles.) An experimental measure of the extent of this scatter is one measure of the resolution.

To study resolution experimentally a well-defined source is needed which can be manipulated as one does a specimen for EM autoradiography. Resolution is conventionally defined in terms of radiation spread around a point source. A well-defined point source which would contain enough radioactivity for statistical analysis is difficult to make experimentally however. Caro (1962) approximated it with biological material. An alternative is to start with a defined extended source, which can be related to a point source mathematically. Such an extended source which is simple to make experimentally is the line source illustrated in Fig. 2. It can be considered as a linear collection of point sources. Such a source was used by Bachmann, Salpeter and Salpeter (1968); Salpeter, Bachmann and Salpeter (1969); and Salpeter and Salpeter (1971) to get quantitative statements and values regarding resolution in EM autoradiography.

The distribution of developed grains with increasing distance from a source can be described in different ways. Two examples are integrated distributions (total grains) (Fig. 3a) and density distributions (grains/unit area) (Fig. 3b). In an integrated distribution, all the grains are summed consecutively up to a predetermined "cut-off" distance from the source. (What constitutes a meaningful "cut-off" distance is discussed in footnote 2.) In such a distribution, the number of grains increases rapidly at first and then levels off. In a density distribution, grains are collected within discrete distance increments from the source and divided by the area included

---

2 For HD (see p. 118) to be a meaningful measure of resolution, grains must be counted far enough from the source so that increasing the distance does not significantly alter the HD. (A slight qualification of this statement for high energy isotopes is discussed by Salpeter and Salpeter, 1971.) Essentially the "cut off" distance must be several times the HD value. The rationale for such a cut off distance can be seen from the integrated distributions (Figs. 6, 7 and 19), which rise rapidly at first and then start to level off at about 3 to 4 HD. In fact, if the cut off distance is less than 2 HD, the measured HD would be approximately onehalf the cut off distance. This appears to be the problem with the resolution values of Miura and Mizuhira (1965). Similarly, Whur, Herscovics and Leblond (1969) have recently claimed that 95—98% of all grains would fall within 2250 Å of a tritium point source. As evidence they point to a histogram (Fig. 8 in Caro, 1962) in which grains are counted only up to 3000 Å. It is obvious that had this histogram presented grains only up to 2250 Å, then 2250 Å would contain 100% of all the grains, but this fact would make us no wiser regarding the resolution of the specimen.

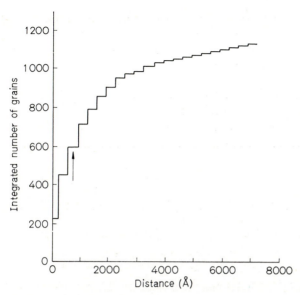

Fig. 3a. Experimental integrated distribution. Developed grains were added consecutively with distance from the line source. (An equal width and area was included for each distance unit.) Such distributions were used to measure resolution (HD) which in this case was 800 Å (arrow). 500 Å section, Kodak NTE emulsion, Dektol developer

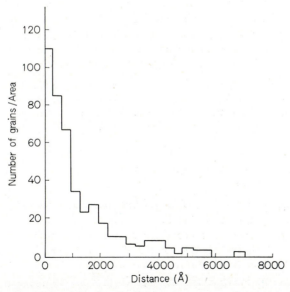

Fig. 3b. Experimental density distribution for same specimen as in Fig. 3a

within each such increment. Such a distribution falls off rapidly with distance from the source. These distributions provide much relevant information. For example, the integrated distribution shows how far from the source one has to go to collect a given fraction of total grains due to the source, while the density distribution shows at what distance the grain density falls to a given fraction of its value over the source.

Table 1. Experimental values for Half Distance (HD)[a]

Emulsion	Specimen Section (Å)	Development	HD in Å H^3	C^{14}
Kodak NTE (pale gold interference)	500 1000—1200	Dektol	800 1000	—[b] 2000
Kodak NTE (purple interference)	500 1000—1200	Dektol	1000 1250	—[b] 2500
Ilford L4 (purple interference)	500	p-phenylenediamine MX	1300 1450	—[b] 1800
	1000—1200	p-phenylenediamine MX D19 Gold-EAS	1400 1600 1450 1450	—[b] 2300 —[b] —[b]
	5000[c]	D19	3500	7000
Ilford L4 (red II interference, 2800 Å)	1000—1200	MX	—[b]	2850

[a] Data from Salpeter et al. (1969); Salpeter and Salpeter (1971); and Salpeter and Szabo (1972). Discrepancy between MX and other developers is believed to be in part due to "dose dependence". (See Salpeter and Szabo, 1972.) To eliminate a misuse of the term HD that has already crept into the literature, we want to emphasize that HD by definition, refers to the 50 % distance for a line source only. The distribution of developed grains differs for different sources and so does the distance within which 50 % of the grains lie. In each case that distance can be expressed as a multiple of HD. By limiting the term HD to refer only to the line source, it retains a unique value for each specimen, and can thus serve as a normalizing unit.
[b] Not measured.
[c] From light microscope autoradiography (Budd and Salpeter, in preparation).

From integrated distributions of the line source we obtained the distance within which 50 % of total grains fell. This was done for a variety of specimens varying in parameters that effect resolution (i.e. section and emulsion thicknesses and silver halide crystal and developed grain sizes). These experimental values (called HD) are reported in Table 1. HD is thus one measure of resolution (i.e. the distance from a *line* source within which 50% of the grains fall).

In addition to being a measure of resolution, HD has proved to be very useful in autoradiographic analyses by providing a normalized "unit of distance" applicable to all specimens independent of their resolution. To explain: Fig. 4 shows examples of density distributions for the line source in several specimens differing in resolution. The spread varies, but the shapes of all the distributions are the same. If each curve is replotted using as a measure of distance not Angstrom units, but units of its own HD (i.e. distance in Angstrom units divided by the HD of the specimen as given in Table 1), they all become coincident. Fig. 5 shows such a coincident distribution for tritium. The statement is however also true for higher energy isotopes (Salpeter and Salpeter, 1971). This "universal" curve happens to be a simple arithmetic function which can be easily restated for a point source. These functions were predicted from a simplified geometric model by Bachmann and Salpeter (1965) (see also Bachmann, Salpeter and Salpeter, 1968). For a mathematical statement, see appendix in Salpeter, Bachmann and Salpeter (1969).

Once one had a normalizing unit of distance (i.e. HD) which eliminated from the developed grain distributions the variability due to resolution, and once one had the function that fitted the actual distribution for a point source, an extrapolation to a variety of well-defined extended sources was possible. This was done by considering each extended source merely as a collection of point sources whose individual contributions were summed. Patience was the limiting factor in the number of different extended structures so treated. We decided that most labeled structures that are likely to be encountered in biological material can be approximated by either circles or bands or a combination of the two (e.g. ovals). A point is a circle of infinitely small radius and a line a band of infinitely small thickness. In addition we considered two situations: the first, where the radioactivity is concentrated all in a linear source at the edge (or circumference) of the labeled structures (called hollow circles or hollow bands), and the second, where it is distributed uniformly throughout the structure (called solid discs or solid bands). For a complete discussion of the families of curves and how they were derived, see Salpeter, Bachmann and Salpeter (1969).

In the present discussion we would like to emphasize that these curves are not in themselves a method of analysis. They merely provide information regarding developed grain distributions which can then be used in a variety of ways in analysing autoradiograms.

For instance, the families of universal curves (Fig. 6—9) illustrate the fact that the spread of developed grains is influenced by the size and shape of the labeled structure. Figs. 6 and 7 give examples of integrated distributions for two solid discs and two solid bands of radius (half-width) 1 HD and 4 HD respectively. The main conclusion from Figs. 6 and 7 is that a lower percentage of total grains due to the source will lie over the

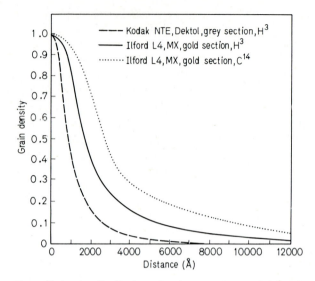

Fig. 4. Experimental density distributions for three specimens differing in resolution. Densities were normalized to be 1 over the source. Note that the shapes of the three distributions are identical although the extent of spread is different

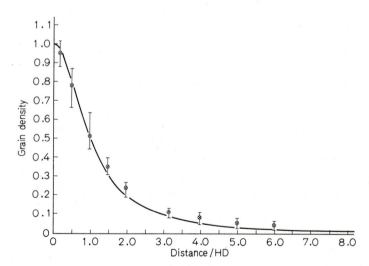

Fig. 5. "Universal" density distribution ($\cos^2 \theta$) for a line source showing the fit to averaged experimental distributions from numerous specimens differing in resolution, after each was replotted in distance units of its own HD (Table 1)

source if it is small than if it is large. Also a higher fraction of the grains due to the source will lie over band sources than over circular sources.

Figs. 8 and 9 give examples of density distributions for the two types of circular sources. We see that for a solid disc (Fig. 9) although the radio-activity by definition is uniformly distributed, the grain density is not. It rises with distance into the disc and falls with distance outside. (All the density distributions are arbitrarily normalized to be one over the circumference.) We also see that even for a hollow circle (Fig. 8), if that circle radius is smaller than 1 HD, the grain density rises with distance into the

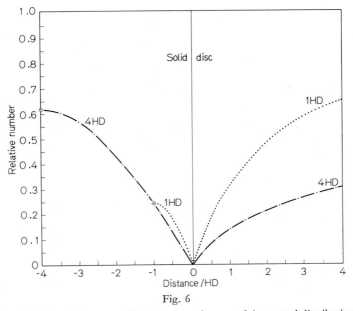

Fig. 6

Fig. 6 and 7. (Fig. 7 is on p. 122) Samples of expected integrated distributions for uniformly labeled sources differing in size and shape. The total number of grains due to a source is normalized to 1. For convenience, all curves start at the edge or circumference of the source (0 on x axis). Negative values are consecutive distances into the source and positive values are outside the source. By definition, all grains over the source plus those outside it add up to 1. The sources in Fig. 6 are two uniformly labeled circles (solid discs), one 4 HD and the other 1 HD in radius; and in Fig. 7 are two uniformly labeled bands, again one 4 HD and the other 1 HD in "half width". From these figures we see that a higher fraction of total grains lies over the larger structure than over the smaller one and similarly a larger fraction lies over the band than over the circular structure. The fraction over the source for a 4 HD disc is 61%, whereas for the 1 HD disc it is only 23%, for the 4 HD solid band 77% of the grains are over the source and for the 1 HD solid band it is 46%. By subtraction from 100% the total per cent outside the source can be calculated. One can obtain values for any unit distance inside or outside the source from the curves. For a complete set of integrated distributions, see Figs. 13—15 in Salpeter, Bachmann and Salpeter (1969)

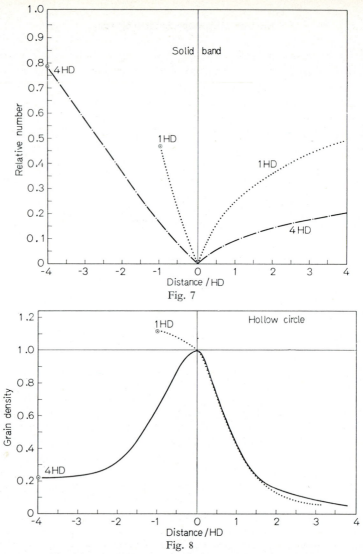

Fig. 7

Fig. 8

Figs. 8 and 9. (Fig. 9 is on p. 123) Examples of universal density distributions for circular sources, 4 or 1 HD in radius. Fig. 8 shows the distribution for hollow circles, i.e. line source at the circum-ference; and Fig. 9 the distribution for solid discs, i.e. uniformly labeled circular source. The density is normalized to be 1 at the edge (circumference). These figures again illus-trate the discrepancy between the distribution of label and that of developed grains. There is no label outside the sources, yet there is considerable grain density, especially within 1 HD. Furthermore, although the label is uniformly distributed in Fig. 9, the grain density rises with distance into the disc to a maximum in the center. (This is due to a higher cross scatter of radioactivity in the center than at the edge of a disc.) Similarly cross scatter from the labeled circumference of a small hollow circle (1 HD) causes a peak grain density over the unlabeled interior. For a complete family of such curves, see Figs. 9—12 in SALPETER, BACHMANN and SALPETER (1969)

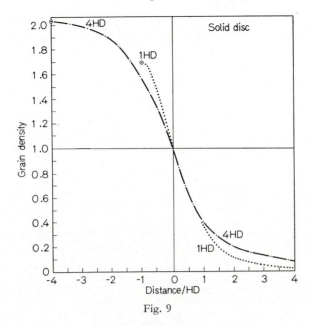

Fig. 9

structure, although by definition the radioactivity is limited to a line source at the circumference. HD is thus very directly related to resolving distance, i.e. one cannot distinguish between a hollow circle and a solid disc if the structure is less than 1 HD in radius. (For a complete description, see Fig. 16 of SALPETER et al., 1969)

Bearing in mind the general relationships between grain distributions and radioactive structures discussed above, we can proceed to the practical question of how to use this information in understanding autoradiograms.

C. Analysis of Autoradiograms

We have frequently been asked, "Does the resolution of EM auto-radiography allow one to determine whether very small vesicles (i.e. < 1 HD) are labeled?" The answer is clearly, "yes", if on the average they are far enough apart. Analysis is limited both by the dimensions of the radioactive compartments and by their distribution in the cell. Small structures approach being point sources, and, as the integrated distributions indicate, the relevant information regarding whether they are labeled or not will come primarily from developed grains which lie outside them. (The larger the structure, the more information will come from grains which lie over the structure.)

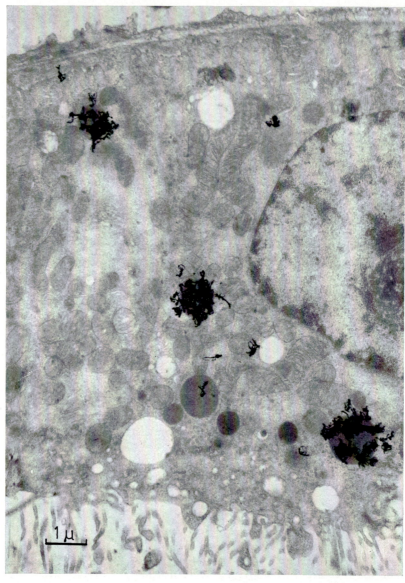

Fig. 10. Proximal tubule cells from rat kidney one hour after start of I^{125} microperfusion. Label primarily over dense cytoplasmic bodies determined to be lysosomes. Ilford L4 emulsion, D19 developer. 12,000 ×. (Fig. 13 from A. B. MAUNSBACH, 1966. By permission.)

I. Qualitative Assessment

The earliest analyses of EM autoradiograms were qualitative. Results were reported in statements such as, "It appears as if the label is primarily in ...".

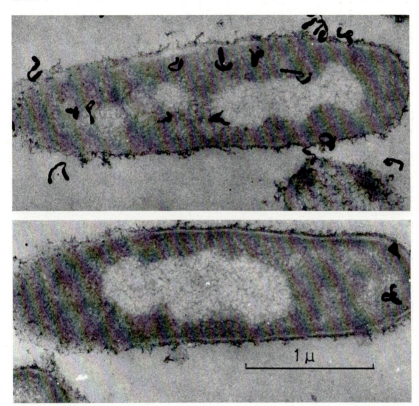

Fig. 11. Escherichia coli incubated with tritiated amino acids. A: Normal uninfected bacterium. B: Bacterium infected with R17. Incorporation of amino acids is greatly depressed although no finestructural changes are seen. Gevaert 307 emulsion, D19 developer. 40,000 ×. (Fig. 2 from GRANBOULAN and FRANKLIN, 1966. By permission)

Such assessments are not necessarily inadequate. In certain situations, this is all one needs. If the radioactive structure is large relative to the resolution (e.g. nuclei or large secretory granules), such qualitative statements contain correct meaningful information either as a starting point for further analysis or as end points in themselves (see for example, Fig. 10).

A problem which lends itself very appropriately to qualitative analysis is one which tries to relate the general finestructure of a cell to its activity at the time of study. An example of such a study is that of GRANBOULAN and FRANKLIN (1966). The results are illustrated in Fig. 11. Changes in metabolic activity and finestructure of bacteria infected with R 17 were tested. Although the finestructure of infected and non-infected bacteria was identical, the metabolic activity clearly was not. A similar approach is

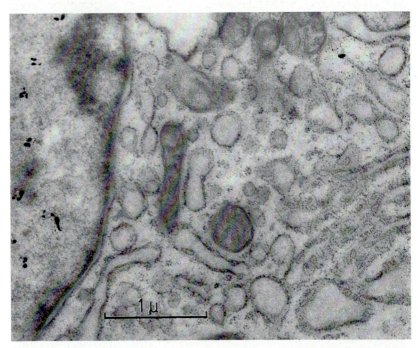

Fig. 12. Mesenchymatous cell in blastema of adult newt, Triturus, labeled with H³-thymidine. The label was used to identify the rapidly dividing cells. They were found to have a more elaborate rough ER than is usually found in undifferentiated populations. 1000 Å section, Kodak NTE emulsion, Dektol developer. 30,000 ×

illustrated in Fig. 12. In the blastema of the regenerating limb of the adult newt, most mesenchymatous cells were found to be rich in rough endoplasmic reticulum (SALPETER and SINGER, 1960 and 1962). This finding was at variance with earlier reports that undifferentiated and rapidly proliferating cells are devoid of rough ER. H³-thymidine was injected to test which cells in the adult blastema were the rapidly proliferating population. The distribution of rough ER was identical in the labeled and unlabeled cells.

The number of problems that can be answered by qualitative assessment is limited however, and qualitative analyses can easily lead to erroneous conclusions. For example, in our early attempts to localize esterases at

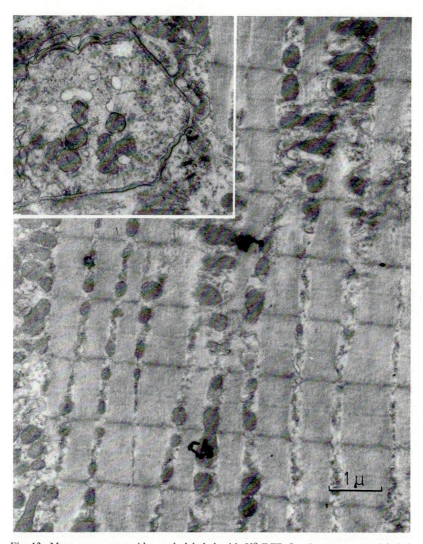

Fig. 13. Mouse sternomastoid muscle labeled with H³-DFP (by the sequence unlabeled DFP-2 PAM-H³-DFP) to reveal acetylcholinesterase sites. Judged qualitatively the number of grains over the large volume of muscle appeared not to differ significantly from the number at the endplate. Quantitative analysis showed more than 40 times higher density of label at the post junctional region, however. (See SALPETER, 1967, 1969). 1000 Å section, Ilford L4 emulsion, Microdol X developer. 15,000 ×

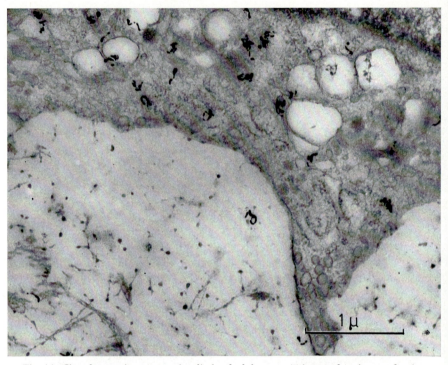

Fig. 14. Chondrocyte in regenerating limb of adult newt, Triturus, four hours after injection of H^3-proline. Developed grains are associated with vacuoles, vesicles, mitochondria and ground substance. No sensible qualitative assessment of its distribution can be made. 500 Å section, Kodak NTE emulsion, Dektol developer. 30,000 ×. (From Salpeter, 1968)

motor endplates by using H^3-DFP, most of the autoradiographic specimens had the appearance illustrated in Fig. 13. Endplates had either no grains or at most two grains near them, and numerous grains were seen distributed over the muscle. This general appearance almost led to the abandonment of the project. The relatively small volume of endplate compared to that of muscle however suggested that a quantitative assessment might lead to a different conclusion. After systematically scanning numerous grids and photographing all endplates (whether labeled or not) and then randomly photographing an equal area of muscle, it was indeed found that there was a 40 times higher grain density (grains/area) at the subsynaptic region of the endplate than in the muscle (Salpeter, 1967, 1969).

Similarly, in the study of secretory pathways, qualitative assessments have proved inadequate. See for example discussions by Ross and Benditt (1965) and Salpeter (1968), and illustration in Fig 14.

II. Quantitative Analyses

1. "Simple Grain Density" Analysis

By simple grain density we mean the number of developed grains lying over a structure, divided by the area occupied by this structure (i.e. grains per unit area). This method of analysis is both simple and frequently used.

Since simple grain density is the first quantitative method to be discussed, many of our comments will be applicable to the other methods as well. In all quantitation one has first to locate developed grains with respect to certain cellular compartments. This involves a decision regarding the location of the grain and the extent or size of the compartment tabulated. Most developers used in EM autoradiography give developed grains of considerable size and complexity. With Ilford L 4 emulsion, these can be as large as about 3000 Å in diameter, considerably above the resolution (HD) of the technique. The fact that the resolution can be better than the diameter of the developed grain appears contradictory and many questions are raised regarding it. It should be remembered however that the size of the developed grain is only one of several factors that effect resolution, and that the radius of the grain, not its diameter, is the important dimension. (For a precise statement of the relative importance of the various factors in resolution, see BACHMANN and SALPETER, 1965.) Aside from its effect on resolution, the extended size of most developed grains also presents a dilemma in localization. Some people have used the convention that if any part of a developed grain lies over a particular structure, the grain should be ascribed to it. This is not satisfactory since not all the grains are the same size. (If an extended localization is desired that goal is better served by the "probability circle" method to be discussed below.) It is simpler and statistically valid to designate the location of a developed grain by some well-defined point. We use the center of the smallest circle which can circumscribe the entire grain[3]. (Since grains were located in this manner in determining HD and the universal curves, analyses using this localization can be compared with the families of expected distributions.)

Once the number of grains located over a given organelle (cellular compartment) is tabulated, the simple density method requires these data

---

3 One procedure for locating this midpoint consists of placing over the developed grain a clear plastic mask with a series of concentric circles of increasing radii drawn around a central small hole. Once the smallest circle that can circumscribe the grain is fitted around it, the center can be punched through the autoradiograph with a teasing needle. The punched center then becomes the location of the grain for all further considerations. The size and shape of the developed grain no longer matter. Alternatively one can use the midpoint of a line drawn between the two ends of a grain. Since the two ends of a grain are not always clearly seen in the general tangle of silver, we prefer the first method.

to be divided by the area of the organelle. The need for area correction in analysis was first stressed by Ross and Benditt (1965). A simple way to obtain the area of a compartment is to place a grid over the autoradiographs so that the grid points are random relative to the finestructure of the cells

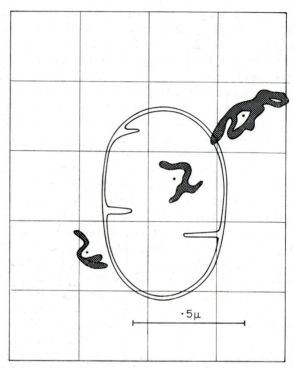

Fig. 15. Schematic of mitochondrion in an autoradiogram at 60,000 ×. A grid (calibrated to 1 intersection point/. 1 μ^2) is placed over the autoradiogram. If the simple density analysis is used for assessing mitochondrial label, all grain centers and grid intersections lying over a mitochondrion are collected. (For adequate statistical sampling numerous autoradiograms have to be tabulated in a similar manner and the data pooled.) The illustrated mitochondrion has 1 grain and 5 points, or 1 grain/.5 μ^2

(see Fig. 15). The number of points which fall over different organelles gives the relative area occupied by these organelles. (If the grid is calibrated as to points per μ^2, area corrections can be made in units of μ^2.) Determinations of area can also be made by cutting out and weighing the different structures of interest. (This procedure can similarly lead to an absolute calibration if one knows the weight of 1 μ^2 of autoradiograph at a given magnification.) If the thickness of the section was measured, the data can also be expressed as grains per unit volume.

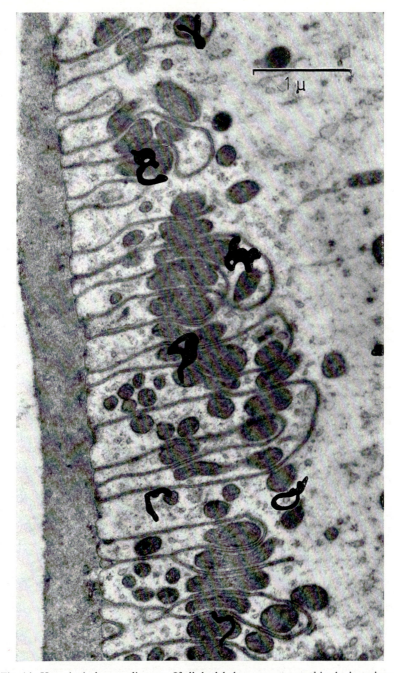

Fig. 16. Hypothetical autoradiogram. If all the label was concentrated in the invaginated plasma membrane, scattered developed grains would give the erroneous impression of labeled mitochondria. 30,000 ×

Several precautions in using the simple density method, some of which are also applicable to the alternative methods of analysis, should be pointed out. In the first place, the fact of choosing the compartments to be analyzed and those to be omitted may introduce a certain bias (i.e. a judgment is made as to the likely localization of label). Even in very complete analyses, in which grain density is tabulated for all the defineable organelles, the cytoplasmic ground substance, small vesicles, membranes, etc., are frequently omitted. In addition to loss in information regarding the untabulated areas, erroneous conclusions may be drawn from the data one does collect, due to radiation spread from the untabulated structures. For example, an unlabeled organelle, surrounded by radioactive ground cytoplasm, would have a considerable grain density merely due to radiation spread from its surroundings. The eventuality of a radioactive rim of ground cytoplasm around a defined organelle must also be considered. Such circumstances are best detected by the complete "density distribution" method, but may also be revealed by the probability circle method, both of which will be discussed below. An estimate of the average label in the ground cytoplasm can be obtained using the simple density method by tabulating grains and points in the ground cytoplasm more than 2-3 HD from the nearest organelle. At this distance the effect due to radiation spread from any hot organelles is minimal (see Figs. 8 and 9).

Radiation spread from one class of organelle to another can also be a problem. Such would be the case if a small radioactive structure permeates a cluster of nonradioactive organelles which are being tabulated. Imagine a radioactive membrane surrounded by mitochrondria (e.g. Fig. 16). A simple grain density tabulation for different organelles would make the mitochondria appear significantly labeled. A similar difficulty would arise if a very radioactive organelle was anatomically coupled with another which has much less or no radioactivity. With simple grain density tabulation, the unlabeled neighbors would invariably have a higher grain density than average tissue background. Similarly a less radioactive neighbor has a higher net gain due to radiation spread than a more radioactive one. Such a situation existed for instance in the study by Faeder and Salpeter (1970) where a radioactive axon is invariably surrounded by a much more radioactive sheath cell (Fig. 17).

There is no simple remedy for these eventualities except to be aware of them. The examples illustrate what we have been repeatedly emphasizing, i.e. that no single prescribed method of analysis substitutes completely for common sense. Whatever the method of analysis, one should always return to the autoradiograms and estimate to what extent radiation spread could have influenced the results. For instance to test whether the radioactive membrane meandering among the mitochondria (Fig. 16) is the cause of the high grain density, the mitochondrial population could be separated into two groups, one near the membrane and the second at a distance. A comparison of the grain densities of the populations would give a meaningful hint. An alternative is to test the hypothesis that the membrane is the radioactive source by using the grain density distribution method to be discussed below.

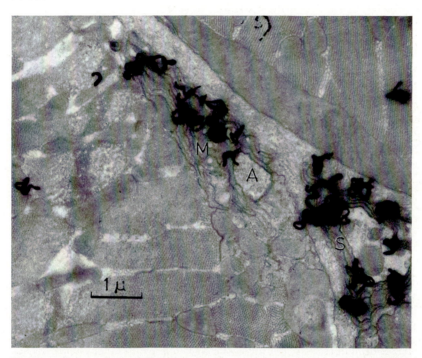

Fig. 17. Overexposed autoradiogram of neuromuscular junction in the cockroach Gromphadorhina. The nerve was stimulated for 1 hour prior to incubation in H^3-glutamate. Heavy label is seen in sheath cells and post-junctional muscle. A net radiation spread into the axon is therefore to be expected. 1000 Å section, Ilford L 4 emulsion, Microdol X developer, 15,000 ×. (From FAEDER and SALPETER, 1970)

The biggest problem which arises when comparing simple grain densities can be predicted from the integrated distributions (see Figs. 6 and 7). These distributions indicate that the relative number of total grains due to a labeled structure which will fall over this source depends on its size. In Fig. 6 this relative number represents 0.23 for the circular source of 1 HD radius and 0.61 for the one of 4 HD radius. Since in a simple density assessment one tabulates only grains over a structure and divides by the area of that structure, an error is introduced when comparing organelles of different sizes. Let us assume that two structures, one being 4 HD and the other 1 HD in radius, are of equal specific activity (i.e. each has the identical number of decays per unit volume of tissue). If that radioactivity produced two developed grains per μ^2 in the exposure time used then, if there were no radiation spread, the relative number over the source would be 1, and the average grain density over the source would be:

$$\frac{1 \times 2 \times \text{area of organelle}}{\text{area of organelle}} = 2.$$

Using the relative numbers given above, the 4 HD disc has a grain density of $0.61 \times 2 = 1.22$ and the 1 HD disc a grain density of $0.23 \times 2 = 0.46$ (a ratio of 1/0.38). Yet, by definition, they have the same density of label. The larger the structure the closer the condition of no radiation spread (i.e. a relative number of 1) is approached. Consequently, if a small structure has a higher grain density than a large one, it is indeed considerably more radioactive.

The integrated distributions can be used in making corrections for radiation spread. Let us assume that a 4 HD structure has a grain density of 0.3 grains/μ^2. If the true grain density (i.e. without radiation spread) is x, and there are no nearby radioactive neighbors to complicate the picture, then we can simply say that $0.61 x = 0.3$ and thus $x = 0.5$ grains/μ^2. Similarly if a 1 HD structure had a grain density of 0.3 grains/μ^2, its true density would be 1.3 grains/μ^2. Since, in many autoradiograms, radiation cross fire from neighboring organelles (< 4 HD distance) introduces a degree of uncertainty, corrected grain densities must be stated with an appropriate level of restraint.

In conclusion then, the simple density method provides the simplest statement related to absolute concentration of radioactivity (i.e. specific activity) of the tissue. Net changes in an organelle with time reflect true fluxes of radioactivity, not merely changes relative to other compartments. If the need for a few corrections is kept in mind, such as those for radiation spread, simple density provides probably the most useful first step in the analysis of numerous autoradiograms.

2. "Per cent" Analysis

A common procedure expresses data in terms of the per cent of total grains associated with any given organelle. Such data can be converted to a per cent density by dividing the per cent of total grains by the per cent of total area occupied by that organelle. (These percentages are not to be confused with per cent of total grains due to a given source which was discussed in relation to the integrated distributions.) If the distribution of grains over the entire tissue is purely random, then per cent density (per cent grains/per cent area) should have a value of 1. Any value above 1 indicates a level of label above average and any value below 1, a level of label below average. The basic advantage of the per cent tabulation over a simple grain density is that it states the label of each organelle relative to the general distribution of label. However if, in the simple density method, the number of grains and points tabulated for each category is given, the per cent information can easily be calculated. If, on the other hand, only per cent data (either only per cent grains or the per cent density) are presented, the data can be very misleading. An illustration of what may happen when data are presented merely in terms of percentages follows. Let us assume that we are studying intracellular transport and release of protein. An animal is given H^3-leucine and the organ of interest prepared for EM autoradiography, either 5 minutes or 30 minutes later. Random

Table 2. Autoradiographic data by "percent" analysis method

Tissue Compartments	Time after Injection (min)	Grains Number	% of Total
Rough ER	5	800	80
	30	560	56
Golgi	5	30	3
	30	270	27
Other	5	170	17
	30	170	17
Total of all	5	1000	100
Compartments	30	1000	100

autoradiograms are photographed and grains tabulated in relation to three compartments, e.g. rough ER, Golgi apparatus, and "other". For high statistical reliability, 1000 grains are tabulated for each time group. Furthermore, it is established that the three organelles occupy the same relative areas in both groups of autoradiograms. The number of grains related to each compartment is expressed as per cent of total grains as in Table 2. The conclusion is drawn that there is a shift in radioactivity with time from rough ER to Golgi.

This interpretation is valid however only if we have a closed system, i.e. only if the total radioactivity for the three compartments remains the same. If any leaves the cell or enters from another "pool" the conclusion no longer holds.

To illustrate, let us assume merely that some of the radioactivity has already left the cell by 30 minutes. To collect the 1000 grains, a larger number of autoradiograms is therefore tabulated. (As before, the ratio of areas occupied by the three compartments remains constant.) Let us now look at the complete data as presented in Table 3.

The only additional information presented in Table 3 is the actual area tabulated in the 2 time groups. The per cent data are the same as in Table 2, but the conclusions are no longer so obvious. Since the grain density in the Golgi is the same at the two times, the actual specific activity (concentration of radioactivity) remains the same. The concentration of radioactivity in the rough ER drops drastically however. (Since we are here comparing each organelle with itself at different times, the problem of different relative radiation spread from a large and a small organelle does not pertain.) Three equally tenable alternative conclusions can be drawn from the data in Table 3: a) all the radioactivity from the rough ER passes through the Golgi; b) part of it passes through the Golgi; c) none of it passes through the Golgi.

Table 3. Expanded data from same autoradiograms as in Table 2

Tissue Compartments	Time after Injection (min)	Grains Number	% of Total	Area (μ^2)	Density $(Grains/\mu^2)$
Rough ER	5	800	80	2000	.4
	30	560	56	18000	.03
Golgi	5	30	3	300	.1
	30	270	27	2700	.1
Other	5	170	17	3600	.05
	30	170	17	32400	.005
Total of all	5	1000	100	5900	.17
Compartments	30	1000	100	53100	.019

Alternative (a) can only be true with this data however, if the turnover rate in the Golgi is considerably higher than it is in the rough ER, as the Golgi would have to handle the larger volume of radioactive material coming from the ER without increasing in specific activity. Yet (b) and (c) are equally possible. The autoradiographic analysis here presented is therefore inadequate to decide among (a) (b) and (c) and a more sophisticated approach is required. (Similar problems in interpretation would exist if the system were fed unequally from an additional outside "pool".)

Thus, the per cent method has the advantage of giving a quick overall view of the relative distribution of radioactivity at any one time, but unless one is able to demonstrate a closed system, the data may be misleading when comparisons are made for different times.

3. "Probability Circle" Analysis

Bachmann and Salpeter (1965) (see also Bachmann, Salpeter and Salpeter, 1968) suggested that if resolution is considered in terms of the distance from a point source within which 50% of the developed grains fall, one obtained a fringe benefit in analysis. The distance from a point source

within which 50% of the developed grains fell is also the distance from a
grain with a 50% probability of containing a point source. Thus, drawing
a probability circle around each grain could aid the search for point sources
by drawing attention to the structures within this circle. (This was an early
statement of what was later expanded by the families of integrated curves.
These curves gave the distances related not only to 50% probability but to

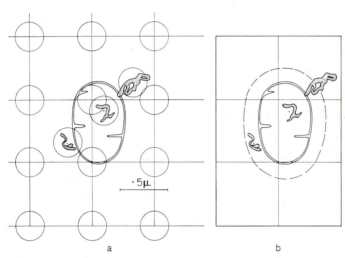

Fig. 18. Schematic of same mitochondrion as in Fig. 15, but at 30,000 × to be analyzed
by the probability circle method. The overlying grid has a larger spacing of 1 grid point/
0.3 μ^2. In (a) circles of 1 HD radius are drawn around each developed grain and each
grid intersection point. By the probability circle method, there are 2 grains and 2 points
whose circles/overlap the mitochondrion. In (b) an annulus equal in width to the radius
of the probability circle is drawn around the mitochondrion. Essentially an identical
result is obtained whether one counts all grains and points whose probability circles
overlap the mitochondrion, or whether one counts all grains and points lying within the
mitochondrion plus its annulus (i.e. 2 grains and 2 points; density = 1 grain/0.3 μ^2)

any probability, and not only for a point source but for a variety of sources.)
From the integrated curve for a point source, Fig. 19, we see that the 50%
probability for a point source is 1.7 HD. One needs to be reminded that
1 HD is the 50% probability for a *line* source only.

The idea of the probability circle has become the basis of one method
for analyzing autoradiograms (SALPETER, 1968), with detailed prescriptions
introduced by WILLIAMS (1969) and NADLER (1971). Basically what the
probability circle analysis involves is illustrated in Fig. 18a. A circle of
given size (given probability to contain the source) is drawn around every

grain center. A compartment (organelle) is considered to be associated with a grain if any part of the circle falls over that compartment. (Partial credit for circles which fall over more than one structure can be assessed to gain additional information.) A large number of circled grains may however be associated with a structure merely because that structure occupies much of the area of the autoradiograph. Area corrections are thus necessary to establish the extent to which randomly distributed probability circles of the same size would overlap the organelle. This can be accomplished by using a grid similar to the one used for area determination in the simple density method, drawing circles around all the grid points and associating a structure with a circled point in the same manner as a structure is associated with a circled grain. A simple statement of the data for each organelle then is the ratio of circled grains/circled points. This essentially becomes a grain density, in which the tabulation includes not only grains and points falling over an organelle (as in the simple density) but also those falling outside the organelle in a zone equal in width to the radius of the probability circle. (See Fig. 18 b.)

By including what lies within a given distance from a tabulated organelle (i.e. possible source) one is in fact collecting a certain fraction of the grains which have fallen outside the source due to radiation spread. As we can also see, especially from the integrated curves, the smaller the source, the larger the fraction of grains falling outside it and the more important this type of analysis becomes. With organelles in the order of 1 HD, a simple density determination is just not sufficient. In a study of H^3-proline incorporation into cartilage cells, Salpeter (1968) used a proability circle analysis primarily to include the possible contribution of small (~ 400 Å) vesicles to the overall labeling pattern. An alternative to the probability circle method for small structures is the "density distribution" method discussed below.

It should be pointed out that although the term "probability" circle is used, the probability itself never enters again into the analysis or interpretation of the data. For this reason, although Bachmann and Salpeter (1969) and Williams (1969) consider an 2500 Åcircle (on an autoradiograph of a 1000 Å section, Ilford L 4 emulsion and MX development) to be approximately a 50% probability circle, and Nadler (1971) considers it to be approximately a 95% probability circle, their analyses are not affected by this drastic difference. The discrepancy matters seriously however in that it reflects the difference in our conclusions regarding the resolution of the autoradiographic technique. Since Nadler bases his statement of 95% probability on as yet unpublished data, no assessment as to its validity can be made at the present time.

The data obtained by the probability circle method differ from that of the simple density method in the following respects. First, for the same

number of autoradiograms, the probability method results in collecting
more grains due to the source, and thus has a higher statistical accuracy.
This is again most important for small structures where a larger fraction
of grains will lie outside them. Secondly, it provides an overlap zone at the
interface between an organelle and the ground cytoplasm or between two
neighboring structures. Circled grains and points which overlap two or
more structures can be tabulated separately from those exclusively over a
single compartment. A comparison of these two densities can lead to

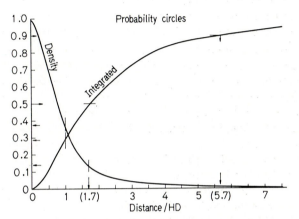

Fig. 19. Density and integrated distributions for a point source. One can see that a 1 HD
circle has a 29% probability of containing the source. At that distance the grain density
is 0.38. A 50% probability circle has a radius of 1.7 HD and a 90% probability circle has
one of 5.7 HD. The grain density (grains/area) in the range from 1.7 HD—5.7 HD is less
than 1. Thus using probability circles above 50% would both lose structural specificity
and dilute the grain density to a meaningless low

significant information regarding the relative contribution of different
compartments at the interface. WILLIAMS (1969) has effectively utilized this
aspect to assess the relative importance of such neighbors. The two densities
represent, in effect, averaged values of two discrete regions in the universal
density distributions, one over the structure, and one at the edge. The
universal curves can thus be used with these data to indicate whether any
contribution at the edge is coming from an outside source, and the relative
importance of the two sources. Finally, in the simple density method the
density value is the average grain density only over the source (i.e. negative
values on the x axis), whereas in the probability circle method we include
in the average the density outside the source (up to the radius of the
probability circle). Since the density drops rapidly in this range (Figs. 8

and 9), it can immediately be seen that as the size of the probability circle is increased, the average density will decrease. This is why using a circle of high probability (e.g. 90%), which on first thought should be superior to a 50% circle, is self-defeating. In Fig. 19 we see that for a point source a 30% probability circle is just over 1 HD, a 50% circle is about 1.7 HD, a 90% circle is just under 5.7 HD, and a 95% circle is well over 7.5 HD. However, the grain density for a point source is already below 0.2 by a distance of 1.5 HD from the source. Thus not only would a 90% probability circle (5.7 HD radius) be so large as to lose any selectivity regarding underlying structures (every organelle would be enclosed by every circled grain) but also the grain density would not reflect the label over any given structure but only a general average tissue label.

The dilution of grain density in the probability circle method, furthermore, results in an even greater difference between a small and a large structure of equal specific activity, than is the case for the simple density method. Assume a probability circle of 1 HD radius (i.e. $\sim 26\%$ probability). From Fig. 6 we see that the relative number of grains over a source of 4 HD radius is 0.61, and that within a 1 HD rim outside, it is 0.13, giving a total of 0.74; and that the relative number of grains over a source of 1 HD radius is 0.23 and, within a rim of 1 HD outside, it is 0.28, giving a total of 0.51. If these discs have the same radioactivity as considered in the discussion of the simple density method, which without radiation spread would have a simple density of 2, then the circled density for the 4 HD disc is 0.94 and for the 1 HD disc is 0.26 (a ratio of 1/0.28 compared to the ratio of 1/0.38 for the simple density, p. 134).[4]

What then is the advantage of the probability circle method. As we said before, it is primarily that small sources, which would be overlooked by the simple density method, can be revealed. Furthermore, when used appropriately, it allows a simple spot check regarding the relative label in

---

4 These densities are derived as follows: Since the number of developed grains due to the decays per unit area of the source is 2, then, without radiation spread, the total grains collected are still $1 \times 2 \times$ area of organelle. The circled density then equals

$$\frac{1 \times 2 \times \text{area of organelle}}{\text{area of (organelle + rim)}}.$$

(This density approaches 2 for large structures.) Since the relative number of grains collected for the encircled 4 HD disc is 0.74, the density for this circled disc is

$$\frac{0.74 \times 2 \times \pi(4)^2}{\pi(4+1)^2} = 0.94.$$

For the circled disc of 1 HD radius it is $\dfrac{0.51 \times 2 \times \pi(1)^2}{\pi(1+1)^2} = 0.26.$

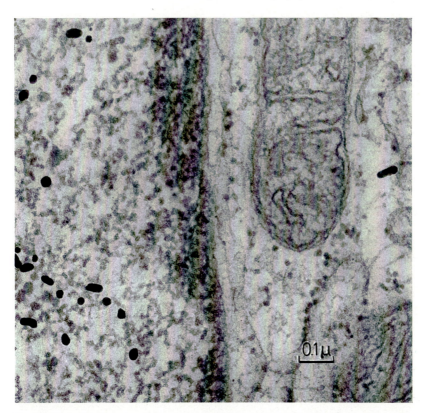

Fig. 20. Mesenchymatous cell in adult newt, Triturus, labeled with H³-thymidine. The nucleus is obviously labeled. Occasionally a developed grain is seen in the cytoplasm. The density distribution method is best suited to answer whether these cytoplasmic grains are due to radiation spread from the nucleus or due to a low level of cytoplasmic label. 500 Å section, Kodak NTE emulsion, Gold latensification-Elon ascorbic developer. 100,000 ×

the cellular ground cytoplasm. In our opinion, WILLIAMS (1969) has utilized the probability circle method to yield these two advantages to a larger extent than NADLER (1971) has.[5]

---

5 We cannot go into detail on the variants introduced by either WILLIAMS or NAD-LER. In our opinion WILLIAMS has the more sophisticated treatment. NADLER seems to discard to a large extent the major advantage of using the probability circle as opposed to the less cumbersome simple density method. Unlike WILLIAMS, he does not consider ground cytoplasm. Furthermore, he excludes all organelles which do not have a certain percentage of "exclusive" grains (i.e. grains whose entire probability circle lies over the organelle). This, by definition, stacks the cards against the small structures (i.e. those which are smaller than the probability circle in radius). Yet it is primarily for the small structures that it is worthwhile using the probability circle method.

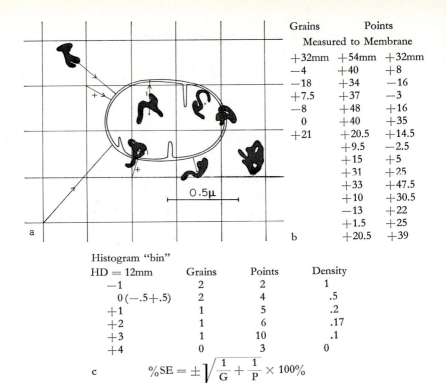

Grains	Points	
	Measured to Membrane	
+32mm	+54mm	+32mm
−4	+40	+8
−18	+34	−16
+7.5	+37	−3
−8	+48	+16
0	+40	+35
+21	+20.5	+14.5
	+9.5	−2.5
	+15	+5
	+31	+25
	+33	+47.5
	+10	+30.5
	−13	+22
	+1.5	+25
	+20.5	+39

a   b

Histogram "bin"

HD = 12mm	Grains	Points	Density
−1	2	2	1
0 (−.5+.5)	2	4	.5
+1	1	5	.2
+2	1	6	.17
+3	1	10	.1
+4	0	3	0

c

$$\%\,SE = \pm\sqrt{\frac{1}{G} + \frac{1}{P}} \times 100\%$$

Fig. 21. a Schematic of mitochondrion in autoradiogram, to be analyzed by density distribution method. The questions asked are whether the mitochondrion is labeled, what the distribution of the label is, and whether the grains outside the mitochondrion are due to radiation spread from it. The shortest distance from each grain and grid intersection point is measured to the outer limiting membrane. The actual measurements are tabulated in b. (These measurements were made at 75,000 × before the figure was reduced for publication.) In c the raw data are retabulated into histogram bins, and grains divided by points to obtain density. The sample size here is obviously not high enough to have any validity but is used merely for illustration (% Sampling Error $= \sqrt{1/\text{grains} + 1/\text{points}} \times 100\%$). d Experimental density scale (y axis) is normalized to be 1 for the bin that stradles the origin. (Normalizing is accomplished by dividing each density value by the value of the 0 bin.) This experimental density histogram fits the expected distribution for a solid disc of 1.5 HD radius (superimposed curve). The data are also compatable with the hypothesis that the grains outside the mitochondrion are due to radiation spread and not to ground cytoplasm label. e If an adequately large sample of grains andpoints had been pooled from many autoradiograms, one should get an experimental histogram (depicted in solid lines) to fit the above hypothesis. If the alternate histogram (broken lines) were obtained, such a hypothesis would have to be rejected. One could conclude however, that the mitochondrion has a higher label than its surroundings but that some label also exists outside. Furthermore, the general label also falls off. It thus might pay to test a few alternate hypotheses to better define the source. In fact, the "dashed" histogram fits a distribution expected from a "disc" with the edge not at the outer limiting membrane of the mitochondrion but at 2 HD outside it. The testing of this alternate hypothesis is left to the reader. (Hint: renormalize the dashed histogram, setting the density of the +2 HD column to 1, and compare the resultant histogram to that expected for a solid disc of $3^1/_2$ HD radius, using the 4 HD curve.)

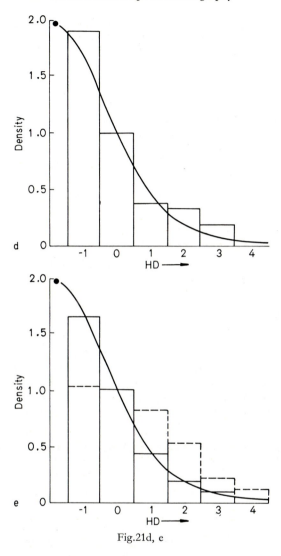

Fig.21d, e

4. "Density Distribution" Analysis

We have tried to show that the families of density and integrated curves can be used to help assess autoradiographic data, independent of the method of analysis used. Ultimately EM autoradiographic analyses involve a form of hypothesis testing. This is done implicitly in all the methods. By choosing the compartments which are tabulated one is testing whether these structures are radioactive, and whether they are more or less so than some other structures. Occasionally one has reason to ask more specific questions:

e.g., what is the distribution of label in organelle "A"? Is organelle "A" the only labeled structure in the vicinity? Is a high grain density in organelle "B" actually due to radioactivity in "B" or is it due to radiation spread from its neighbor "A"? For example, in Fig. 20 one may ask if the developed grain in the cytoplasm is due to a low level of label there or if it is due to radiation spread from the nucleus. To answer such specific questions more of the information contained in the universal curves must be used than is required in any of the other methods discussed above. One needs to compare full experimental grain density histograms tabulated in relation to a given organelle, with those expected for different kinds of sources. The goodness of fit can be used to 1) identify or reject a structure as a source, 2) determine the distribution of label within this source, or 3) formulate new hypotheses for subsequent testing.

Let us see how this works. In Fig. 21 we see again a labeled mitochondrion. The hypothesis being tested is whether or not the mitochondria in our tissue are labeled. To tabulate the experimental grain density distribution, the shortest distance from each grain and grid intersection, or random point is measured to the outer limiting membrane of the mitochondrion with these exceptions: 1) no grains or points need be tabulated which are more than 4 HD from the mitochondrion since the grain density is already very low at this distance; 2) if there is more than one mitochondrion in the autoradiograph, the measurement should be to the nearest one; 3) if the grain is closer than 2 (or preferably 3) HD to more than one organelle, it should be excluded to avoid excessively high densities due to radiation spread. Points as well as grains must be excluded to make an area correction for the zone in which no grains are being tabulated.

Each measurement is designated as (—) or (+) depending on whether it is over the organelle or outside it[6]. Once an adequate sample is obtained,

---

6 As in the case of all the methods of analysis, to get adequate statistical sampling, large numbers of autoradiograms have to be scanned and the information from all the organelles of a given category pooled. The assumption is made that all the organelles (e. g. mitochondria) come from one homogeneous population as far as radioactivity is concerned. If there is any doubt about this assumption and multiple populations (due to shape, size or other factors) are suspected, these sub-populations must be tested separately. If no difference shows up, the data can be pooled. Adequacy of statistical sampling involves several factors, e.g. number of animals used, number of tissue blocks sampled and number of grains and points tabulated. An estimate of the sampling error for the grains and points is easy to make. Each of these follows Poisson statistics but is independent of the other. The per cent sampling error (or standard error) for a sample size N is $\pm \frac{1}{\sqrt{N}} \times 100\%$; that for a ratio of two samples, e. g. grains and points (G and P), is

$$\pm \sqrt{\frac{1}{G} + \frac{1}{P}} \times 100\%.$$

For a more detailed discussion, see Technical Considerations, Salpeter and Bachmann (1972).

the data should be collated within histogram "bins" of unit distance from the source. We conventionally use a 1 HD unit as our distance bins and have one such unit straddle the edge (i.e. we collect all grains and points within —0.5 to +0.5 HD; then those within —0.5 to —1.5 HD and

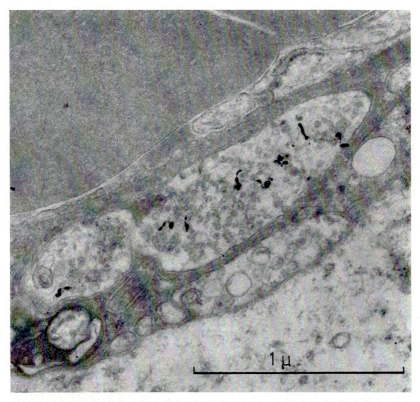

Fig. 22. H^3-norepinephrine bound in the sympathetic nerve terminal of the mouse pineal. From BUDD and SALPETER (1969)

those within +0.5 to +1.5 HD, etc.). Once thus collated, the number of grains per bin is divided by the number of points in that bin and thus a density distribution is obtained. Since the edge or circumference of an organelle is easy to define, expected distributions were arbitrarily normalized to be 1 at the edge. One should therefore normalize the experimental histograms using the same origin. To do this the densities of each bin are divided by the density in the zero bin (i.e., the bin from —0.5 to +0.5 HD).

Since we know the HD of our specimens we can also ascertain the average radius of our organelle in HD units, (~ 1.5 HD for the mitochondrion in Fig. 21). (Oval structures can be considered circular with a radius equal to the average of the two axes. If the axes are very different,

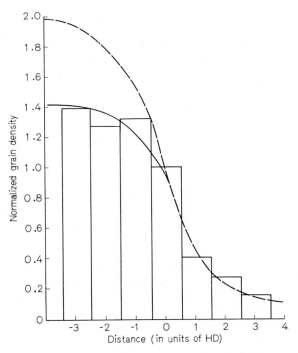

Fig. 23. Experimental grain density histogram around sympathetic nerve endings as illustrated in Fig. 22. The superimposed dashed curve corresponds to the distribution expected for a circular disc equal in radius to the average nerve terminal. The grain distribution outside the nerve fits that expected solely from radiation spread. Inside the nerve, however, the expected grain density for such a disc rises to a higher peak than does the experimental density. This indicates a non-uniform distribution of label, with a higher concentration at the periphery. From Budd and Salpeter (1969)

one can treat the organelle as a band structure of half width equal to half the smaller axis.) Given the radius in HD units we can see which of the curves in the various families fits the experimental distribution. In Fig. 21 d the experimental histogram has a reasonable fit to the distribution expected for a uniformly labeled disc. The statistical sample used in constructing this histogram is obviously too small for any deviation from the expected distribution to carry any significance. For a more meaningful interpretation, several hundred grains from numerous autoradiograms would have to be pooled. Fig. 21e gives two possible alternative histograms, assuming such a statistically valid sample.

To illustrate further how this type of analysis works see Figs. 22 to 26. Fig. 22 (Budd and Salpeter, 1969) is a typical autoradiogram of H³-

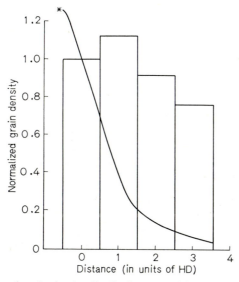

Fig. 24. Experimental grain density distribution around large (~ 1000 Å) dense core granules. The superimposed curve represents the expected density for the condition that only these granules are labeled. This hypothesis has to be rejected

norepinephrine bound in the sympathetic nerve terminal of the mouse pineal. The nerve is obviously the most heavily labeled structure in the field. In numerous autoradiographs of this material, there are occasional grains also over pinealocytes or endothelial cells. To test whether they represent extraneuronal binding sites, we tabulated a grain density distribution with respect to the limiting membrane of the nerve terminals. This distribution is depicted in Fig. 23. It fits the hypothesis that the developed grains outside the nerve are due to radiation spread from the nerve. The number of extraneuronal binding sites, if any, were thus below the sensitivity of detection in our system. The data do not fit the hypothesis however that the radioactivity is uniformly distributed throughout the nerve. The best fit is to a composite curve for a distribution of radioactivity in which there is a higher concentration of label at the circumference than in the center.

Sympathetic nerve terminals contain two populations of dense core granules, large (1000 Å) and smaller (500 Å) ones. The next hypothesis

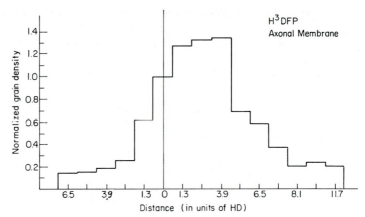

Fig. 25. Experimental grain density distribution on either side of the axonal membrane in mouse endplates labeled with H³-DFP (Fig. 13). The distribution does not fit any of the universal curves but indicates that the label is localized in a diffuse zone on the muscle side of the axonal membrane. The finestructure of the endplate suggested that this zone may be related to the junctional folds. When the grain distribution was replotted around the junctional fold membranes the histogram illustrated in Fig. 26 was obtained

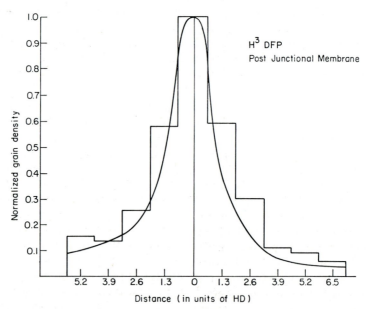

Fig. 26. Experimental grain density distribution around the post-junctional membrane in a mouse endplate labeled with H³-DFP. The data fit expected distribution for a line source with the curvature characteristic of the endplate. From Salpeter (1969)

tested was that the large dense core granules have a higher label than their surroundings. A grain density distribution relative to the outer limiting membrane of such granules was constructed. Fig. 24 compares this experimental distribution with that expected from a circular source of egual radius (0.5 HD). The hypothesis of preferential label clearly has to be rejected.

The next example illustrates how another experimental histogram led to the rejection of one hypothesis, and to the formulation of an alternative one which was subsequently accepted. From autoradiographs of the mouse endplate labeled with H^3-DFP as in Fig. 6 (SALPETER, 1967, 1969), an experimental histogram was tabulated in relation to the axonal membrane, Fig. 25. This distribution did not fit any of the expected distributions but indicated that the radioactivity is located in a broad band towards the muscle side of the endplate. This band appeared to coincide with the junctional fold region of the endplate. Fig. 26 shows the histogram which resulted when the experimental grain density distribution was retabulated in relation to the junctional membranes. The fit to the distribution expected from a line source was such that we could not reject the hypothesis that the junctional folds are the source of the radioactivity. Other studies which further illustrate the uses of the family of curves and the density distribution method of analysis are: ISRAEL, SALPETER and STEWARD (1968); PLATTNER et al. (1970); HEDLEY-WHYTE et al. (1969); GAMBETTI et al. (1972). LENTZ (1972). The four methods for quantitative analysis discussed here represent those currently used in different laboratories. Hopefully the information on expected grain distributions available from the families of universal curves will lead to further refinements and increased sophistication in extracting information from E.M. autoradiograms.

D. Conversion of Developed Grain Data to Information on Radioactivity

Ultimately quantitation in EM autoradiography should extend information about developed grains to information about the radioactivity in the tissue. This conversion is not difficult. First, the sensitivity of the technique has to be established, i.e. the probability that a radioactive decay will lead to a developed grain. For this purpose a radioactive source comparable in dimensions to an EM autoradiographic specimen is needed. It must be of known specific activity and its thickness must be measurable. When coated with a calibrated emulsion layer, exposed and developed under controlled and reproducible conditions, the number of developed grains per unit area of emulsion can be related to the calculated number of decays which occurred in the source. Such values and their variances have been reported (Caro and Schnos, 1965; Bachmann and Salpeter, 1967; Wisse and Tates, 1968; Vrensen, 1970; Salpeter and Szabo, 1972) and are reproducible within 20 to 30%.

Given the number of developed grains collected per compartment, the sensitivity value of the particular technique being used and the specific activity of the administered material, one can easily calculate how many molecules (M) (both radioactive and non-radioactive) of the administered material have been retained per unit area (or volume) of the tissue after processing for EM autoradiography[7]. A simple formula for obtaining M is:

$$M = \frac{Gd}{E} \cdot \frac{SD}{A} \tag{1}$$

where

G = grain density, i.e. the number of developed grains per unit area (or volume) of the tissue;

d = the inverse of sensitivity, i.e. the number of decays correlated with one developed grain;

E = exposure time in minutes (1.44×10^3 minutes = 1 day);

S = specific activity of administered material (Curies/mmole). (Specific activity represents decays for total material and not only for the radioactive molecules.);

D = decays/minute/Curie (2.22×10^{12}). The same unit of time must be used for D as for E and S.

A = Avogadro's number (6.02×10^{20} molecules/mmole). The same units must be used for A as for S.

The main limitation of a conversion from grain density to molar concentration is neither the reproducibility of the EM autoradiographic technique nor the accuracy of the sensitivity values. It is rather the accuracy with which one can define the source and can collect all the grains due to that source, including those dispersed by radiation spread i.e. the accuracy of obtaining G. Here again the families of curves are useful in two different ways. First, if the data were analyzed by the simple density method, then one can, from the size of the compartment and the integrated curves, estimate the per cent of total grains that lie outside the source and therefore are not included in the tabulated data. One can then correct for the radiation spread and include these grains in the value for G, or one can at least indicate the extent to which "G" may be in error. For an example of such a treatment, see Faeder and Salpeter (1970); Salpeter and Faeder (1971).

---

[7] In analyzed autoradiograms we can get information only regarding the retention by the tissue of exogenously applied material. Neither the extraction of radioactive material during processing nor the endogenous pools are discussed here since we believe them to be problems unrelated to analysis of autoradiograms.

The second approach to the collection of total grains before converting to molar concentration is by the use of the density distribution analysis. Once an experimental histogram is shown to fit a given source and the source is thus defined, the total grains in the histogram should be divided by the volume (area) of that source and the resultant grain density used as G in formula (1). See for example, Salpeter (1969).

References

Bachmann, L., Salpeter, M. M.: Autoradiography with the Electron Microscope: A quantitative evaluation. Lab. Invest. **14**, 1041—1053 (1965)

Bachmann, L., Salpeter, M. M.: Absolute Sensitivity of Electron Microscope Autoradiography. J. Cell Biol. **33**, 229—305 (1967).

Bachmann, L., Salpeter, M. M.: Electronenmikroskopische Autoradiographie. In Methodensammlung der Electronenmikroskopie. Eds.: G. Schimmel and W. Vogell. Stuttgart: Wiss. Verlagsges. 1972.

Bachmann, L., Salpeter, M. M., Salpeter, E. E.: Das Auflösungsvermögen elektronenmikroskopischer Autoradiographien. Histochemie **15**, 234—250 (1968).

Budd, G. C.: Recent Developments in Light and Electron Microscope Radioautography. Int. Rev. Cytol. **31**, 21—56 (1971).

Budd, G. C., Salpeter, M. M.: Distribution of Labeled Norepinephrine within Sympathetic Nerve Terminals Studied with Electron Microscope Radioautography. J. Cell Biol. **41**, 21—32 (1969).

Caro, L. G.: High Resolution Autoradiography. II. The problem of resolution. J. Cell Biol. **15**, 189—199 (1962).

Caro, L. G.: Progress in High Resolution Autoradiography. In Progress in Biophysics and Molecular Biology. Pergamon Press. **16**, 171—190 (1966).

Caro, L. G.: A Common Source of Difficulty in High Resolution Radioautography. J. Cell Biol. **41**, 918—919 (1969).

Caro, L. G., Schnos, M.: Tritium and phosphorus-32 in high resolution autoradiography. Science **149**, 60—62 (1965).

Faeder, I. R., Salpeter, M. M.: Glutamate Uptake by a Stimulated Insect Nerve Muscle Preparation. J. Cell Biol. **46**, 300—307 (1970).

Gambetti, P., Autilio-Gambetti, L. A., Gonatas, N. K., Shafer, B.: Protein Synthesis in Synaptosomal Fractions. Ultrastructural Radioautographic Study. J. Cell Biol. **52**, 526—535 (1972).

Granboulan, N., Franklin, R. M.: High Resolution Autoradiography of Escherichia Coli Cells Infected with Bacteriophage R 17. J. Bact. **91**, 849—857 (1966).

Hedley-Whyte, E. T., Rawlins, F. A., Salpeter, M. M., Uzman, B. G.: Distribution of Cholesterol-1, 2-H^3 during Maturation of Mouse Peripheral Nerve. Lab. Invest. **21**, NO. 6, 536—547 (1969).

Israel, H. W., Salpeter, M. M., Steward, F. C.: The Incorporation of Radioactive Proline into Cultured Cells: Interpretation based on Radioautography and Electron Microscopy. J. Cell Biol. **39**, 698—715 (1968).

Jacob, J.: The practice and application of electron microscope autoradiography. Int. Rev. Cytol. **30**, 91—181 (1971).

Lentz, T. L.: Distribution of leucine—^{3}H during axoplasmic transport within regenerating neurons as determined by electron microscope radioautography. J. Cell Biol. **52**, 719—732 (1972).

Maunsbach, A. B.: Absorption of I^{125} Labeled Homologous Albumin by Rat Kidney Proximal Tubule Cells. J. Ultrastruct. Res. **15**, 197—241 (1966).

Miura, T., Mizuhira, U.: Determination of autoradiographic resolution by H³- or P³²-labeled RNA phages in electron microscopy. J. El. Micros. (Japan) **14**, 327—336 (1965).

Nadler, N. J.: Interpretation of Grain Counts in Electron Microscope Radioautography. J. Cell Biol. **49**, 877—882 (1971).

Pelc, S. R.: Theory of Electron Autoradiography. J. roy. micr. Soc. **81**, 131—139 (1963).

Plattner, H., Salpeter, M. M., Saltzgaber, J., Schatz, G.: Promitochondria of Anaerobically Grown Yeast. IV. Conversion into Respiring Mitochondria. Proc. nat. Acad. Sci. (Wash.) **66**, 1252—1259 (1970).

Reimer, L.: Elektronenmikroskopische Untersuchungs- und Präparationsmethoden, 556—568. Berlin-Heidelberg-New York: Springer 1966.

Ross, R., Benditt, E. P.: Wound healing and collagen formation. V. Quantitative electron microscope radioautographic observations of proline-H³ utilization of fibroblasts, J. Cell Biol. **27**, 83—106 (1965).

Salpeter. M. M.: General Area of Autoradiography at the Electron Microscope Level. In Methods in Cell Physiology, **II**, 229—253. Ed.: D. Prescott. New York: Academic Press 1966.

Salpeter, M. M.: Electron Microscope Autoradiography as a Quantitative Tool in Enzyme Cytochemistry: The Distribution of Acetylcholinesterase at Motor Endplates of a Vertebrate Twitch Muscle. J. Cell Biol. **32**, 379—389 (1967).

Salpeter, M. M.: H³-Proline Incorporation into Cartilage: Electron Microscope Autoradiographic Observations. J. Morph. **124**, 387—421 (1968).

Salpeter, M. M.: Electron Microscope Radioautography as a Quantitative Tool in Enzyme Cytochemistry. II. The Distribution of DFP-Reactive Sites at Motor Endplates of a Vertebrate Twitch Muscle. J. Cell Biol. **42**, 122—134 (1969).

Salpeter, M. M., Bachmann, L.: Electron Microscope Autoradiography. In Principles and Techniques of Electron Microscopy: Biological Applications. Vol II. 221—278 Ed.: M. A. Hayat. New York: Van Nostrand Reinhold 1972.

Salpeter, M. M., Bachmann, L., Salpeter, E. E.: Resolution in Electron Microscope Radioautography. J. Cell Biol. **41**, 1—20 (1969).

Salpeter, M. M., Faeder, I. R.: The Role of Sheath Cells in Glutamate Uptake by Insect Nerve Muscle Preparations. Prog. in Brain Res. **34**, 103—114 (1971).

Salpeter, M. M., Salpeter, E. E.: Resolution in Electron Microscope Radioautography. II. Carbon¹⁴. J. Cell Biol. **50** NO. 2, 324—332 (1971).

Salpeter, M. M., Singer, M.: The Fine Structure of Mesenchymatous Cells in the Regenerating Forelimb of the Adult Newt, *Triturus*. Develop. Biol. **2**, 516—534 (1960).

Salpeter, M. M., Singer, M.: The Fine Structure of Mesenchymatous Cells in Regenerating Limbs of Larval and Adult *Triturus*. 00—12, Fifth Int. Cong. for El. Micr. New York: Academic Press 1962.

Salpeter, M. M., Szabo, M.: Sensitivity in Electron Microscope Radioautography using Ilford L4 Emulsion: The Effect of Radiation Dose. J. Histochem. Cytochem. **20** NO. 6, 425—434 (1972).

Vrensen, G. F. J. M.: Some New Aspects of Efficiency of Electron Microscope Autoradiography with Tritium. J. Histochem. Cytochem. **18** NO. 4, 278—290 (1970).

Whur, P., Herscovics, A., Leblond, C. P.: Radioautographic visualization of the incorporation of galactose-³H and mannose-³H by rat thyroid in vitro in relation to the stages of thyroglobulin synthesis. J. Cell Biol. **43**, 289—311 (1969).

Williams, M. A.: The Assessment of Electron Microscope Autoradiographs. Adv. Opt. and Electr. Micros. Eds.: R. Barer and V. E. Coslett. Academic Press **3**, 219—272 (1969).

Wisse, E., Tates, A. D.: A gold latensification elon ascorbic acid developer for Ilford L4 emulsion. 4th Eur. Conf. on Elec. Micr., Rome, 465—466 (1968).

Scanning Electron Microscope Techniques in Biology

Thomas L. Hayes

A. Introduction

In contrast to the conventional electron microscope, the scanning electron microscope's utility does not generally lie in the area of ultra high resolution. Rather, it is the process of information transfer which is particularly aided by the scanning microscope.

This paper will attempt to describe some of the techniques of scanning electron microscopy (SEM) that have utilized this information transfer capability in biomedical research. The references are examples from the literature which happen to be familiar and significant to the author rather than a complete listing of the biological applications. For a more extensive bibliography of the SEM, the reader is referred to the excellent publications of WELLS (1967, 1968, 1969, 1970, 1971) or CARR (1971).

While the conventional electron microscope has extended our resolution very significantly, there are limitations on its ability to transfer chemical information and three dimensional images of complete biological units. The scanning electron microscope, although generally possessing lower resolving power than the conventional electron microscope, can help to fill some of these information gaps.

I. General Principles of Operation

The scanning electron microscope departs from the traditional spatial focusing method of image formation. Most images (light microscope, conventional electron microscope, telescope, eyeglasses, human eye) are formed by focusing radiation after it has left the specimen in such a way as to produce the necessary one-to-one correspondence between points on the specimen and points on the image. The scanning electron microscope utilizes instead a technique of painting the image through a time sequence of points, a technique similar to that used in television imaging (OATLEY et al., 1965).

The light microscope and the conventional electron microscope are related by the basic similarity of their imaging forming process. Many interesting points can illustrate the development of the conventional electron microscope as an extension of light microscope theory (COSSLETT, 1966). The SEM, however, is more related to the electron microprobe or to television imaging where all focusing takes place prior to interaction with the specimen and the image points are addressed in time.

```
QIKGVVBHLFREOBTROXGQTJDBKWXFEIOVTQOL
ADYAZJCIVVJXFICWLFCWDBSDRFMIVTMBROCA
VSZNIYXPRJSUCFFBTZVOJBJZINQYLKKVWOJJZ
DNFICNPXDJQDPVEBZLBOBHNBDHKNVIZCXYTB
ZNWHWBSCANNINGERMZTFFTYDFVUQVPBIOTR
BTRNAFCLKDDHXSERXEZMZJAJHAXKDTGYHZS
XRZLTKDELECTRONPDOBFFTRNQVAXWKNGZYX
DRKOXMQZQJUHZEKPUQSBRXGIQSKGFNIAKKS
DCLSIGOIMICROSCOPYTZEPULPLZNLMEIHECSEC
UQEGJFBZJHVMFAXCDDLHVPQYVVDVNCJGULY
ILEBPUFHNNFYWEWRJVNFCNELLISMYERSQWUF
CFEMKGSMLETAQVTQBLJHCTVBPWQRIQVFQHV
ZKPSQITOPLVQSGGQHUMVLBLLDOQRBGCZUWR
```

```
QIKGVVBHLFREOBTROXGQTJDBKWXFEIOVTQOL
ADYAZJCIVVJXFICWLFCWDBSDRFMIVTMBROCA
VSZNIYXPRJSUCFFBTZVOJBJZINQYLKKVWOJJZ
DNFICNPXDJQDPVEBZLBOBHNBDHKNVIZCXYTB
ZNWHWBSCANNINGERMZTFFTYDFVUQVPBIOTR
BTRNAFCLKDDHXSERXEZMZJAJHAXKDTGYHZS
XRZLTKDELECTRONPDOBFFTRNQVAXWKNGZYX
DRKOXMQZQJUHZEKPUQSBRXGIQSKGFNIAKKS
DCLSIGOIMICROSCOPYTZEPULPLZNLMEIHECSEC
UQEGJFBZJHVMFAXCDDLHVPQYVVDVNCJGULY
ILEBPUFHNNFYWEWRJVNFCNELLISMYERSQWUF
CFEMKGSMLETAQVTQBLJHCTVBPWQRIQVFQHV
ZKPSQITOPLVQSGGQHUMVLBLLDOQRBGCZUWR
```

DBL 723-5185

Fig. 1. Information coding. The amount of information transferred by an image can depend on qualities other than resolution. Two images of an identical, nearly random, alphabet sequence are shown at the same resolution. The ability to perceive the message in the lower depends more on qualities such as contrast and tone than on any parameter related to resolution or detail

The two basically different methods of imaging have been distinguished in the two instruments which produce pictures at a great distance. One long distance imaging instrument is a telescope and the other is television. It might be useful to also make this distinction in the imaging of small objects by referring to the light and conventional electron instruments as microscopes and to the SEM as microvision (HAYES, 1971a).

The SEM might be described as consisting of two systems: the probing system into which the specimen is placed and the display system which forms the visual image (EVERHART and HAYES, 1972). In the probing system, electrons leave a hot filament or other source and are focused by means of electron lenses to a very fine point on the surface of the specimen. As the electrons interact with the material of the specimen at this point, they induce a variety of radiation such as characteristic X-rays, infrared, visible light, and streams of electrons.

These radiations leaving the surface of the specimen in a variety of directions can be counted and the measure of the number of photons or electrons of a specific type would reflect a particular quality of the specimen in terms of our knowledge of the production of that type of radiation by the primary bombarding electron beam. The direction in which a particular radiation leaves is immaterial as long as that radiation is counted.

The measure of the amount of radiation leaving a point on the specimen at any instant can be used to modulate the brightness of the beam of the display cathode ray tube as it rests at the corresponding point on the image forming screen of this display tube. When the radiation from the point on the specimen is high, the point on the cathode ray tube would be bright.

If the probing beam is now moved to an adjacent spot on the specimen, information from that point can be used to modulate the adjoining point on the display cathode ray tube. In a similar fashion, a third and fourth point can be added in sequence until all points of the specimen have been covered in a regular array. The one-to-one- correspondence necessary for the formation of an image occurs because of the single position of the probing beam corresponding to a single position of the display cathode ray tube beam at each instant in time.

In practice, the points follow one another with great rapidity so that the image of each point becomes an image of a line and the line, in turn, can move down the screen so rapidly that, to the eye, a complete image is formed. Such a process of the rapid sweep of an electron beam to produce a complete picture is familiar to us in television imaging.

If it is necessary to build up a better statistical sample of each point (EVERHART, 1970), a slower scan rate is selected and photographic film can be used to record each point and to integrate the entire picture over a longer period of time than can be integrated in a flicker-free fashion by the brain. Thus either through the direct visualization of a rapidly moving spot or by the collection over a long period of time of many spots on photographic film, the end effect is an integrated picture of the specimen being studied.

In the familiar lens imaging system, (e.g. light microscope, conventional EM), or in the mirror electron microscope (BOK et al., 1971). the information signal leaving the specimen is a vector quantity containing directional (localization) as well as amplitude and phase (information) components. In the scanning system, the information signal is a scaler quantity. All localization is a function of time and only the amplitude component is contained in the video signal. At each instant the scanning system needs only to measure the number of photons or electrons leaving the spot on the sample. It does not need to determine their direction.

The radiation leaving a single spot on the specimen at any instant is collected and then this signal is amplified and used to modulate the brightness of the cathode ray tube at that instant. There can be a complete separation in kind between the radiation used as the probing beam and the radiation that is collected and counted as the informational signal.

The two electron beams; one sweeping over the specimen, and the other sweeping over the face of the display cathode ray tube are in synchrony. They start at the top of their sweeps at the same instant, sweep through the same number of lines and end up at the bottom of their arrays at the same instant (HAYES, 1970a).

While the two beams sweep in synchrony over the same number of lines, the area of the array of lines (or raster) is not the same size on the specimen and on the face of the diyplay cathode ray tube. In fact, the magnification of the scanning electron microscope is produced by having the display raster very much larger than the size of the synchronous raster sweeping the specimen. Magnification is defined as the ratio of the linear size of the display raster to the size of the specimen raster. If, for example, the beam probing the specimen describes an array of lines one millimeter by one millimeter, and this information is related to points on the display cathode ray tube occupying a raster of ten centimeters by ten centimeters, the magnification of the system would be one hundred times. The useful range of magnification is dictated by the resolving power of the instrument (upper limit) and by the lens design and placement of the specimen (lower limit). The usual range of operation in scanning electron microscopy is of the order of ten times at the low end to fifty thousand times at the upper end. The range of magnification available to the scanning instrument puts it in a position between the light microscope and the conventional electron microscope, and in fact, the scanning instrument can act as a bridge between the other two.

II. A Comparison of Resolution

The resolution of a scanning electron microscope varies widely depending on the particular radiation being used as the information signal. The most commonly used signal is secondary electrons which produces the familiar three dimensional topographic image. Operating in this mode, currently available SEMs offer a resolution of about one hundred angstroms.

If transmitted electrons are utilized as the signal, resolutions below 5 Å have been achieved (CREWE et al., 1970; CREWE, 1971). If cathodoluminescence (visible light produced by electron bombardment) is used as the information signal, resolution may be no better than 1000 Å (PEASE and HAYES, 1966b, 1967).

Often the kind of information required in the image dictates the mode of operation and thus the available resolution. If, for example, the image must display the three dimansional relationship of a bulk specimen, the ultra-thin sections required for high resolution transmission scanning are prohibited. Or again, if the biochemical information of cathodolumine-scence is needed, the somewhat higher resolution secondary electron image will not do. The type of information required in the image may take precedence over resolution and resolution as an end in itself becomes less important (HAYES, 1971b).

However, in any mode of operation the highest resolution is always sought (NIXON, 1968), and a consideration of some of the factors effecting resolution might be useful. The conventional electron microscope can achieve resolutions well below 5 Å. In spite of this remarkable achievement, it takes only a simple calculation to show that even when operated with this high degree of precision, the conventional electron microscope is utilizing only a small part of the resolution that is available based on the wavelength of the electrons alone.

The main factor which limits the resolving power of the conventional electron microscopes is not the wavelength-sensitive defraction, but rather spherical aberration of the electron lens (COSSLETT, 1966). In his elegant treatment of the factors effecting resolution in the conventional electron microscope, ZEITLER has considered factors which range from the design parameters of the instrument itself to the physiological and psychological aspects of human vision (ZEITLER, 1969). It is pointed out that the conventional electron microscope today operates at a resolution limit in the 10 Å range that the goal of research is pushing this resolution towards a 1 Å atomic resolving power.

The scanning electron microscope, when utilizing its most common mode of signal collection, can exhibit, at best, an order of magnitude less in resolving power, (100 Å). In a particular mode of operation, (scanning transmission electron microscopy), the SEM can, in fact, achieve high resolution (CREWE, 1971) equal to that of the conventional electron microscope, but for the general operating conditions, the SEM ranks considerably lower in resolution than the conventional electron microscope.

The factors which control the resolution of the SEM in its most commonly used mode of secondary electron signal can be grouped into those factors associated with: 1) the limitations of a demagnified electron beam, 2) the limitations imposed by the interaction volume generated when the probing electron beam strikes the specimen, and 3) the limitations which are connected to design parameters dictated by the specimen conditions necessary for signal extraction (EVERHART et al., 1959; OATLEY et al., 1965; THORNTON, P. R., 1968; BROERS, 1970).

Using general optical principles, Langmuir, in a classic paper, showed that the current density in a focused beam of electrons has an upper limit (LANGMUIR, 1937). If we express this maximum current density obtainable in terms of a minimum diameter of the probing beam of electrons even in a system free of aberrations, we would have a minimum diameter of the spot as expressed in equation number 1.

$$d_0^2 = \frac{4i}{0.6\pi J_c \alpha^2} \frac{kT}{eV}. \tag{1}$$

In this equation, d_0 is the Gaussian probe diameter, J_c is the emission current density at the cathode, e is the electronic charge, eV is the electron energy, i is the probe current, α is the semiangle of convergence of the electron probe, k is Boltzmann's constant, T is the absolute temperature of the cathode.

In addition to this minimum diameter imposed by the basic electron optical considerations, there are also lens aberrations of the type found in the conventional electron microscope. These generally are listed under spherical aberration, chromatic aberration and astigmatism. There is also a defraction limiting term involved in the determination of the final minimum diameter available. If we describe the diameter of a disk of confusion associated with each of these aberrations or limitations, we find that they are functions of the semiangle of convergence α, the wavelength of the electrons used λ, and the fractional energy spread of the beam ΔV. These relationships can be expressed in the following equations (OATLEY et al., 1965):

$$d_s = \frac{1}{2} C_s \alpha^3. \tag{2a}$$

$$d_c = C_c \frac{\Delta V}{V} \alpha. \tag{2b}$$

$$d_f = \frac{1.22\lambda}{\alpha}. \tag{2c}$$

In equation 2, d_s, d_c, and d_f are the diameters of the least disk of confusion associated with spherical aberration, chromatic aberration and diffraction limits and C_s and C_c are the spherical aberration and chromatic aberration coefficients. Both C_s and C_c for magnetic objective lenses, have values comparable with their focal length. Astigmatism is a correctable aberration and will be treated in considerable detail in the section dealing with instrument operation.

In order to find the total probe diameter d, arising from the basic electron optical considerations expressed in equation 1 and from spherical aberration, chromatic aberration and diffraction as expressed in equation 2, the usual procedure (OATLEY et al., 1965) is to add the respective diameters in quadrature:

$$d^2 = d_0^2 + d_s^2 + d_c^2 + d_f^2$$

$$d^2 = \frac{4i}{0.6\pi J_c \, \alpha^2} \frac{kT}{eV} + 1.5\lambda^2 \frac{1}{\alpha^2} + \frac{1}{4} C_s^2 \, \alpha^6 + C_c^2 \left(\frac{\Delta V}{V}\right)^2 \alpha^2 \qquad (3)$$

Utilizing the above equation with certain reasonable assumptions, we can find an optimum value for α, and from this determine the resolving power for the conventional electron microscope as compared with the scanning electron microscope operated in the secondary electron signal mode (HAYES and PEASE, 1968). Such a calculation as carried out by PEASE and others indicates that the resolution, $d/2$, of the conventional electron microscope lies below 10 Å, but for the scanning electron microscope, chromatic aberration is appreciable and a current of 10^{-12} amp. (about the minimum required for scanning microscopy) can be focused into a spot no smaller than 100 Å in diameter utilizing conventional sources (PEASE and NIXON, 1965). A part of the difference between the two instruments can be attributed to a higher spherical aberration coefficient and a lower operating voltage connected with the operation of a scanning electron microscope as compared to the conventional instrument. The higher spherical aberration coefficient results from the need to collect the low energy secondary electrons which emerge from the specimen surface. Such collection can only take place in field-free space and requires, therefore, that the focal length of the scanning electron microscope final lens is considerably longer than the lenses associated with the objective lens of a conventional electron microscope (about 1 centimeter in the SEM compared to 2 millimeters for a CEM). The larger spherical aberration and chromatic aberration coefficients limit the spot diameter (PEASE and NIXON, 1965).

The lower SEM accelerating voltage, V, is suggested by considering the interaction volume that occurs when the electron beam strikes the specimen. If the signal being utilized in the scanning electron microscope actually originates from all parts of this interaction volume, it is the diameter of this pear-shaped interaction zone that determines the resolution of the instrument rather than the diameter of the incident electron beams. In the commonly used secondary electron signal mode of operation, the very soft secondary electrons are collected only from the volume very close to the surface and very near to the axis of the incident electron beam. This in spite of the fact that the incident beam can penetrate on the order of microns into the specimen, producing secondary electrons throughout this rather large volume. Due to their very low energy, the secondary electrons can only escape to be collected when they lie near the surface. This results in an effective collection diameter of about 100 Å for the secondary electron mode. If characteristic X-rays or visible light were being counted as the video information signal, the effective collection diameter would be considerably larger. Even in the case of secondary electrons, it is advantageous to limit the penetration of the electron beam by reducing the accelerating voltage.

The resolution of the SEM is determined by the probe diameter and by the factors related to the interaction between the probing electrons and the specimen. The relationship between resolution and contrast in terms of the important concept of signal to noise ratio has been presented in several of the papers by EVERHART and collaborators (EVERHART et al., 1959; EVERHART, 1968; EVERHART, 1970).

Utilizing what might be considered a standard scanning electron microscope (tungsten hairpin cathode, conventional oil diffusion pump vacuum stystem, and secondary electron information signal), PEASE was able to obtain resolution comparable to the theoretical limit placed on electron probe size and current as determined by the aberrations of the lens, electron noise and contrast levels (PEASE and NIXON, 1965). The electron probe size achieved was approximately 50 Å and a resolution in secondary electron mode of 100 Å was demonstrated.

If we refer to Eq. (1), it is clear that advances in resolution could be achieved by utilizing a high intensity source in the electron gun. If the current density at the cathode can be increased, a smaller probe diameter with sufficient intensity to produce a picture in a reasonable time would be possible. A new high resolution design utilizing a lanthanum hexaboride cathode was presented by BROERS in 1969. In addition to the high intensity gun, the effects of vibration, stray field and specimen contamination were reduced. Utilizing this specially designed instrument, BROERS was able to demonstrate a minimum probe diameter of approximately 30 Å and a point-to-point resolution of approximately 50 Å in the secondary electron mode (BROERS, 1969).

The approach of ALBERT CREWE's group in Chicago has been directed towards high intensity guns coupled with an ultrahigh vacuum system and the possibility of multipolar lenses; a combination that might be capable of resolving individual atoms (CREWE, 1966). The success of this program to date has been quite spectacular and, utilizing transmitted electrons as the information signal, CREWE has reported the visualization of individual uranium atoms (CREWE et al., 1970).

The SEM operated in the transmission scanning mode has been analyzed by applying the reciprocity theorem of optics, with the conclusion that this mode is capable of producing the same contrast mechanism effects as the conventional electron microscope (ZEITLER, 1971; CREWE, 1971). This work pointed out some unique advantages of scanning transmission electron microscopy as compared to the conventional electron microscope; notably in the method of information presentation and the improvement of signal to noise ratio. Since the information is presented sequentially as an electronic signal, processing of the information is made considerably easier and because the scanning transmission electron microscope can utilize all electrons interacting with the specimen rather than being restricted to elastically scattered electrons as is the usual case of the conventional electron microscope, we have an improvement possible in the signal to noise ratio.

Certain restrictions are also associated with scanning electron microscopy operated in the transmission mode. Since the signal being utilized consists of electrons transmitted through the specimen, the thickness of the specimen is reduced to the dimension familiar in conventional electron microscopy. Thus some of the depth information that can be valuable when studying a bulk specimen is not available in the scanning transmission mode. Also, the high intensity guns, in general, require considerably better vacuum conditions than the conventional hairpin filament. These conditions while improving specimen contamination, can also contribute to delays and inconvenience in specimen exchange and place restrictions on specimen outgassing.

The SEM as operated in its most general form (PEASE, 1971), currently has a usable resolution of approximately 1 order of magnitude less than a conventional electron microscope. There is available, however, the scanning transmission mode (CREWE, 1971; KIMOTO et al., 1969) which allows resolutions comparable to that of the conventional electron microscope to be obtained with certain restrictions as to the type of specimen and the operating conditions that must be used.

III. Comparison of Information Transfer

In comparing the information available with scanning electron microscopy, we will find that while the SEM has a lower resolving power, it can offer some additional modes of information transfer (HAYES and PEASE, 1969). Such information can be described as analytic and as subjective, or experiential.

1. Analytic Information

When the probing electron beam interacts with the material of the specimen, several types of radiation are produced and, in addition, alterations in the electrical properties of the specimen occur. Any of these radiations or alterations, if they occur essentially instantaneously, can be utilized to modulate the display cathode ray tube and thus paint out a picture of the specimen. Each of these pictures will contain information that is related to the way in which the secondary radiation has been produced. Among the radiations which have been utilized as a video signal are:

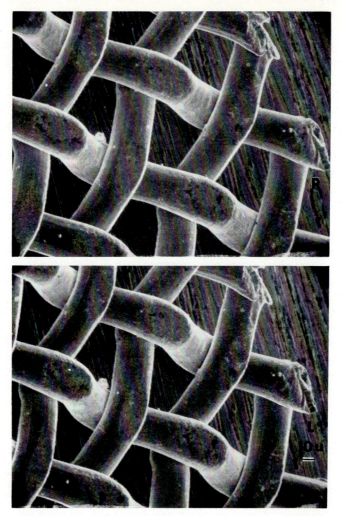

Fig. 2. Stereoscopic viewing. By adding the code of binocular depth and shape perception to the shading and overlap codes usually seen with single scanning electron micrographs, an enhancement of subjective contact as well as resolution of analytic ambiguities can be achieved. These two stereo-pairs ($\times$ 270) represent two views of a two-hundred mesh woven metal grid used in the preparation of electron microscope samples. There are approximately two hundred wires to the inch. Our understanding of the system is enhanced by the addition of the stereoscopic imaging which is added to an already vivid characterization by means of the shading contrast seen in individual members of the pair of pictures. — In order to aid in the viewing of stereo-pairs presented in this chapter it is suggested that the reader construct a small, inexpensive prism (see Appendix 1). The pair of pictures as viewed through the prism with the right eye will be displaced vertically downward as compared to the image viewed simultaneously by the left eye which looks at the picture directly without any optical device. By changing the distance between the prism and the page, the picture intended for the right eye (R) will merge with that for the left (the "R" will overlap the "L") allowing assimilation of stereoscopic information. A non-stereoscopic image for comparison will be seen above and below the center, superimposed image when the page is viewed in this way

visible light (PEASE and HAYES, 1966b; PEASE and HAYES, 1967; MANGER and BESSIS, 1970; KRINSLEY and HYDE, 1971; REMOND et al., 1970; MUIR et al., 1971), characteristic X-rays (RUSS, 1971; TOUSIMIS, 1969; JOHARI, 1971b; MUIR et al., 1971), induced specimen current (EVERHART et al., 1964; EVERHART, 1966) and backscattered electrons in addition to the most commonly used signal, the low energy or secondary electron radiation.

The SEM, by utilizing these diverse radiations, can form images which identify chemical bonds, elemental composition, electrical properties, and topography of the specimens. Thus in terms of the types of interaction that can be recorded in the image, the SEM is somewhat more versatile than the conventional electron microscope.

There is also some advantage in the ability of the SEM to view bulk specimens. If geometric analysis is required in three dimensions, the rather difficult procedure of reconstruction from serial sections (STACKPOLE et al., 1971) is necessary with the conventional electron microscope since each specimen observed in this instrument must be thin enough to transmit the electrons used to form the image.

Since many of the three dimensional geometries are of considerable importance in biology, the additional information available through the ability of the SEM to sample depth and shape becomes important. This is particularly true in the areas of the nonmetric, nonprojective, enumerative geometries encompassed by the field of topology (RASHEVSKY, 1961).

2. Subjective or Experiential Information Transfer

The possibility of complementing objective, analytic information transfer with subjective, experiential aspects of image-observer interaction can be investigated using SEM techniques since several of these techniques mimic in their functional character the mode by which the individual experiences the world around him (HAYES et al., 1969). Thus, the SEM can be utilized to extend our senses as well as to accumulate analytic data. Such an ability accounts for much of the attraction that scanning electron micrographs possess for scientists and laymen alike. In addition to being an attractive method of information presentation, such subjective contact can allow a more fundamental investigation of the usefulness of non-analytic information transfer as a technique in the imaging of biological systems.

B. Specimen Preparation

The requirements of specimen preparation for scanning electron microscopy are no less rigorous and demanding than those of conventional electron microscopy or any other form of microscopic investigation

(Boyde and Wood, 1969; Pfefferkorn, 1969; Pease, 1964; Baker, 1958). The investigator using the scanning electron microscope often finds the majority of his time spent in specimen preparation when compared to time spent in actual viewing with the microscope. Specimen preparation begins with the selection of appropriate tissue material and proceeds through the necessary fixation, dehydration and conductivity considerations.

I. Selection of Tissue

One of the most critical decisions to be made in any scanning electron microscope study is to determine whether or not there is a reasonable chance that scanning electron microscopy can yield useful information about the subject. This decision depends not only on the material itself, but also on the kind of information that is required about the specimen. In many cases, it will be more appropriate to investigate the material using conventional electron microscopy, light microscopy, or electron micro-probe rather than the scanning electron microscope. Limits of resolution, the fact that most of the techniques will reveal only the outermost surface of the specimens, and the severe environmental conditions that the specimen must experience in the column of the instrument, are factors that should be kept in mind. However, a large variety of biological materials of many types have been examined successfully (Carr, 1971; Boyde and Williams, 1968; Barber and Boyde, 1968; Pease and Hayes, 1966a; Pease and Hayes, 1968a; Smith et al., 1971; Thornhill et al., 1965; Hayes et al., 1966; Golomb and Bahr, 1971).

1. Natural Surfaces

Since the probing electron beam penetrates only a few microns into the specimen and in certain operational modes such as secondary electron, the signal emerges from an even thinner layer of the specimen, it is generally a surface that we are investigating in the scanning electron microscope. If a biological system contains information distributed on a surface that is in contact with either air or a solution, the preparation for scanning electron microscopy is relatively simple and does not require dissection or sectioning techniques. Such surfaces, however, must be considered quite carefully with respect to fixation and dehydration in order to prevent the introduction of unsuspected artifacts. At times the type of information that is required will remain invariant under certain deformations (Hayes, 1972), in particular, certain topological properties of shape will remain invariant under many types of deformation. For most purposes, however, the surface integrity must be maintained without distortion and the proper handling of these surfaces through fixation and dehydration as outlined below is essential (Boyde and Wood, 1969).

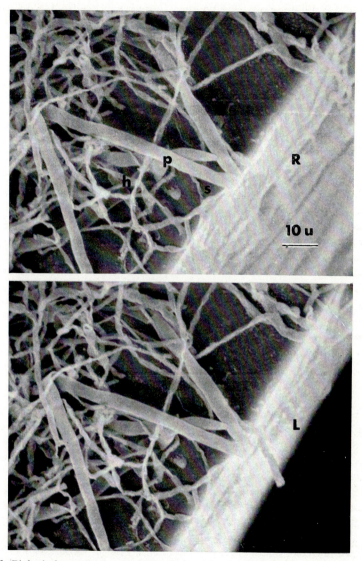

Fig. 3. Biological organization in depth. Stereo-pairs ($\times$ 1000) showing the growth of organisms on a soft contact lens (SPENCER, et al.). When viewed stereoscopically, the organism (Aspergillus fumagatus) exists as strands reaching high off the surface of the lens material. Note that strands on the surface (s) can be distinguished from strands (h) high above the surface. Also note nearly vertical post (p) difficult to recognize in single picture (MATAS, B. R., SPENCER W. H., HAYES, T. L.: SEM of hydrophilic contact lenses. Arch. Ophthal. **88**, 287—295, (1972).)

The eye might be taken as an example of a biological system where considerable information is displayed on surfaces which are relatively accessible for scanning electron microscope viewing. The outer and inner surface of the cornea, the surface of the lens, the rather intricate filter system of the trabecular meshwork and even parts of the retinal complex can be prepared with little but simple gross dissection required. These surfaces of the eye are both cellular and noncellular and their study can yield information which can be correlated with conventional electron and light microscopy (SPENCER et al., 1968; SPENCER and HAYES, 1970; BLUMKE and MORGENROTH, 1967; KUWABARA, 1970; MATAS et al., 1971).

Leaf surfaces (HESLOP-HARRISON and HESLOP-HARRISON, 1969), insect cuticle morphology (SOKOLOFF et al., 1967; HARTMAN and HAYES, 1971), and paleobiological samples (ERBEN, 1970), are other examples of specimens where information can be obtained directly with a minimum of dissection or sectioning.

Systems composed of individual cells offer surfaces which are readily accessible for scanning electron microscopy. Investigations of blood cells in both normal and pathological states (BESSIS and LESSIN, 1970; KAYDEN and BESSIS, 1970; CLARKE and SALISBURY, 1967; SALISBURY and CLARKE, 1967; WARFEL and ELBERG, 1970; MICHAELIS et al., 1971), tissue cultured cells (SHEIE and DAHLEN, 1969; BOYDE et al., 1968), and of protozoa (SMALL and MARSZALEK, 1969; HORRIDGE and TAMM, 1969) are examples in this category.

2. Dissected Material

Most biological material exists in three dimensional solid arrays which do not offer the open surfaces discussed in the section above. For such material quite elaborate dissection techniques are necessary in order to reveal the parts of the specimen which are of interest. LEWIS, in his work on neuronal tissues has developed several of these ultradissection and tissue clearing techniques (LEWIS et al., 1969a; LEWIS et al., 1969b; LEWIS, 1971; HILLMAN and LEWIS, 1971). Extending the work of THOMAS in conventional electron microscope preparations (THOMAS, 1969), LEWIS has also applied techniques of low-temperature ashing in the preparation of scanning electron microscope specimens. This procedure can help to clear the specimen of unwanted tissue components that would tend to obstruct the view in the scanning electron microscope and can be augmented by the introduction of nonvolatile components through the process of staining.

It has been suggested that a combination of dissecting and ripping of tissue following aldehyde-perfused and fixed tissue is a very suitable method for revealing internal surfaces (BOYDE and WOOD, 1969).

The cytological techniques used to separate cell components can be utilized for the preparation of SEM material. Isolated stained chromosomes have been studied and their SEM morphology correlated with the light microscope image (SHEID and TRAUT, 1971).

3. Sectioned Tissue

It is possible to utilize a standard microtome to prepare sections of biological material where the surfaces to be investigated lie deep within a solid organ or tissue. These sections can be viewed in the SEM after certain preparative techniques and with proper care in interpretation (ELIAS, 1971) can provide useful structural information. The preparative techniques of fixation, dehydration, and embedding often will not preserve ultrastructure in terms familiar to conventional electron microscopists but may retain some useful information at the cellular and very large polymer level. By investigating such sections utilizing the SEM, we can take advantage of the increased resolving power when compared with the light microscope and the ability of the scanning electron microscope to visualize the sides or vertical surface of these sections (MCDONALD et al., 1967). Such sectioning technique has found application in the study of healing wounds (FORRESTER et al., 1969a; FORRESTER et al., 1969b), certain parts of the eye (SPENCER et al., 1968), central nervous system (MCDONALD and HAYES, 1967; BOYDE and WOOD, 1969) and heart muscle (POH et al., 1971).

The technique in general, consists of cutting paraffin sections (or utilizing slides previously prepared for light microscopy, (POH et al., 1971) after following the general procedures utilized in light microscope histological preparations. The sections are cleared in a solvent such as xylene, dried, metal coated and viewed in the SEM.

4. Living Specimens

A few organisms can retain sufficient water to survive the rather extreme environmental conditions (CROWE and COOPER, 1971) of vacuum and radiation encountered in the column of the SEM. Such specimens, while generally very much reduced in any physiological activity while being viewed, recover after removal from the column and present a unique type of specimen for study (PEASE et al., 1966; HUMPHREYS et al., 1967; HARTMAN and HAYES, 1971; PEASE and HAYES, 1968a). The preparative procedure for these specimens is minimal. In general, no fixation, dehydration or coating with conductive material is required. The elimination of preparative steps and the rather strict criteria of survival means that fidelity is probably quite good in these preparations.

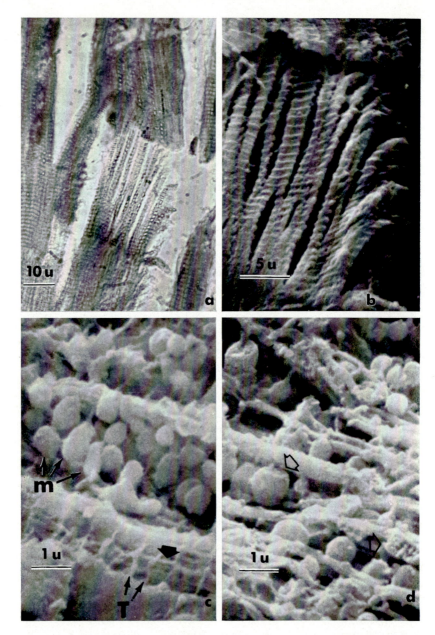

Fig. 4. Sectioned material. Thick sectioned material prepared for light microscopy can occasionally be a useful subject for scanning electron microscopy. Such material can be compared directly in the two instruments, and information gained through the interaction of visible light with specific parts of the specimen can be used to make identification

5. Ion Etching

Ion etching, pioneered by BOYDE and STEWART represents an additional technique for revealing underlying surfaces (BOYDE and STEWART, 1962; STEWART and BOYDE, 1962; BOYDE, 1967). In this procedure, a beam of ions such as argon is utilized to blast away the surface of the tissue in order to reveal the underlying material. It is necessary to try and distinguish between structures that have been revealed as a result of removal of the outer layers of material from those structures related to the ion etching process itself. BAKER has considered some of the factors important to an interpretation of ion etched material, particularly in reference to the images of ion etched red blood cells (BAKER, 1969).

6. Freeze-Etching Techniques

Because of the obvious relationship in terms of surface investigation, the higher resolution freeze-etching technique utilizing conventional electron microscopy must always be considered as an alternative to the secondary electron mode scanning electron micrograph. The freeze-etching technique has proven to be of great usefulness in the investigation of surfaces where the qualities might be described as rough, but not too rough. That is, if the surface projections are relatively small (such as the particles that have been characterized on the surface of membranes), the freeze-etching technique can be of great value. Many aspects of freeze-etching technique have recently been reviewed by KOEHLER (KOEHLER, 1968).

The addition of complementary replicas techniques where both surfaces of a fracture can be compared (STEERE, 1971a) and the further enhancement of depth information by stereo-pair micrographs (STEERE, 1971b) has

---

of the scanning electron micrograph morphology more certain. Fig. 4a ($\times 900$) shows a section of heart muscle photographed utilizing polarized light that allows the bire-fringent areas associated with the muscle striations to be identified. Fig. 4b ($\times 2,700$) shows the same section in the scanning electron microscope and identification of individual bands can be made by assigning specific landmark points to the two micrographs. In this fashion it can be shown that the ridges seen on the surface of the myofibril in the scanning electron microscope image are associated with the isotropic regions of the myofibril, thus they cross the myofibril at the level of the Z band and correspond to the position of the transverse T tubule system as described in the conventional electron microscope. Figs. 4c ($\times 11,000$) and 4d ($\times 11,000$) show mitochondria within rabbit heart muscle cells. The material was paraffin embedded and sectioned for light microscope examination prior to clearing and viewing in the scanning electron microscope. The mitochondria can be seen as closely packed bunches of oblong particles (m), lying between longitudinal fibrils (closed arrow). Occasionally, there is an indication of a mitochondrion that has been cut open to reveal the internal cristae structure (3 open arrows)

measurably improved our ability to interpret surface and shape characteristics by this technique. The depth of the replica is quite striking and the high resolution possible (30 Å) is bound to impress the scanning electron microscopist who has struggled to achieve a 100 Å resolution.

Depth and shape characteristics of the replicas can also be seen by scanning electron microscopy. It is possible to correlate the appearance of identical areas using SEM and CEM (McAlear et al., 1967). As will be discussed in further detail below, the freezing process is at times difficult to carry out without introducing artifacts of structure and the interpretation of replicated frozen surfaces can sometimes be a challenging task. In general, however, the freeze-etching technique offers an extension of resolution and an alternative method of imaging underlying surfaces that can be exposed in at least a semi-controlled fashion. Cryofracture without replication can also be a useful method for revealing specimen interiors for the scanning electron microscope (Haggis, 1970; Lim, 1971). Frozen material may also be examined at low temperatures without dehydration (Echlin, 1971).

II. Fixation

1. Ultrastructure Fixatives

Although there are significant differences in terms of the embedding matrix and required resolvable detail between the SEM and conventional electron microscope preparations, the most often used fixatives for scanning electron microscopy are those adapted from conventional electron microscope techniques. For example, osmium tetroxide and OsO_4—$HgCl_2$ fixative (Parducz, 1967) has been used very successfully to fix protozoa for SEM examination (Horridge and Tamm, 1969; Small and Marszalek, 1969). A formaldehyde-glutaraldehyde fixative of high osmolality first suggested for use in conventional electron microscopy by Karnovsky (Karnovsky, 1965) was utilized in a scanning electron microscopy study of the structure of avian lung (Nowell et al., 1970) and has found general application in our laboratory as a SEM fixative.

The fixation methods utilized in conventional electron microscopy are discussed thoroughly by Pease (Pease, 1964) and certain of the specialized applications of fixation of biological samples for scanning electron microscope examination have been investigated (Arnold et al., 1971; Arenberg, 1971; Boyde and Wood, 1969; Marzalek and Small, 1969).

2. Light Microscope Fixatives

Occasionally the fixative of choice is more related to standard light microscope fixation than to the specialized ultrastructural fixatives used in conventional electron microscopy. PARDUCZ's fixative (PARDUCZ, 1967) was originally developed for instantaneous fixation of the ciliates. BOYDE has found that this fixative is very suitable when used to harden the free surfaces of soft tissues, isolated cells, or cultures (BOYDE and WOOD, 1969).

The cross linking or precipitating properties of fixatives (BAKER, 1958) can be utilized to produce the desired degree of elasticity or brittleness in a tissue prior to dissection. BOYDE has found that glutaraldehyde and formaldehyde make tissues tough, while PARDUCZ's fixative makes them very hard and brittle (BOYDE and WOOD, 1969).

Finally, it may be useful to examine material which had originally been prepared for light microscopy (POH et al., 1971). In this case, the preparative procedures are those associated with the standard light microscope preparation and as a result, the degree of fidelity is not comparable to the ultrastructural fixation and imbedding common in conventional electron microscopy. However, the advantages of having access to historical material prepared by standard light microscope pathology techniques may outweigh considerations of ideal fixation.

III. Dehydration and Drying

Dehydration, usually through a graded series of alcohols, is a familiar process in both light microscope and conventional electron microscope techniques. This dehydration is a necessary prerequisite to imbedding in the non-aqueous matrices utilized for the sections which are the standard specimens in these techniques. In scanning electron microscopy it is often necessary to carry out such dehydration for the purposes of freeze-drying from a non-polar solvent, or for substitution of a liquid which can be conveniently removed through critical-point drying.

In general, however, the specimens for the scanning electron microscope do not consist of sections, and therefore are not surrounded by any supporting matrix. The need in scanning electron microscopy preparation is to remove the water completely rather than to substitute an embedding material for the water originally present. The techniques utilized for this removal of the water have generally been derived from the dehydration techniques utilized in conventional electron microscopy and have drawn heavily on the technology of freeze and critical point drying techniques. These will be discussed below.

1. Freeze-Drying

Freeze-drying has probably been the most often used technique to remove water from scanning electron microscope preparations with a minimum of artifactural change. Although freeze-drying has sometimes

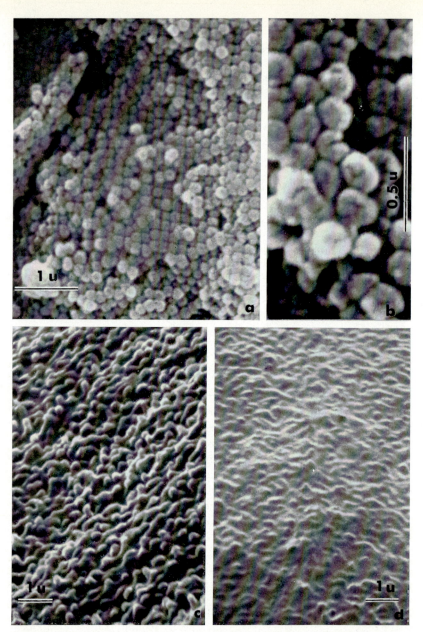

Fig. 5. Standard specimens and drying artifacts. Precious opal can be utilized as a convenient standard for the assessment of resolution and stability of the scanning electron microscope. In addition to the array of spherical units visible at a moderate degree of magnification (Fig. 5a, ×17,000), a very fine internal structure associated with each individual sphere can be seen at higher magnifications (Fig. 5b, ×50,000). The spherical particles also offer a very useful geometry for the correction of astigmatism (see also Fig. 8). The effects of air drying as compared to freeze drying can be seen in the lower pair of photographs (×19,000) showing the corneal epithelium of the rabbit (MATAS et al., 1971). In Fig. 5c the microvilli are quite well preserved following freeze-drying, while in Fig. 5d the surface has been greatly flattened and detail has been lost following the air drying process

been referred to as an "artifact-free" method, it is as prone to difficulty in this direction as other ultrastructure preparative techniques (KOEHLER, 1968; BOYDE and WOOD, 1969). The artifacts of preparation may arise during the freezing process or during the sublimation of the ice under the low temperature, high vacuum conditions of this technique.

After considerable experimentation, BOYDE and WOOD have concluded that the tissue surface is least distorted following freeze-drying from a non-polar solvent, e.g. Amyl acetate can be substituted for the tissue water. Freeze-drying from water is a relatively slow process requiring days or even weeks of vacuum evaporation at low temperatures (—70 °C). By substituting the non-polar solvent, the freeze-drying time can be shortened to a matter of under one hour (BOYDE and WOOD, 1969). A second advantage is that the non-polar materials freeze more in the form of an amorphous glass than of a typical ice crystal. Since ice-crystal formation is one of the most serious problems associated with freeze-drying, substitution of organic liquids for the water is a definite advantage. At the same time the possible chemical alterations produced by changing the chemical environment from water to a non-polar solvent must be considered a drawback.

Several commercial freeze-drying units are available. It is advantageous that the unit be equipped with an electronically cooled stage where the temperature of the specimen can be monitored continuously and a specimen chamber that is free of contamination produced by outgassing of chamber components or back-streaming of oil vapor from the mechanical pump. Since the specimen temperature is often maintained in the —70 °C to —80 °C range it is also useful to have a cold trap at liquid nitrogen temperatures to assist in reducing the contamination on the surface of the material during freeze-drying.

Upon completion of the freeze-drying (approximately 3 days for specimens of 1—3 mm in thickness, BOYDE and WOOD, 1969), the specimen is warmed to slightly above room temperature before air is readmitted to the vacuum chamber. Such warming is necessary to prevent condensation which might re-hydrate the specimen.

The specimen can now be stabilized by coating with a thin layer of carbon, which will help to maintain the dried state for specimens which are hydroscopic. Most specimens also will be coated with a thin layer of conducting heavy metal by evaporation techniques (see below). BOYDE has developed a vacuum station in which both freeze-drying and the evaporative steps assoiated with the carbon and heavy-metal layers can be carried out in a single chamber during one vacuum cycle. Such a procedure eliminates the danger of rehydration, but the very long times associated with freeze-drying means that the evaporation unit will be tied up for long periods of time and not available for other uses.

2. Critical Point Drying

A second method for removing the water from the specimen without the damaging effects of surface tension is the critical-point technique developed by Anderson (Anderson, 1951). In a two-phase system such as the liquid water in equilibrium with the gaseous water vapor over the specimen, there is a point in terms of the environmental temperature and pressure applied to these two phases where the liquid phase will become indistinguishable from the vapor or gas phase. At this point the liquid will be converted to the gas which can then be removed from the system without the surface tension effects associated with air drying. The values of temperature and pressure, where this transition occurs, is called the "critical point."

Water is not a practical material for the critical-point process because of its extremely high pressure and temperature values; therefore, the water of the specimen is replaced; first, by an intermediate liquid and then by the transitional liquid having the critical point within the range of design characteristic of a laboratory apparatus. Anderson utilized ethyl alcohol as the dehydrating agent, amyl acetate as the intermediate liquid, and carbon dioxide as the transitional liquid. Boyd has utilized this method in the construction of a critical-point apparatus for scanning electron microscopy. A detailed description of the apparatus and its use is contained in his article describing preparative techniques (Boyd and Wood, 1969). He found generally satisfactory results but noted a 10% bulk shrinkage artifact with this method, as measured in certain tissue samples.

The lower critical point associated with the fluorocarbons, "freons", has been utilized by Cohen in the construction of a critical-point apparatus using readily available commercial components and capable of rapid and efficient operation (Cohen et al., 1968).

At least three commercial models of critical-point apparatus which operate with CO_2, freon, or both, are now on the market.

A comparison of freeze-drying and critical-point drying as preparative techniques for scanning electron microscopy is demonstrated in two papers dealing with the structure of protozoa. The freeze-drying technique was utilized in one (Small and Marzalek, 1969), while the other utilized the critical-point technique for drying similar osmium fixed protozoan specimens (Horridge and Tamm, 1969). The two methods can be used together in order to identify freezing and chemically induced artifacts.

3. Air Drying

The surface tension effects associated with drying in air in general distort and obscure the details of the specimen. Occasionally however, this surface tension flattening can be an advantage (e.g. in the disclosure of internal granules of a cell which would not be revealed if the spherical form of the cell were maintained through freeze drying or critical point drying, McDonald and Hayes, 1969). Cell junctions may be emphasized by air drying (Spencer and Hayes, 1970), and the A, I, and Z bands in striated muscle fibers can also be accentuated by this technique. Boyd and co-workers have demonstrated that sub-surface features such as nuclei and mitochondria can also be revealed (Boyde et al., 1968; Boyde et al., 1969). Air drying is useful when certain kinds of information are to be extracted from the specimen preparation (e.g. comparative form, Roth, 1971) but in general it cannot be substituted for the more faithful methods of freeze drying and critical-point drying.

IV. Improving Conductivity

Since it is generally desirable to maintain the electrical potential of the specimen surface at ground level, the specimens for scanning electron microscopy are usually coated with a thin layer of conducting material by vacuum evaporation techniques prior to viewing in the instrument. Most biological specimens in the dehydrated state conduct electricity rather poorly and this layer prevents the build-up of beam induced charge and the resulting artifacts. Living material, where the presence of the hydrated system seems to prevent an accumulation of charge (Pease et al., 1966) is an exception, usually requiring no metal coating. The interpretation of the bright areas and image distortions produced by charging has been discussed by several authors (Boyde and Wood, 1969; Pawley, 1972a, c).

1. Metal Evaporation

Most frequently the conducting layer consists of a layer of carbon, which provides for a well covered surface, followed by an outer layer of heavy metal, usually gold, that has been evaporated on the sample in order to improve the secondary electron coefficient and to limit somewhat the interaction volume. In order to achieve uniform coverage it is essential to rotate the specimen quite rapidly. The carbon layer (not more than 200 Å thick) is tough and stabilizes the specimen while providing a good substrate to which the gold layer is firmly adherent. Carbon is easy to evaporate, causes little specimen heating, and is easily scattered by the residual air molecules to form a continuous layer over the specimen. Gold is very easy to evaporate from a tungsten filament, has a reasonably small granularity by SEM standards and has a high secondary electron emission coefficient (Boyde and Wood, 1969).

At very high scanning electron microscope magnifications the granulation of the gold film is sometimes observable and may become a problem. In order to reduce the granularity the metal coatings utilized in high resolution conventional electron microscopy can be used for scanning

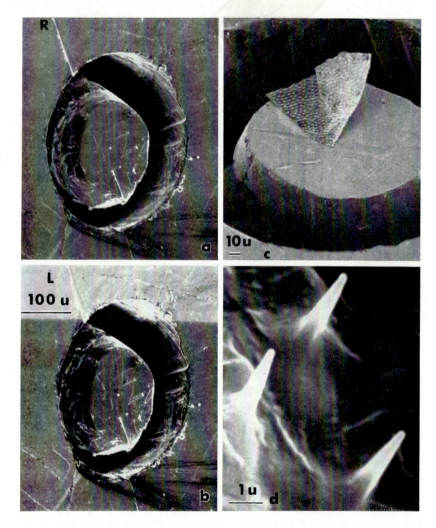

Fig. 6. Charging artifact (J. PAWLEY, 1972c). Fig. 6a, b (× 90) is a stereo-pair showing the micro-stub used for holding specimens at a fixed potential. Again, a prism can be used to superimpose "R" (right eye) with "L" (left eye) for stereo viewing. The small stub is electrically isolated and can be set at a potential different from that of the surrounding surface. In this way, many of the charging artifacts can be produced under controlled conditions. Fig. 6c (× 300) shows a small fragment of the wing of the flour beetle (Tribolium confusum) attached to the micro-stub and Fig. 6d (×9,000) illustrates the setae (bristles) on this wing as the specimen is held at a potential of minus 90 volts. Binocular addition of grey levels can also be illustrated by simultaneously stereo viewing the bright (6b) and dark (6a) band at the top of the pair

electron microscope preparations. Very small granularity and relative ease of evaporation from tungsten is possible with a platinum-palladium alloy or with chromium.

Specimen damage sustained during the metal evaporation procedure due to heating is probably quite small. KOEHLER has discussed the temperature rise associated with radiant heating and the heat associated with phase changes experienced by the condensing metal on the specimen surface. For most evaporation procedures it would seem that the temperature increment at most is one or two degrees Centigrade which is far below that necessary to produce volatilization of the components of the dried tissue sample (KOEHLER, 1968).

2. Conducting Sprays and Solutions

For certain studies it is possible to spray an antistatic agent onto the specimen as an alternative to metal evaporation (SIKORSKY et al., 1968). The antistatic agent, (e.g. Duron) can be applied very conveniently and without exposing the sample to the vacuum of an evaporator. It tends, however, to cover small surface detail and therefore is most useful at the lower magnifications.

A liquid such as a fatty acid mixture can also be applied to the specimen surface in order to improve conduction under electron bombardment. Again, residual material can obscure fine detail and introduce unwanted artifact. In one of the early studies of mammalian tissue, soaking of tissue specimens in solutions of metal salts was tried with only moderate success (JAQUES et al., 1965).

C. Viewing Techniques

The actual operation of a scanning electron microscope for biological investigation can be carried out by the investigator himself or through the use of a specialized technical operator. The wear and tear on an instrument is, of course, increased by having a multitude of different operators, some of whom have relatively little experience in the operation of intricate electronic apparatus. However, if the operator is utilized there is always the difficulty of the biologist working "over his shoulder." The technical operator is more proficient in the operation of the instrument but may lack the biological background in the specialty of the investigator. This background often seems important in the interpretation of images and in the selection of the field, magnification, viewing aspect, etc.

Often the scanning electron microscope represents a joint investment of several departments or divisions which may include the physical sciences, engineering, biology, and the medical sciences. Instrument design, which in the early days of the commercial instrument was directed rather exclusively toward the engineering and physical requirements, has been extended to include instrumentation designed more directly for the needs of the biological science investigator. Several companies have instruments which are directed toward the operation by the biological investigator himself, and if price considerations can be improved, it would be hoped that small scanning electron microscopes will be available, not only within the biological departments but even, perhaps, in the individual laboratories of the biological investigators.

As the operator faces the scanning electron microscope, it is apparent that there are many variable operating parameters that must be considered in the production of the image. In order to feel comfortable with all of the variables involved, it is necessary to sit down and actually operate the instrument. During this learning period some of the following suggestions may prove useful.

I. Standard Specimens

If we are to make reasonable progress in our attempt to integrate the various operating parameters it is very useful to have a specimen which is stable and contains morphological detail that can be used to test the operation of the instrument.

Standard specimens which are biological in nature have the advantage in that their density, surface texture, and chemical composition is more directly related to those samples which will probably be under investigation. MANNING has shown that tobacco mosaic virus paracrystals and serum lipoprotein macromolecules might be used as standard specimens (MANNING et al., 1968)

In general, however, non-living material offers a much greater specimen stability, and can be used for repeat testing over long periods of time. Mineral samples can be prepared with surface structures that test the maximum resolving power of the instrument, and with a judicious choice of density and elemental composition we can still approximate the composition of biological material.

We have found that a sample of precious opal, that has been etched with hydrofluoric acid and then coated with a conducting layer by evaporative techniques (JONES, 1969) is a very satisfactory standard specimen. Such a specimen remains stable over many months and contains structure at two size levels: one corresponding to the moderate range of magnifications of a scanning electron microscope and a fine structure that is near the resolution limit of our present instruments. Opal has been investigated by light microscope and conventional electron microscope techniques (JONES et al.,

1964; Darragh et al., 1966; Sanders, 1964) and thus has the additional advantage of presenting a well-established structure that has been identified by several imaging methods.

II. Signal Monitor

If we are to make the necessary quantitative appraisal of the performance of the scanning electron microscope it is essential that we have a method for monitoring the information signals (Pawley et al., 1969). Most commercial instruments have this capability to a more or less satisfactory degree. The signal monitor acts as an exposure meter for the photographic recording of the image and the signal displayed should represent as nearly as possible the amount of light at each point on the display cathode ray tube. If the monitor displays this signal, both the brightness and the contrast levels can be read directly, and no subjective estimates or dependence on prior experience is required for efficient photographic recording. Completely automatic photographic systems are also now available.

By observing a single line of the scan (utilizing a vertical defeat circuit) and sharpening the peaks of the display a best focus position can be determined. Such focusing is particularly useful at very high magnifications where the required low specimen current for maximum resolution often means that the image will be quite reduced in intensity.

The monitor can also help in the analysis of signal distortions produced by stray fields, electronic failure, or mechanical vibration. Identification of the difficulty is often aided by an ability to determine both the frequency and the amplitude of the imposed distortions.

III. Accelerating Voltage

The operator generally has a considerable range of accelerating potentials from which to select, accelerating potentials from 1 kV to 50 kV being most often utilized. It has been suggested that for the observation of biological material accelerating voltages no higher than 10 kV be used. Adequate resolution, at least for certain microscopes, may only be obtained at a higher accelerating potential and some experimentation is usually in order. It is probably best for the operator to try several potentials, keeping in mind that such things as increased interaction volume and the charging effects produced require careful interpretation. If this is kept in mind, it seems that for certain samples the increased signal brightness produced by slight charging effects at higher voltages might actually aid in the visualization of some detail.

IV. Specimen Current

The specimen current selected by the operator should be as low as possible while still enabling the formation of a relatively noise-free image. The ability to measure specimen current is quite useful and the micro-micro-ammeter usually used for this purpose is often consulted. The probe diameter is controlled in part by the specimen current (see Eq. (1)), and for the usual high resolution work currents in the range of 1 or 2×10^{-11} A are desirable. For survey work at somewhat lower magnifications specimen currents in the 10^{-10} A range are often useful, since they provide a considerable increase in the signal and therefore improve noise considerations.

For operational modes other than secondary electron collection it may be necessary to increase the specimen current even further. Cathodoluminescence and characteristic X-ray images can be produced in general only with a considerable increase in the specimen current. Resolution in these modes is generally determined by considerations other than probe diameter, and specimen currents as high as 10^{-8} or 10^{-7} amps are commonly utilized.

V. Contrast; Photo-multiplier

In operating a scanning microscope the contrast of the image is often controlled by regulating the voltage of the photomultiplier tube. Contrast within the range of recording film can be set subjectively, utilizing the display cathode ray tube, or can be determined by measuring the peak height of the signal monitor display. If a process of re-photographing is to be used, the possibility of increasing contrast through successive photographic reproduction should be considered. The selection of a contrast level somewhat below that which might appear optimum on the display cathode ray tube is suggested. Extreme contrast in the image is often a result of charging and is best corrected by improving the conduction of the sample (see B.IV above).

VI. Scan Rate

Frame speeds ranging from hundreds of seconds to TV rates (1/30th of a second or faster) can be utilized in scanning electron microscopy.

The slower scan rates are used when it is necessary to extract statistically significant information for each of the picture elements utilized in a photographic recording of the image. Typically a frame speed (or exposure time) of about 40 seconds has been used. However, a 40-second frame cannot be integrated by the brain and for direct viewing and for the observation of dynamic techniques the much faster sweep speeds are called for.

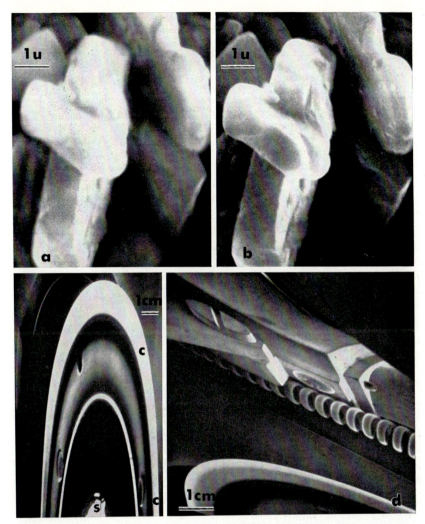

Fig. 7. Specimen current and resolution. Specimen current is an important parameter determining, in part, the available resolution of the SEM image. Figs. 7a ($\times$ 9,000) and 7b ($\times$ 9,000) compare the resolution at specimen currents of 1×10^{-10} amps and 0.7×10^{-11} amps respectively. For maximum resolution, the specimen current must be reduced to the lowest value consistent with a relatively noise free image. In order to determine some of the quantitative results of variations in specimen current, beam current, and collected secondary electron current, it can sometimes be useful to utilize a small stage mounted electromagnet (J. PAWLEY) to deflect the beam directly onto the scintillator. Characteristics of the raster are maintained quite well as shown by the imaging possibilities of such a deflected beam. In Fig. 7c the scintillator (s) and surrounding collector shield (c) can be seen at a magnification of less than one, ($\times 0.5$). In Fig. 7d ($\times 1$) a detail of the stage mount can be recognized after deflection of the beam by the stage mounted magnet. Calibration of the scintillator-photomultiplier can be carried out by such direct deflection of the incident beam to the collector after first measuring the beam current with a Faraday cup specimen

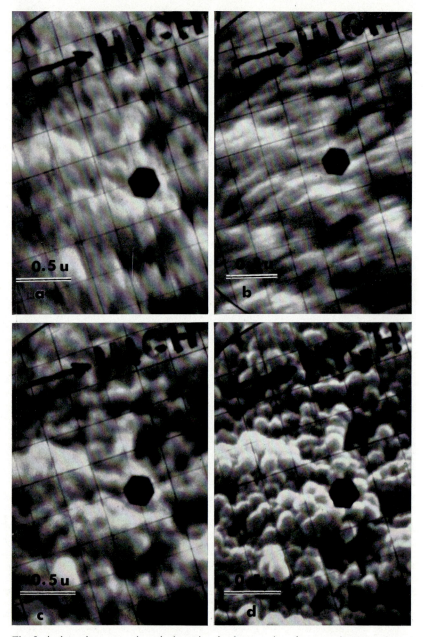

Fig. 8. Astigmatism correction. Astigmatism in the scanning electron microscope image is a common limitation to resolution. The operator can correct this aberration by applying an astigmati m that is equal but opposite to the inherent astigmatism found in the instrument at that time. A rectilinear grid inscribed on a transparent plastic disk is

Television scan rates (KIMOTO et al., 1969) are very useful for focusing and for observing astigmatism and movement of the specimen. Survey work, that precedes a more detailed high magnification study, also benefits from flicker-free display. TV scan rate helps to correlate adjustment of the focus control with the instantaneous image; eliminating confusion with the long-persistance image used for visual observation of lower scan rates.

In order to assess specimen movement in a smooth and accurate fashion it is necessary that the frame speed approach standard TV rates and the use of a micromanipulator in the stage of a scanning electron microscope can only be accomplished if the images can display movement. For this type of micro-dissection it is essential to be able to observe the specimen and the micro-dissection needles in the TV scan rate mode.

VII. Astigmatism Correction

The process of astigmatism correction, possibly more than any other operator function, will determine the resolution of the final image.

The most effective way to deal with astigmatism is to prevent its occurence. Astigmatism often is the result of contamination in the column of the microscope, in particular, contamination of the final aperture. The best resolution is obtained when the column is absolutely clean and a minimum astigmatism correction need be applied. However, in day to day operation a small astigmatism correction is generally necessary.

Any astigmatism correction contains two parameters: one associated with the magnitude of the correction and the other associated with the direction in which this correctional field will be applied. The fact that two variables are concerned in astigmatism correction makes it somewhat more difficult to manipulate than a single variable adjustment such as focus. The purpose of the astigmatism correction is to introduce into the electron optical system an artificial astigmatism which will be equal in magnitude but opposite in direction to the astigmatism that is inherent in the microscope at the moment.

---

placed over the face of the display cathode ray tube and is rotated in order to locate and mark the direction of astigmatism during the correction procedure. Astigmatism means that different directions focus at different points. We can see in Figs. 8a and 8b that the images are smeared in one direction but sharp at 90 degrees to that direction and that the directions reverse between Fig. 8a (low focus) and 8b (high focus). If we attempt to take an intermediate focus position, we find that there is no position which will give good resolution. Fig. 8c represents an attempt at a best focus, non-corrected image. Fig. 8d shows the beneficial effect of astigmatism correction. Now all directions focus at the same point and the resolution of the micrograph is considerably improved
(Fig. 8a, b, c, d, opal, all × 30,000)

The astigmatic system will focus one direction in the specimen plane at a different focal plane than the orthogonal specimen direction. By determining the direction of low and high focus the azimuthal characteristic of the astigmatism of the microscope column can be determined. It is sometimes helpful in this respect to utilize a plastic overlay that can be rotated and set to record the direction of the astigmatism. The next step is to locate (with the azimuthal control of the stigmator) the proper setting which will produce an astigmatism that is opposite in direction to that found in the instrument. To do this the magnitude control of the stigmator is increased until the introduced astigmatism is considerably larger than that present in the instrument so that the direction can be calibrated independently of any astigmatism that is present in the instrument.

Once the direction of astigmatism correction has been determined it is next necessary to introduce the proper magnitude that will exactly cancel the inherent astigmatism. Some idea of the magnitude of astimatism present can be gained by observing the difference in focus steps between the low focus and high focus directions. Most often the magnitude of the astigmatism correction is set while observing the image at rapid scan rate, increasing the magnitude of the astigmator from zero until the best image is obtained. If the magnitude of the introduced astigmatism is too low, the astigmatism direction will be that of the original determination; if it is too high the direction will be that of the introduced correctional astigmatic field.

VIII. Final Aperture Size

The selection of a final aperture is dictated by the magnification, resolution and by the depth of focus that is required. For very high resolution work, a large aperture is appropriate since high magnification observations are limited in terms of height variations. At low magnifications a large depth of focus is often a prime requirement and for this purpose we should use a smaller final aperture (large objective aperture is possibly 200 microns and a small aperture might be 50 microns, BOYDE and WOOD, 1969). As noted above, the cleanliness and centering of the final aperture is essential if astigmatism of the system is to be low. It is a considerable advantage of instrument design if the final aperture diameter can be selected and positioned conveniently.

IX. Viewing Aspect

The aspect of the specimen shown in the secondary electron image of a scanning electron microscope is a projection of the specimen on a plane perpendicular to the probing beam. If the specimen has structure that is intricate in the sense that certain parts are hidden in this projection, it will be necessary to rotate and tilt the specimen in order to see all parts of it.

A goniometer stage of high mechanical stability and good range in the three translational directions as well as rotation and tilt is an important component of the microscope. In addition to increasing the number of surfaces which can be visualized, an accurate tilting stage allows the production of stereo-pairs (BOYDE, 1970). Special techniques for manipulating the specimen might also be considered (BYWATER and BUCKLEY, 1970).

If the stage is tilted through an angle of 8 to 10 degrees between successive photographs, the two photomicrographs will represent information that can be compared to the two viewing aspects of human binocular vision. Such stereo-pairs allow analytic processing that cannot be achieved through any single photograph (WELLS, 1960; BOYDE, 1970). The stereo-pair images can be recombined by several techniques, analytic and subjective (see also section E.II below) and represent a very large increase in both analytic and experiential contact with the specimen.

X. Micromanipulation

PAWLEY has designed and constructed a micromanipulator that can be used within the column of a SEM (PAWLEY and HAYES, 1971; PAWLEY, 1972a; PAWLEY, 1972b). This micromanipulator utilizes piezo-electric crystals to move two small needles in an independent fashion over very small distances. These needles can be positioned to within 0.1 microns and their response time is very short allowing for joy stick control and good coordination with the hands and eye of the operator. The micromanipulator has proven valuable where dissection or tearing of the specimen is required and has also been used to introduce a microprobe that can be utilized in the study of the charging artifact (PAWLEY, 1972c).

D. Signal Processing

I. Differentiation

In order to compress the signal range and to emphasize characteristics as a function of the rate of change of the signal rather than the signal level itself, several applications of electronic differentiation have been utilized (CREWE et al., 1969; HEINREICH et al., 1970). Differentiation can be utilized as a mathematical tool related to the "lighting direction" as seen by the observer. If the derivative function of an object is considered, we find that the signal itself may approximate a square pulse, but that the derivative of the signal would consist of two peaks, one positive and one negative. The combination of a positive and negative derivative signal when utilized

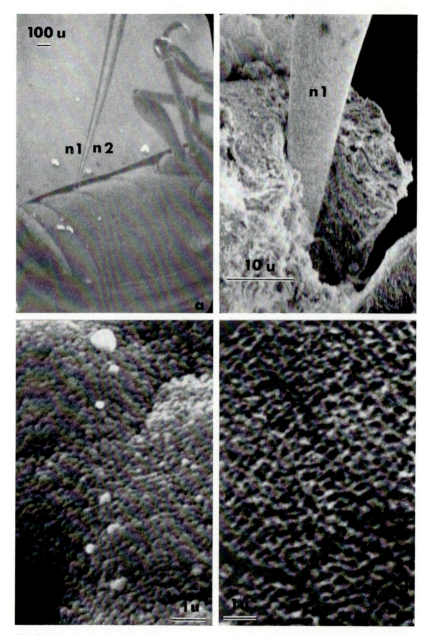

Fig. 9. Micromanipulation (J. PAWLEY). Fig. 9 demonstrates micromanipulation within the column of the scanning electron microscope. Fig. 9a (× 36) shows two independently controlled needles (n 1, n 2) in contact with the surface of the small insect Tribolium confusum. The needles can operate as fine microelectrodes for the controlled application

to modulate the display cathode ray tube, results in "lighting" the object from one side. Thus, where the derivative is positive, the sample would appear bright, while on the opposite side of the sample, there would be a dark area corresponding to a shadow. A mixture of normal signal and differential signal can sometimes be useful in improving the transfer of shape and depth information aiding in identification.

II. Deflection Modulation

EVERHART, WITTRY, and others have developed signal deflection techniques which aid in analysis and can generate striking three-dimensional displays (EVERHART, 1966; WITTRY and VAN COUVERING, 1967). The original signal may have no depth codes (i.e., the specimen-induced current) but may be more effective if displayed or transformed into a spatial mode. Deflection modulation utilizes the information signal to deflect the beam of the display cathode ray tube rather than to change its intensity as is the usual case in intensity modulated images. This process converts the signal strength into apparent height of a solid three dimensional object. Relationships between signal intensities can then be measured in terms of this deflection and at times additional insights into the signal image characteristics are possible.

Deflection modulation has also been utilized in connection with the secondary electron signal in order to emphasize surface texture. It should be noted, however, that in contrast to secondary electron depth coding which is basically a function of the angle between the beam and the specimen at each point, the deflection modulation "height" is an arbitrary function and is completely at the disposal of the operator. To say, therefore, that deflection modulation increases the resolution of height is misleading. Resolution in height (Z direction) of the specimen will not be effected by this method of presentation (modulation) of the image. It is quite true that deflection modulation may make certain height relationships over the surface of the sample more apparent and more easily correlated in the mind

---

of electrical potential and can be quite useful for localized grounding of nonconducting specimens. Fig. 9b ($\times$ 1,800) shows the more normal use of one of the dissection needles (n 1) to separate tissue while it is being viewed in the scanning electron microscope. This picture was made by photographing the face of a TV scan rate display tube. The lower photographs (Fig. 9c $\times$ 9,000, 9d $\times$ 9,000) illustrate two complementary surfaces that were obtained by separating layers of tissue from the stomach of the frog. Such complementary surfaces are somewhat analogous to those achieved through certain freeze-etching techniques. In Fig. 9d (right), the small villi of the surface are represented as holes in the facing surface of the observer, but the basic distances that can be separated in terms of Z resolution are not effected by the generation of an artificial texture through deflection modulation.

III. Color Modulation

In the previous section we have considered utilizing the information signal to deflect the beam of the display cathode ray tube. It is also possible to utilize signal strength to determine the color of each point of the image. In such a color modulation system (PAWLEY et al., 1969) a given signal intensity is converted to a particular color. In color modulation, the assignment of a particular hue or color to a particular signal intensity is at the disposal of the operator, but once this assignment is made it is the signal itself as it is generated at the specimen surface that will determine the color pattern of the final picture (HAYES et al., 1969).

Color modulation can be useful in depicting iso-intensity areas of the image and can also help to accentuate gray level steps that are too small to be regognized without such color coding. By contrasting color between two adjacent gray levels, patterns of signal distribution can be recognized more easily (EVERHART and HAYES, 1972).

Color has also been utilized to code two or more separate information signals collected simultaneously for image display. This process, generally referred to as color coding to distinguish it from the color modulation described above, has been utilized for example, to combine the cathodo-luminescent and secondary electron images in a single picture.

IV. Computer Processing

Certain of the image processing techniques found to be useful in light and conventional electron microscopy (FRANK, 1972; BURKE et al., 1971) and several aspects of transfer theory (BUDINGER, 1971) are applicable to SEM image evaluation. Because the information from the SEM is presented as an electronic, time sequence signal, direct processing by computer is a possibility. In addition to processing the image, the computer may be utilized to control the microscope and generate the raster for the scanning beam. McDONALD has reported a computer SEM system utilizing an IBM 1800 computer which processes the video signal and also generates the raster (MAC DONALD, N. C., 1968). Pattern recognition programs may be utilized to analyze information from the micrographs (LIPKIN and ROSENFELD, 1970) and stereometric analysis has also been carried out using computational methods (DORFLER and RUSS, 1970; BRAGGINS et al., 1971).

E. Recording Techniques

Recording techniques serve to provide a permanent picture that can be reviewed and analyzed and also serve to integrate an image which may be too slow for physiological response time. For specimens where no motion takes place, recording media are generally films or Polaroid positives and for the recording of dynamic phenomena, video tape is the commonly used recording media.

I. Photographic Integration

1. Polaroid Film

Since many of the photographs taken with the SEM are for the purpose of allowing the observer to see an integrated high resolution image, Polaroid positive roll film (Type 42) is often chosen. Since intensity of the display cathode ray tube is not a problem, very fast film is not required. When an image needs to be processed for publication, display, teaching purposes, etc., the positive Polaroid print can be quite satisfactorily re-photographed and the negative produced in this process can be stored for future use. The fraction of photographs, in our experience, that need be rephotographed is small, and the use of positive film is somewhat more economical than the production of negatives directly from each SEM recording through the use of Polaroid PN (positive-negative) film. However, if re-photographing is not convenient or if the percentage of the images where negatives are required is quite high, the PN film can be the preferred method.

Polaroid Polacolor film has been utilized to record color modulation and color coding images (PAWLEY et al., 1969) utilizing the visible light spectrum from a standard CRT display tube. Probably of more general use, in the sense of color conversion, are photographic processes which allow the color conversion to take place away from the microscope itself. This separation provides more latitude in the production of the color of the final color print and also frees the microscope for other uses during the dark-room color conversion.

2. 35 Millimeter Standard Roll Film

The savings in cost of film are considerable if a 35 mm standard roll film is used instead of Polaroid. The obvious loss is the immediate contact with the high resolution image which is provided with Polaroid. If experimental work has already been carried out which will make the selection of image forming parameters quite certain, then it is possible to switch over to a recording mode using 35 mm film and to develop and print a larger number of pictures later at a much reduced cost per picture.

II. Stereo-Pairs

The use of two images taken at two different aspects (tilt angle) greatly enhances both the analytical possibilities and the experiential contact with the specimen. While such a process requires an additional expenditure in time and effort in order to obtain a second picture of the specimen, the results are most rewarding and stereo-pair production has become an essential part of scanning electron microscope operation.

1. Resolution of Analytic Ambiguities

Quantitative, three-dimensional analysis can be carried out by computer techniques and by semi-quantitative modeling techniques (Boyde, 1970; Wells, 1960). In addition to height measurements, it is also possible to utilize stereo-pairs to resolve questions concerning the general shape of the specimen and certain nonmetric properties such as intersection, overlay and general topological properties.

2. Enhancement of Experiential Contact

A single image produced by the secondary electron signal in a SEM utilizes depth codes of shading and overlap to impart to the observer the shape and depth characteristics necessary for experiential contact. By utilizing a second image taken at an aspect that would correspond to the position of the second eye, we can provide the information through the depth code of binocular vision. It is quite clear to anyone who has utilized a good stereoscopic viewer that the enhancement in contact, in the feeling of "being there," is considerable when this additional code is superimposed on the other two. If we are to utilize our intuitive response to systems, it is valuable to make our contact with the image as broad as possible in terms of mechanisms for viewing our environment. The addition of a stereo viewer close to the scanning electron microscope console allows the operator to add this code for depth and shape in a convenient manner.

Individual stereo viewing equipment ranges in price from $ 10.00 to several hundred dollars and, in general, the choice of the equipment depends somewhat on the range of use anticipated. For some individuals it is possible to assimilate a stereo-pair with a minimum of assistance from optical equipment; however, for the average viewer a good quality stereo viewer (perhaps costing $ 200.00) is worthwhile if fatigue and discomfort is to be avoided and if we are to guarantee a maximum stereoscopic participation by all individuals.

3. Methods of Stereo-Pair Presentation

The presentation of stereo-pairs to more than the single viewer is usually accomplished by either slide projection or by some form of printed image. In both cases there are considerable technical difficulties present. If we consider first the presentation of stereo-pairs to a large audience by means of projection, we find that a special projector is generally needed and that the audience must also be equipped with special glasses. Today, most commonly the coding necessary to separate the image for left and right eye is carried out through means of polarized light. However, other methods have been used, notably color coding (red, green) with the same requirement for a coding projector and a decoding pair of glasses for each member of the audience. The alignment of the projected images is some-times quite delicate and for many projection rooms it is not possible to have the alignment satisfactory for all members of the audience. It is also necessary that a special screen (lenticular screen) be used to maintain the distinction in direction of polarization of the two images if polarized light is used.

One modification of this projection technique is to superimpose two color images on a single slide utilizing the appropriate stereoscopic viewer for proper mounting. The requirement for analyzing glasses for the audience still exists (NEMANIC, 1971) (see Appendix 2).

In terms of printing stereo-pairs in a book or journal, we are again faced with rather formidable difficulties. Several techniques have been utilized but each of these techniques have certain drawbacks. Perhaps the simplest technique for printing stereo-pairs is directed towards the some-what selected audience that can assimilate a stereo-pair simply by looking at the two images (perhaps with the aid of a separating card between them). Such a method is very simple but obviously suffers because of the rather small audience that can be expected to achieve success with this method.

Scientific American, in their January, 1954 issue, used a mirror image printed as one of the stereo-pairs. Synthesis of the two images by the reader is achieved by placing a small mirror vertically to the page in between the two pictures.

It is also possible to print the two members of the stereo-pairs side by side and suggest to the viewer that he utilize a stereoscopic viewer to make the synthesis. If a standard design could be set for publication of the stereo-pair with respect to both size and position on the page and to the selection of a standard inexpensive stereoscopic viewer this method would become more accessible.

A fourth method of printing stereo-pairs for large circulation is based on a superimposition of images as was discussed above with respect to the projection of slides. The two images can be superimposed and coded in two colors which will later be separated by the reader by means of glasses or filters. Such a presentation has been quite successful in some instances and the publishers have even considered it useful to include an inexpensive pair of viewing glasses as a part of the book or journal.

The superposition technique can also be utilized in a fashion such that the sorting of the images is accomplished by an overlay layer rather than by requiring glasses to be worn by the reader. This technique generally makes use of small lenticular screen overlays which act as cylindrical lenses and present one of the offset images to one eye and one to the other. Such devices have been utilized to a considerably extent in the production of so-called 3 D postcards. This technique, while very expensive, has been utilized for the presentation of stereoscopic information of scientific value (HARTE and RUPLEY, 1968).

It would seem that each of these methods is rather restrictive either in terms of the auxilliary equipment needed by the reader or in terms of cost. A really workable, inexpensive method for the presentation of stereopairs to large audiences through the media of books or journals is still needed. The vertically mounted system used in this chapter and detailed in Appendix 1 might be a step in this direction.

III. TV Tape

Rapid scan rates are useful in preliminary viewing of the stationary specimens and are essential for the observation of dynamic systems (KIMOTO et al., 1969; PAWLEY and HAYES, 1971). It is sometimes useful to record these dynamic events and for this purpose the most appropriate vehicle is television tape recording. However, the difficulties of presentation of this dynamic information either in published or lecture form are considerable. As a first approach one can simply photograph the face of the TV screen in order to present one or more static samplings of the TV image. Since the function here is solely one of recording for future review or presentation rather than a dual function of recording and also integrating the image for assimilation by the observer, the types of photographic procedures are quite standard and in general would not benefit as much from the rapid processing capabilities offered by Polaroid film. We have found a Speed Graphic with a $4'' \times 5''$ back quite satisfactory for this purpose.

The TV tape recording system associated with the SEM need not be expensive and it is possible to utilize nearly any of the commercially available tape recorders. However, the various brands of tape recorders are not compatible and some care must be taken to insure compatibility between the recorded tapes and the play-back recorders available at meetings or lectures.

Projection facilities in terms of adequate TV monitors throughout the audience are often lacking at scientific meetings and while there are devices which project an image from TV tape onto a screen, they are limited in terms of brightness and mangification and in our experience, have limited applicability. Such TV tape projectors may be considered only when the audience is not too large and when a sufficient number of TV tape presentations will be made to justify their cost.

A more convenient mode of presentation is to have the TV tape material transcibed onto standard 16 mm movie film. The transcription process is rather expensive, but once the material is on 16 mm film, presentation becomes no problem.

The static printing in journal or book form of material from TV tape fails to convey the essential quality of movement. Perhaps it will be possible in the future to distribute the tapes through a kind of "journal of video tape" but for the moment, journal presentations must rely on descriptions provided by authors and a limited number of single frame, still pictures.

F. Information Assimilation by the Observer

The end point in our chain of information transfer which started at the specimen is the transfer of information from the image to the observer. Such information transfer might be considered as taking place along two channels. The first channel relies on the objective, analytic processes associated with our systems of ideas (physics, chemistry, biology, mathematics). It is this analytic process that is by far the most productive and well known scientific channel. However, it has not been without its famous critics. KIERKEGAARD in particular has pointed out the limitations of objective reflection (KIERKEGAARD, 1848). We might explore then, the possibilities of adding subjective or experiential channels of information transfer to the more classical analytic, objective methods. First, however, let us consider the analytic methods that are available to us through SEM (JOHARI, 1971a).

I. Analytic Information Processing

The information available in the image from the SEM might be divided on the basis of its physical characteristics into geometric information, chemical information and information related to the electrical properties of the specimen.

1. Geometric Information

a) Metric Geometry

Metric geometry deals with properties of length, angle and area and the functional relationships between these properties. This geometry is historically the oldest and still serves as the first point of contact between an observer and the analytic properties of the object being investigated.

The metric properties of a sample are invariant under translational changes but are quite sensitive to other transformations such as projection or deformation. Any study of the metric properties of a sample requires preparative methods that insure the fidelity of the angles, distances and areas involved. Also, since metric geometry by definition depends on the measured value, the resolution of the instrument is of considerable importance since it places a limit on the accuracy of measurement.

Perhaps our most common form of analysis is the familiar operation of determining geometric size and shape by measuring the lengths, angles and areas associated with the plane and solid figures represented in the scanning electron micrograph.

Metric geometry is not the only geometry available for study with respect to the spatial characteristics of the specimen. Historically, projective geometry properties are next in order to be considered. Projective properties are somewhat more general than the metric geometry properties in that they are invariant not only to translation but also to projection. Under projective transformation, the properties of distance, angle and area pertinent to metric geometry do not remain constant, but there are other properties (e.g. linearity) which do remain invariant.

b) Topologic Geometry

There is a still more general set of properties which remain invariant under a wide variety of transformations. These are referred to as the topological properties and constitute a non-metric, non-projective, enumerative geometry. Such properties (e.g. genus) because of their invariance to a wide range of transformations, might be considered the more basic spatial properties of the specimen (RASHEVSKY, 1954, 1961).

The technique of studying systems by introducing controlled transformations has been most valuable in physics and mathematics and can also be applied to some extent to imaging studies (HAYES, 1972). In contrast to the usual goal of maintaining fidelity, in these studies transformations (artifacts) are purposely introduced in the hope that by sifting through specimen properties as a function of their invariance under certain transformations we may be able to extend our analytical techniques.

In general, the study of topological properties requires the ability to be able to sample information in three dimensional depth. In terms of available microscopes, the light microscope at high resolution is limited to optical "sectioning" as a result of its very narrow depth of field and the CEM by the physical restrictions of a very thin specimen. The SEM combines a very large depth of focus with the ability to look at bulk samples. Such a sampling of intricate shapes often leads to a better appreciation of certain of the topological characteristics and to some extent can justify a loss in the metric geometry analytic capabilities as a result of the lowered resolution of the SEM (HAYES, 1970b).

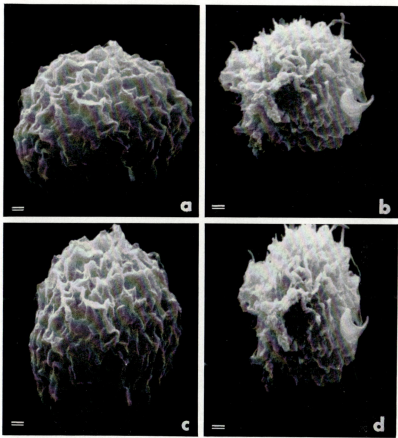

Fig. 10. Geometric transformations. Projective or topologic transformations must be assessed with some care in the interpretation of the images seen in the scanning electron microscope. Figs. 10a ($\times$ 3,300) and 10b ($\times$ 3,300) are "uncorrected" images of spherical cells taken at 45 degree specimen tilt. Figs. 10c and 10d are the corresponding "corrected" images after applying commercially available electronic modification of the raster to take into account the foreshortening of planar measurements taken at 45° tilt. Since the aspect presented in any scanning electron microscope image is that of observation from the electron gun (line of sight parallel to the probing beam), it is clear that the measurements on a tilted plane, as projected from this viewpoint are, in fact, a simple function of the angle of the tilt. However, three dimensional objects such as a spherical cell, although resting on the tilting plane, may not be projected according to this same funtion. A sphere, for example, remains circular in projection regardless of the angle of tilt. Thus, in these pairs of images the circular uncorrected image is accurate whereas the electronic tilt correction has introduced serious distortion into the image. The fact that the long axis of all particles are in the direction of the tilt and that the axial ratio can be calculated from the angle of tilt (on the assumption of spheres as true shapes) gives a strong indication that the "corrected" images are products of an artificial distortion. In any application of geometric factors, it is necessary that the proper dimensional qualities are considered if we are to avoid spurious descriptions (1 µ mark shown). (A. H. WARFEL)

2. Chemical Information

The analysis of chemical composition has been a very useful part of light and conventional electron microscopy. Such chemical information can be presented in tabular form such as micro-photometric or electron diffraction data (GLAESER and THOMAS, 1969) or can be presented as an image such as the areas of specific stain uptake shown in the histochemical micrograph. Both numerical and imaged chemical information can be obtained from the SEM and because of the separation of probing radiation (electrons) from image forming signal (x-rays, Auger electrons, visible light, etc.) the SEM offers some unique advantages in this area (JOHARI, 1971b).

a) Characteristic X-Ray Elemental Analysis

The electron microprobe for many years has utilized the characteristic x-rays emitted from a sample bombarded by electrons to identify and quantitate the specimen elemental composition (ANSELL and JUDO, 1969; KNASTON and STEWART, 1969). These x-rays originate from the excited electron shells (K, L. M, etc.) for the specimen atoms and the x-ray wavelength or energy is characteristic for each element. The characteristic x-ray spectra in general cannot be used to determine which chemical bonds exist between these atoms.

For biological applications, the usefulness of an x-ray attachment for the SEM depends largely on the value of elemental analysis to the biological researcher. While there are some examples of biological systems which are very sensitive to the concentration of a particular element, it is more the general rule that biology is more concerned with the chemical relationships between fairly common atoms. It is possible that as we limit the area studied by utilizing the very fine probing beam of the SEM (SUTFIN et al., 1971), we may find locally high concentrations of unusual elements but most often we are dealing with very low concentrations of all but the common light elements. Elemental micro-analysis by the SEM (ANSELL and JUDD, 1969; RUSS, 1971) should be studied with the same attention to questions of sensitivity (elemental concentration which can be measured), matrix interference, specimen geometry and counting statistics that would be necessary before applying the electron microprobe (BROWN et al., 1971; GARDNER and HALL, 1969) or x-ray fluorescence techniques to the problem.

The ability to form an SEM (or microprobe) image using the characteristic x-ray signal is also available and offers an additional way that the observer can understand the relationship of the chemical elements within the specimen (RUSS, 1971; TOUSIMIS, 1969).

b) Auger Spectra

Energies of the Auger electrons ejected from the specimen are also characteristic of the elemental composition but the Auger yield for the light elements $(2 < Z < 15)$ is high compared to the characteristic x-ray emission in this region (MAC DONALD, 1971). An analysis of the Auger spectra also gives an indication of the discrete location of the element below the surface of the specimen. The short range of the Auger electron compared to the corresponding characteristic x-ray allows depth analysis by Auger electron spectroscopy in the 1 to 10 Å range. This very short range also means that the volume analyzed is very much smaller than the volume analyzed by characteristic x-ray techniques. This could be of considerable importance to the biological investigator if certain elements exist in reasonably high concentrations only in very small domains.

c) Cathodoluminescence Analysis

Certain materials will emit visible light when bombarded with electrons This cathodoluminescence can be counted and used as the information signal in the SEM. This mode of operation provides information concerning the chemical composition of the specimen at a more complex level than elemental analysis. Cathodoluminescence is more related to the molecular and solid-state properties of the specimen.

It would be of great advantage in biological research to use a variety of specific cathodoluminescent stains to locate these chemical entities but, although some work in this direction has been partially successful (PEASE and HAYES, 1966 b), the problems of photon yield, poisoning of the stain and electron radiation damage remain a considerable obstacle (PEASE and HAYES, 1967).

d) Energy-loss Spectra

As electrons pass through a thin specimen in the scanning transmission mode of operation, they lose energy in a way that is charcteristic of the composition of the specimen. If the energy loss peaks are analyzed , a great deal of useful information can be extracted, some of which corresponds to the information contained in the analysis of several of the excitation modes. CREWE and co-workers have developed techniques of energy loss spectroscopy (CREWE, 1966) and combined them with very high resolution scanning transmission microscopy to produce chemical contrast in images and quantitative data describing composition (CREWE et al., 1970; CREWE, 1971). The ability to utilize more of the in-elastically scattered electrons as well as elastically scattered electrons improves the image characteristics of this scanning transmission system as compared to conventional electron microscopy. Such advantage may also allow the examination of thick sections (SWIFT and BROWN, 1970) at very much lower accelerating potential than the ultra high voltage conventional electron microscopes used for this purpose.

3. Electrical Properties and Charging

One of the earliest and most prominent areas of application of the SEM is the study of the electrical properties of inorganic materials and electronic circuits. The net number of electrons gained by the specimen (specimen current) or, more commonly, the changes produced in a current flowing in the specimen (induced specimen current) is used as the video signal for the scanning electron micrograph (EVERHART et al., 1964; EVERHART, 1966). Direct applications of these modes of operation to biological samples has been small but the electrical properties of the specimen and their interaction with the electron beam is of considerable importance to biologists as it affects the artifact of charging (PAWLEY, 1972a, c; EVERHART, 1970; BOYDE, 1971). This commonly encountered artifact can only be properly recognized if the observer is aware of the factors that influence secondary and backscattered electron emission, collection efficiency and surface potential.

II. Experiential or Subjective Information Processing

In addition to its capabilities as an analytical tool, the SEM has some obvious advantages in presenting information that can be experienced directly.

1. Models of Perception

If we are to construct instrumentation that can extend our senses, it is useful to consider some models of visual perception that might be applied to the microscopic world under investigation. ARNHEIM has suggested three models of perception based on an interpretation of the relationship between an object and its environment (ARNHEIM, 1969): 1) The first mode of perception does not distinguish between the properties associated with the object and those properties contributed by the environment. It can be represented by the camera or "peep-hole" approach. Each small unit of the image is considered independently of its surroundings and no distinction is attempted between essential characteristics and accidental attributes. In such a perceptive attitude, the principal criteria is fidelity between image and object. 2) The second model of perception suggested by ARNHEIM might be characterized as the absolute scientific approach. In such an approach the object is stripped of all "non-essential" properties and only the abstracted values relating to a predetermined essence of the object are considered. This kind of approach has shown itself to be very valuable as a scientific technique where the action of the environment may be considered as artifactual and influencing the image in a distortive way. 3) The third model of preception uses the environment to reveal properties of the specimen that are not readily apparent in any one static environment. In art, this model might be represented by the Impressionist school of painting. Environmental factors such as lighting are observed under changing conditions in order that the qualities of the object are more clearly revealed. Environment becomes a tool for the study of the object. At various levels the invariance to certain transformations reflect fundamental characteristics of the specimen that would not be apparent under static nonvariant conditions.

Instead of treating all artifacts or deformations as negative or unproductive, we might be able to utilize these same factors as an additional way of probing for the characteristics of the specimen.

2. Limits of Analytical Information Processing

The transfer of information by visual pathways is not limited to analytic processes; an artist through his painting transfers large amounts of information about himself and his view of the world by methods which often resist the analytic process. Many of the same techniques that have been so successful in visual art could be attempted in viewing the microscope world if instrumentation techniques could be developed which allow the individual observer to contact the specimen through channels which are similar to the ways in which he views his own environment.

Often the kinds of things that we learn through such subjective interaction are not easy to describe. Our understanding of the object viewed seems to be increased in a general way but specific statements about the type of learning which we have experienced are difficult to make. About all we can say is that it seems to be valuable to be open to the possibilities of subjective, experiential contact and to be free to make use of feelings and ideas that such contact may inspire.

The analytic, objective approach has protected the observer from unrecognized artifact, image aberrations, misleading display modes and personal bias. But it has also been restrictive in that objectivity and analysis tend to remove the observer as a person from the investigation (KIERKEGAARD, 1848) and in his place substitute a very limited system of ideas.

3. Possibilities of Complementary Subjective and Analytic Investigations

It would seem useful for the observer to be able to move comfortably between the objective and subjective methods. An over-strict adherence to objectivity can be severely limiting to the general understanding and appreciation of the specimen. On the other hand, an over indulgence in subjectivity leaves the observer adrift in a sea of emotions.

It is true that a scanning electron microscopist may be overly impressed with simply "pretty pictures" but there is an equal danger that our respect for beauty as a powerful and respectable criterion in our search for methods of transferring information may be seriously weakened if we pay too much attention to the cries for objectivity heard so often. Perhaps it is not necessary to choose between artistic methods and scientific methods; between subjective and objective approaches (HAYES, 1969, 1971a, 1971b). If we can operate with an objectivity which is free of personal bias and also apply a subjectivity which is independent and unafraid of the criticism of our peers, we can bring to the attack on the problem not only the best of our systems, but the best of ourselves.

G. Conclusion

I. Questions Regarding a Scanning Electron Microscope Program for Biological Study

I would like to pose some questions that might face the biological scientist as he considers the possibilities for scanning electron microscopy in his own field.

1. Is the SEM Really Necessary?

This question can only be answered after carefully considering both the kinds of information that are required from the microscope system and also the possible alternative systems which are available. Often the technique of replication (either directly from the specimen or after freeze-etching procedures) would be the alternative of choice (KOEHLER, 1968; STEERE, 1969a, 1971b). The increased resolution (STEERE, 1971a) and the ability to provide a method for reaching deeper-lying structures while still preserving the integrity of the biological system (KOEHLER, 1967; STEERE, 1969b) makes this technique very attractive.

One should also keep in mind that by using careful microscopic techniques which utilize visible light radiation, a great deal of information can be obtained at resolutions which often are quite adequate for the particular system under investigation. Differential interference microscopy (Nomarski) (ALLEN et al., 1969) and the procedures of phase microscopy may result in a satisfactory acquisition of the needed information (KAYDEN and BESSIS, 1970; BAJER and ALLEN, 1966) from the specimen without the very significant capital outlay associated with scanning electron microscopy. In addition, the particular advantages in terms of a hydrated system and the interaction of visible light with matter may, in fact, allow the light microscope to cover areas not available to SEM. Before resorting to scanning electron microscopy, a rather clear decision is necessary that the particular needs of the experiment cannot be met through the use of the more traditional forms of microscopy (KOEHLER, 1971).

2. Which Instrument?

Several commercial brands and sizes of scanning electron microscopes are now available and each manufacturer has several models or accessory packages. Each of these instruments offers certain unique features and the selection of a particular brand requires generally a rather close contact with the actual operating instrument. It is most often necessary for the investigator to observe his own sample material in several of the instruments in order to try to appreciate which instrument will be best for a particular use. The kind of sample that will be used, the number of different operators that may use the microscope, the facilities for maintenance and repair that are available to the prospective buyer all must be considered when selecting an instrument. Special stage requirements, a variety of signal collection systems (x-ray, cathodoluminescence, etc.), TV scan rate, are also often of prime importance to particular investigations.

In addition to primary scanning electron microscopes, there are also electron microscopes which can operate in both the conventional and the scanning modes. Such instrumentation, particularly with respect to scanning transmission operation (SWIFT and BROWN, 1970), can be quite effective and offer (after a moderate amount of modification), the capability of high resolution conventional electron microscopy coupled with a useful flexibility in terms of both secondary electron image and the transmitted electron signal.

The choice of instruments is not a simple one and requires that the investigator visit the demonstration facilities of several manufacturers; carrying with him his own sample and even preferably operating the instrument himself. It is only by making such a thorough investigation that the advantage of the different instruments to any one program can be accurately determined.

3. What Auxilliary Equipment Might be Needed?

Nearly all SEM investigations are greatly enriched by a broad correlative study in other disciplines (FORRESTER et al., 1969b; LEWIS, 1971). In particular, light microscope and conventional electron microscope facilities provide information which makes the scanning electron micrograph very much more useful (KOEHLER and HAYES, 1969a; KOEHLER and HAYES, 1969b; MCDONALD and HAYES, 1969; MCALLISTER and HADEK, 1970). An SEM isolated from these facilities, particularly in biological research, would be difficult to use to its full potential. The addition of multiple microscope techniques yields a final output which is considerably greater than the sum of its parts and close cooperation between light microscopists, conventional electron microscopists and scanning electron microscope investigators (GEISSINGER, 1971; MCALEAR et al., 1967; MCCALLISTER and HADEK, 1970) is very desirable.

There are also several units of auxilliary SEM equipment that are generally useful: 1) *A binocular, stereoscopic light microscope*. The SEM is often used as an extension of a dissecting or stereoscopic light microscope and the scanning facility should include the very best instrument of this type. The advantages in terms of orienting the observer to the specimen by means of this microscope's relatively large field of view and its use in the preparation and mounting of specimens justifies the acquisition of an excellent light microscope of this type. 2) *Freeze-dryer*. Freeze-drying is probably the most often utilized method of removing the water from biological tissues with a minimum of artifact. The freeze-dryer utilized should have monitoring and control of both temperature and vacuum systems and should dry the sample with a minimum of contamination. 3) *Critical-point dryer*. Critical-point drying represents an excellent alternative method of water removal and drying by this technique is a very valuable check on freezing artifacts possibly introduced during freeze-drying. A critical-point dryer that is capable of utilizing either carbon dioxide or the freons as the transition fluid is desirable. 4) *Vacuum evaporator*. The vacuum evaporator is utilized as the common method of placing a conducting layer on the sample. Considerations of minimum contamination are important and the ability to rotate (and possibly tilt) the sample is necessary if a uniform conducting layer is to be achieved. It is often necessary to evaporate several different kinds of material without breaking the vacuum and therefore multiple electrodes are desirable. 5) *Stereo viewer*. An optical stereo viewer placed near the microscope allows the operator to view stereo-pairs conveniently and has become a very much used part of the SEM laboratory equipment. The viewer should be flexible enough to provide comfortable viewing for different operators. 6) *Signal monitor oscilloscope*. The monitoring of the video signal and the measurement of the final light production on the display cathode ray tube is very valuable in the recording of images and in trouble shooting the microscope. It is generally worthwhile to establish a permanent signal monitoring facility with sufficient capabilities to carry out both of these functions effectively. 7) *TV tape recorder*. For any dynamic study, it is necessary that the information be stored for review and analysis on video tape. The video tape recorder need not be elaborate and a variety of suitable models are on the market. If other video facilities are near by, the question of compatibility between video recorders might be a consideration.

II. Prospects for the Future

Imaging in the secondary electron mode as an aid to recognition of shape and structure will probably remain the central use of the instrument. The resolution in this mode has improved with the addition of high intensity guns and further improvement can be expected at least for specialized modes of operation such as scanning transmission. The x-ray and cathodoluminescent modes have been less well explored in terms of biological applications and their use in an histochemical analysis will become more common in the future.

It would seem that the smaller specialized instruments available at lower cost will prove valuable in many aspects of biological research. Small scanning electron microscopes can now be designed in a cost range comparable to a conventional electron microscope and a very limited scanning system might be constructed in the future for a price comparable to that of a good light microscope. Such simple systems, of course, would necessarily be limited in their performance, but it may be found that the operations that they can perform are the ones needed in many biological investigations. If this proves to be the case, we might find that scanning systems can be located in individual biological laboratories rather than in a central facility, and that they can be available to the individual investigator in a way similar to the light microscope.

We must resist the temptation to say that a new instrument is automatically a better instrument. The uses of the SEM in biology are relatively specific and narrow as compared to the light microscope or even the conventional electron microscope. However, it seems that already the position of scanning electron microscopy, as a useful additional tool for the investigation of biological structure, is assured.

Acknowledgments

The author would like to acknowledge the generous support of the Electronics, Mechanical and Graphic Arts groups at Donner Laboratory and Lawrence Berkeley Laboratory. A special acknowledgment belongs to Dr. JAMES B. PAWLEY who has contributed so much to the SEM program at Donner. His instrimentation design, experimental results and critical comments have been of great value during the preparation of this chapter. My thanks also to Ms. DOROTHY SPRAGUE for her expert help in preparing the manuscript and to Dr. JAMES KOEHLER, University of Washington, for his continued interest in this project.

H. Appendices

I. Optical Aids for the Viewing of Vertically Mounted Stereo-Pairs

1. Lucite prism. A small prism can be held in front of one eye in order to deflect the image of one of the stereo-pairs making this image coincide with the image of the second member of the stereo-pair which is viewed directly with the other eye. The amount of deflection is controlled by

adjusting the distance between the page of the book and the prism. Problems of convergence, inter-ocular distance and picture size are reduced by utilizing vertical rather than horizontal mount of the stereo-pairs.

When the prism to picture distance is properly adjusted, the observer will see three images: a monocular image above and below the coincident stereoscopic image which is in the center.

The prism can be constructed by cutting a wedge of lucite of about 25° pitch and faces of perhaps $2'' \times 2''$. The faces are then polished until a satisfactory clarity is obtained. Such a prism will provide the appropriate deflection when held approximately 12 inches from the page of the book. We have found it convenient to hold the prism about $2''$ in front of the right eye with the apex pointing towards the observer, and the bottom surface parallel to the plane of the book.

2. Microscope slide-corn syrup prism. A simple and inexpensive prism can be constructed using standard $1'' \times 3''$ light microscope slides to form the wedge and liquid corn syrup as the refracting material. The long edges of two slides are held tightly together and placed on the center section of a $5\frac{1}{2}''$ strip of $\frac{1}{2}''$ wide adesive tape. The tape will stick to the bottom edges of the slides and can be picked up and pressed against the outer faces of the slides, sealing the long bottom edge. The top edges of the slides are now separated to a position a little less than $\frac{1}{2}''$ apart forming an open wedge. Holding the slides in this position the strip of tape is brought up the ends of the wedge. The tape now forms the end covers of the wedge.

The wedge is now filled with clear corn syrup (Karo, light) and the top is sealed with a second strip of $\frac{1}{2}''$ adhesive tape.

Such liquid prisms tend to be self-sealing as the corn syrup might be exposed to air and the prisms retain good optical properties over many months time with no appreciable leaking or drying.

The finished prism is held about $2''$ in front of the right eye with the apex pointing down and the distance from prism to page is adjusted until coincidence of the left and right eye image occurs (see above).

3. L. A. MANNHEIM has reported that two small plastic prisms have been incorporated into a pair of glasses suitable for commercial manufacture (MANNHEIM, 1971). The price of these glasses would be in the range of $ 3.00 to $ 6.00 and they are the product of Stereo Vertrieb Nesch, 44 Munster, Enschedeweg 78, West Germany.

4. Two small mirrors (purse mirrors) can also be used to accomplish the vertical image displacement. In contrast to the single prism method, two mirrors are necessary, one for each eye, in order to prevent one eye seeing a mirror image and the other a direct image. The two mirrors are held against the eyebrows at a 45° angle so that the observer looks straight ahead and sees the page of the book directly beneath the mirrors. The mirrors are now rotated slightly producing vertical displacement of one

member of the stereo-pair with respect to the other until the image for the left eye and the image for the right eye coincide to give a stereoscopic display.

II. Projection of Stereo-Pairs by Means of a Superimposed Color-Coded Transparency

($3\frac{1}{4}'' \times 4''$ lantern slide, standard projector and screen).

M. NEMANIC, University of California, Berkeley

1. *Original Micrographs.* Take normal black and whites on No. 107 Polaroid film with 7° of tilt between the pair.
   a) focus using the z-axis control;
   b) center the same area in both micrographs.
   2. *Framing for Superposition.* Use a Polaroid MP-3 Industrial Viewer, a graflex 120 roll film back and a framing table.

   a) place the black and white print in the framing table; center the micrograph by viewing through the camera lens; enlarge the image to fill the format of the roll film back;
   b) place Ectacolor or Ektachrome color film in the back.

   3. *Color Coding of the Two Superimposed Images.*
   a) use Wratten 25a (red) and 58 (green) gelatin filters;
   b) exposure settings:

		Speed	*f*
Ectacolor Type S (ASA 100)	Green:	1/2	4.7
(for negative slides)	Red:	1/8	5.6
High Speed Ektachrome	Green:	1/2	5.6
(ASA 160)	Red:	1/8	8
(for positive slides)			

   c) standardize the use of red or green for the lower angle micrograph;
   d) expose the first micrograph of the stereo-pair with the appropriate filter over the lens (hand-held);
   e) place the second micrograph of the pair in the same frame as the first without advancing film; set speed and shutter opening for the new filter; expose, producing double exposed superposition and then advance the film for the next pair.

   4. *Processing and Mounting.*
   a) standard color processing is carried out by the photo-lab or by a commercial camera shop;
   b) the developed film is mounted between lantern slide glass in the usual way.

5. *Viewing.*

a) red and green acetate, found in an art supplies store, can be used to make simple viewers for large groups of people. Two thicknesses of green and three of red approximate the color and optical density of filters used to make the color composite. Small squares are held in front of each eye while viewing the projection;

b) instruct viewers as to the correct color filter for right and left eye, depending on convention adapted in 3 (c) above.

References

Allen, R. D., David, G. B., Normarski, G.: The Zeiss-Normarski differential interference equipment for transmitted-light microscopy. Z. wiss. Mikr. **69**, 193—221 (1969).

Anderson, T. F.: Techniques for the Preservation of Three-Dimensional Structure in Preparing Specimens for the Electron Microscope. Trans. N.Y. Acad. Sci. **13**, 130—134 (1951).

Ansell, G. S., Judd, G.: Thin Film Electron Probe — Electron Microscope Microanalysis Techniques in Biomaterials Studies. Trans. N.Y. Acad. Sci., Ser. II, **31**, 637—647 (1969).

Aremberg, I. K., Marovitz, W. F., Mac Kenzie, A. P.: Preparative Techniques for the Study of Soft Biologic Tissues in the Scanning Electron Microscope. Trans. Amer. Acad. Ophthal. Otolaryng. **75**, 1333—1345 (1971).

Arnheim, R.: Visual Thinking, 1—345, Univ. of Calif. Press, Berkeley, and Los Angeles 1969.

Arnold, J. D., Berger, A. E., Allison, O. L.: Some Problems of Fixation of Selected Biological Samples for SEM Examination. In: Scanning Electron Microscopy. Ed.: O. Johari, 249—256. I.I.T. Research Institute, Chicago, Illinois 1971.

Bajer, A., Allen, R. D.: Structure and Organization of the Living Mitotic Spindle of Haemanthus Endosperm. Science **151**, 572—574 (1966).

Baker, J. R.: Principles of Biological Microtechnique, 1—357. London: Methuen & Co. 1958.

Baker, R. F.: Ion Etching of Red Blood Cells. In: Red Cell Membrane Structure and Function. Eds.: Jameson, G. A. and Greenwalt, T. J., 13—35. Philadelphia: J. B. Lippincott Co. 1969.

Barber, V. C., Boyde, A.: Scanning Electron Microscopic Studies of Cilia, Arch. Zellforsch. **84**, 269—284 (1968).

Bessis, M., Lessin, L. S.: The Descocyte-Echinocyte Equilibrium of the Normal and Pathologic Red Cell. Blood **36**, 399—403 (1970).

Blumke, S., Morgenroth, K.: The Stereo Ultrastructure of the External and Internal Surface of the Cornea. J. Ultrastruct. Res. **18**, 502—518 (1967).

Bok, A. B., Le Poole, J. B., Roos, J., De Lang, H.: Mirror Electron Microscopy. In: Adv. in Optical and Electron Microscopy. Eds.: R. Barer, and V. E. Cosslett. Vol. 4, 161—383. New York: Academic Press, 1971.

Boyde, A., Stewart, A. D. G.: A Study of the Etching of Dental Tissues with Argon Ion Beams. J. Ultrastruct. Res. **7**, 159—172 (1962).

Boyde, A.: The Development of Enamel Structure. Proc. roy. Soc. Med. **60**, 923—928 (1967).

BOYDE, A., WILLIAMS, J. C. P.: Surface Morphology of Frog Striated Muscle as Prepared for and Examined in the Scanning Electron Microscope. J. Physiol. (Paris) **197**, 10 P—11 P (1968).

BOYDE, A., JAMES, D. W., TRESMAN, R. L., WILLIS, R. A.: Outgrowth from Chick Embryo Spinal Chord in vitro Studied with the Scanning Electron Microscope. Z. Zellforsch. **90**, 1—18 (1968).

BOYDE, A., GRAINGER, F., JAMES, D. W.: Scanning Electron Microscopy of Chick Embryo Fibroblasts in vitro, with Particular Reference to the Movement of Cells Under Others. Z. Zellforsch. **94**, 46—55 (1969).

BOYDE, A., WOOD, C.: Preparation of Animal Tissues for Surface-Scanning Electron Microscopy. J. Microscopy **90**, 221—249 (1969).

BOYDE, A.: Practical Problems and Methods in the Three-Dimensional Analysis of Scanning Electron Microscope Images. In: Scanning Electron Microscopy. Ed.: O. JOHARI, 105—112. I.I.T. Research Institute, Chicago, Illinois 1970.

BOYDE, A.: A Review of Problems of Interpretation of the Scanning Electron Microscope Image with Special Regard to Methods of Specimen Preparation. In: Scanning Electron Microscopy. Ed.: O. JOHARI, 1—8. I.I.T. Research Institute, Chicago, Illinois 1971.

BRAGGINS, D. W., GARDNER, G. M., GIBBARD, D. W.: The Applications of Image Analysis Techniques to Scanning Electron Microscopy. In: Scanning Electron Microscopy. Ed.: O. JOHARI, 393—400. I.I.T. Research Institute, Chicago, Illinois 1971.

BROERS, A. N.: A New High Resolution Reflection Scanning Electron Microscope. Rev. Sci. Inst. **40**, 1040—1045 (1969).

BROERS, A. N.: Factors Affecting Resolution in the Scanning Electron Microscope. In: Scanning Electron Microscopy. Ed.: O. JOHARI, 1—8. I.I.T. Research Institute, Chicago, Illinois 1970.

BROWN, A. C., GERDES, R. J., JOHNSON, J.: Scanning Electron Microscopy and Electron Probe Analysis of Congenital Hair Defects. In: Scanning Electron Microscopy. Ed.: O. JOHARI, 369—376. I.I.T. Research Institute, Chicago, Illinois 1971.

BUDINGER, T. F.: Transfer Function Theory and Image Evaluation in Biology: Applications in Electron Microscopy and Nuclear Medicine. University of California, Berkeley, Lawrence Berkeley Lab. Report, LBL, 565, 232 pp., 1971.

BURKE, J. F., INDEBETOUW, NOMARSKI, G., STROKE, G. W.: White-light Three Dimensional Microscopy Using Multiple-image Storing and Decoding. Nature **231**, 303—306 (1971).

BYWATER, N. E., BUCKLEY, T.: A New Technique for Specimen Manipulation in the Scanning Electron Microscope. J. Microscopy **92**, 113—118 (1970).

CARR, K. E.: Applications of Scanning Electron Microscopy in Biology. In: Int. Rev. Cyt. Eds.: G. H. BOURNE and J. F. DANIELLI, Vol. 30, 183—255. New York: Academic Press 1971.

CLARKE, J. A., SALSBURY, A. J.: Surface Ultramicroscopy of Human Blood Cells. Nature (Lond.) **215**, 402—404 (1967).

COHEN, A. L., MARLOW, D. P., GRANER, G. E.: A Rapid Critical Point Method Using Fluorocarbons ("Freons") as Intermediate and Transitional Fluids. J. Microscopie **7**, 331—342 (1968).

COSSLETT, V. E.: Modern Microscopy, 160 pp. London: G. Bell and Sons, Ltd. 1966.

CREWE, A. V.: Scanning Electron Microscopes: Is High Resolution Possible? Science **154**, 729—738 (1966).

CREWE, A. V., ISAACSON, M., JOHNSON, D.: A Simple Scanning Electron Microscope. Rev. Sci. Instr. **40**, 241—246 (1969).

CREWE, A. V., WALL, J., LANGMORE, J.: Visibility of Single Atoms. Science **168**, 1338—1340 (1970).

CREWE, A. V.: A High Resolution Scanning Electron Microscope. Scientific American **224**, (4) April, 26—35 (1971).

CROWE, J. H., COOPER, A. F.: Cryptobiosis. Scientific American **225**, (6), December, 30—45 (1971).

DARRAGH, P. J., GASKIN, A. J., TERRELL, B. C., SANDERS, J. V.: Origin of Precious Opal. Nature (Lond.) **209**, 13—16 (1966).

DORFLER, G., RUSS, J. C.: A System for Stereometric Analysis with the Scanning Electron Microscope. In: Scanning Electron Microscopy. Ed.: O. JOHARI, 65—72. I.I.T. Research Institute, Chicago, Illinois 1970.

ECHLIN, P.: The Examination of Biological Material at Low Temperatures. In: Scanning Electron Microscopy. Ed.: O. JOHARI, 225—232. I.I.T. Research Institute, Chicago, Illinois 1971.

ELIAS, H.: Three-Dimensional Structure Identified from Single Sections. Science **174**, 993—1000 (1971).

ERBEN, H. K.: Application of the Scanning Electron Microscope in Paleobiology. In: Scanning Electron Microscopy. Ed.: O. JOHARI, 233—239. I.I.T. Research Institute, Chicago, Illinois 1970.

EVERHART, T. E., WELLS, O. C., OATLEY, C. W.: Factors Affecting Contrast and Resolution in the Scanning Electron Microscope. J. Elect. Control **7**, 97—111 (1959).

EVERHART, T. E., WELLS, O. C., MATTA, R. K.: Evaluation of Passivated Integrated Circuits Using the Scanning Electron Microscope. J. Electrochem. Soc. **111**, 929—936 (1964).

EVERHART, T. E.: Deflection-Modulation CRT Display. Proc. Inst. Elect. Electron. Engrs. **54**, 1480—1482 (1966).

EVERHART, T. E.: Reflections on Scanning Electron Microscopy. In: Scanning Electron Microscopy. Ed.: O. JOHARI, 1—12. I.I.T. Research Institute, Chicago, Illinois 1968.

EVERHART, T. E.: Contrast and Resolution in the Scanning Electron Microscope. In: Proc. 3rd Amer. Stereoscan Colloquium, 1—8. Kent Cambridge Scientific, Inc., Morton Grove, Illinois 1970.

EVERHART, T. E., HAYES, T. L.: The Scanning Electron Microscope. Scientific American **226**, No. 1, 54—69 (1972).

FORRESTER, J. C., HAYES, T. L., PEASE, R. F. W., HUNT, T. K.: Scanning Electron Microscopy of Healing Wounds. Nature (Lond.) **221**, 373—374 (1969a).

FORRESTER, J. C., ZEDERFELDT, B. H., HUNT, T. K.: The Tape-Closed Wound — A Bioengineering Analysis. J. surg. Res. **9**, 537—542 (1969b).

FRANK, J.: Computer Processing of Electron Micrographs, in this volume, 1973. Ed.: J. KOEHLER.

GARDNER, D. L., HALL, T. A.: Electron-microprobe Analysis of Sites of Silver Deposition in Avian Bone Stained by the v. Kossa Technique. J. Path. Bact. **98**, 105—109 (1969).

GEISSINGER, H. D.: Correlated Light Optical and Scanning Electron Microscopy of Gram Smears of Bacteria and Parafin Sections of Cardiac Muscle. J. Microscopy **93**, 109—117 (1971).

GLAESER, R. M., THOMAS, G.: Application of Electron Diffraction to Biological Electron Microscopy. Biophys. J. **9**, 1073—1099 (1969).

GOLOMB, H. M., BAHR, G. F.: Scanning Electron Microscope Observations on Surface Structure of Isolated Human Chromosomes. Science **171**, 1024—1026 (1971).

HAGGIS, G. H.: Cryofracture of Biological Material. In: Scanning Electron Microscopy. Ed.: O. JOHARI, 97—104, I.I.T. Research Institute, Chicago, Illinois 1970.

HARTE, R. A., RUPLEY, J. A.: Three-Dimensional Pictures of Molecular Models — Lysozyme. J. biol. Chem. **243**, 1663—1669 (1968).

HARTMAN, H., HAYES, T. L.: Scanning Electron Microscopy of Drosophila. J. Hered. **62**, 41—44 (1971).

HAYES, T. L., PEASE, R. F. W., McDONALD, L. W.: Applications of the Scanning Electron Microscope to Biologic Investigations. Lab. Invest. **15**, 1320—1326 (1966).

HAYES, T. L., PEASE, R. F. W.: The Scanning Electron Microscope: Principles and Applications in Biology and Medicine. In: Advances in Biological and Medical Physics, Vol. 12. Eds.: J. H. LAWRENCE and J. W. GOFMAN, 85—137. New York: Academic Press Inc. 1968.

HAYES, T. L., GLAESER, R. M., PAWLEY, J. B.: Information Content of Electron Microscope Images: Color and Depth Information as Vehicles for Subjective Information Transfer. In: Proc. 27th Ann. Meeting Electron Microscopic Soc. Amer., Arceneaux, C., ed., 410—411. New Orleans: Claitor's Publishing 1969.

HAYES, T. L.: The Scanning Electron Microscope: A High Information Content Image of Biological Systems. Int. Convention, Inst. Elect. Electron. Engrs. Digest, 158—159. I.E.E.E., New York (1969).

HAYES, T. L., PEASE, R. F. W.: The Scanning Electron Microscope: A Non-focused, Multi-information Image. In: Data Extraction and Processing of Optical Images. Ed.: W. E. TOLLES, **157**, 497—509. Annals N.Y. Acad. Sci., 1969.

HAYES, T. L.: The Penetrating Eye — Biomedical Applications of the Scanning Electron Microscope. 21 min. sound 16 mm film, Medical Communications Division, Lilly Research Laboratories, Indianapolis, 1970a.

HAYES, T. L.: Some Comments on the Imaging of Biological Systems. Int. Convention Inst. Elect. Electron. Eng. Digest, 154—155. I.E.E.E., New York (1970b).

HAYES, T. L.: Microvision. In: Scanning Electron Microscopy. Ed.: O. JOHARI, 521—528. I.I.T. Research Institute, Chicago, Illinois 1971a.

HAYES, T. L.: The Scanning Electron Microscope: Multi-Radiation, Non-focused Imaging. In: Advances in Medical Physics. Ed.: J. S. LAUGHLIN, 124—140. 2nd Int. Conference Med. Phys. Inc., Massachusetts General Hospital, Boston, 1971b.

HAYES, T. L.: Invariance Under Transformation: A Useful Concept in the Interpretation of SEM images. In: Scanning Electron Microscopy. Ed.: O. JOHARI. 57—64. I.I.T. Research Institute, Chicago, Illinois 1972.

HEINRICH, K. F. J., FIORI, C., YAKOWITZ, H.: Image-Formation Technique for Scanning Electron Microscopy and Electron Probe Microanalysis. Science **167**, 1129—1131 (1970).

HESLOP-HARRISON, Y., HESLOP-HARRISON, J.: Scanning Electron Microscopy of Leaf Surfaces. In: Scanning Electron Microscopy. Ed.: O. JOHARI, 117—126. I.I.T. Research Institute, Chicago, Illinois 1969.

HILLMAN, D. E., LEWIS, E. R.: Morphological Basis for a Mechanical Linkage in Otolithic Receptor Transduction in the Frog. Science **174**, 416—419 (1971).

HORRIDGE, G. A., TAMM, S. L.: Critical Point Drying for Scanning Electron Microscope Study of Ciliary Motion. Science **163**, 817—818 (1969).

HUMPHREYS, W. J., HAYES, T. L., PEASE, R. F. W.: Transmission and Scanning Electron Microscopy of Cryptobiotic Cytoplasm. In: Proc. 25th Ann. Meeting Electron Microscope Soc. Amer., Arceneaux, C., ed., 50—51. Claitor's Publishing, New Orleans 1967.

JAQUES, W. E., COALSON, J., ZERVINS, A.: Application of the Scanning Electron Microscope to Human Tissues: A Preliminary Study. Exp. Molec. Path. **4**, 576—580 (1965).

JOHARI, O.: SEM and Analytic Possibilities. In: Scanning Electron Microscopy. Ed.: O. JOHARI, 529—536. I.I.T. Research Institute, Chicago, Illinois 1971a.

Johari, O.: Total Materials Characterization with the Scanning Electron Microscope. Research/Development **22**, 12—20 (1971b).

Jones, F.: The Use of Opal as an SEM Standard, Personal Communication (1969).

Jones, J. B., Sanders, J. V., Segnit, E. R.: Structure of Opal. Nature (Lond.) **204**, 990 (1964).

Karnovsky, M. J.: A Formaldehyde-Glutaraldehyde Fixative of High Osmolarity for Use in Electron Microscopy. J. Cell Biol. **27**, 137 A—138 A (1965).

Kayden, H. J., Bessis, M.: Morphology of Normal Erythrocyte and Acanthocyte Using Normarski Optics and the Scanning Electron Microscope. Blood **35**, 427—436 (1970).

Kierkegaard, S.: Concluding Unscientific Postscript, 1848. Translated by David F. Swenson and Walter Lowrie. Princeton University Press 1941.

Kimoto, S., Hashimoto, H., Tekashima, S.: Transmission Scanning Microscope. In: Scanning Electron Microscopy. Ed.: O. Johari, 81—88. I.I.T. Research Institute, Chicago, Illinois 1969.

Kimoto, S., Sato, M., Adachi, T.: TV Scanning Device in a Scanning Electron Microscope and Its Applications. In: Scanning Electron Microscopy. Ed.: O. Johari, 65—72. I.I.T. Research Institute, Chicago, Illinois 1969.

Knaston, D., Stewart, A. D. G.: Composition Analysis in the Stereoscan. In: Scanning Electron Microscopy. Ed.: O. Johari, 465—470. I.I.T .Research Institute, Chicago, Illinois 1969.

Koehler, J. K.: Studies in the Survival of the Rotifer Philodina after Freezing and Thawing. Cryobiology **3**, 392—399 (1967).

Koehler, J. K.: The Technique and Application of Freeze-Etching in Ultrastructure Research. In: Advances in Biological and Medical Physics. Eds.: J. H. Lawrence and J. W. Gofman, Vol. 12, 1—84. New York: Academic Press Inc. 1968.

Koehler, J. K., Hayes, T. L.: The Rotifer Jaw: A Scanning and Transmission Electron Microscope Study, I. The Trophi of Philodina acuticornis odiosa. J. Ultrastruct. Res. **27**, 402—418 (1969a).

Koehler, J. K., Hayes, T. L.: The Rotifer Jaw: A Scanning and Transmission Electron Microscope Study, II. The Trophi of Asplanchna sieboldi. J. Ultrastruct. Res. **27**, 419—434 (1969b).

Koehler, J. K.: Comparative Examination of Biological Samples by SEM, TEM and Other Techniques. In: Scanning Electron Microscopy. Ed.: O. Johari, 241—248. I.I.T. Research Institute, Chicago, Illinois 1971.

Krinsley, D. H., Hyde, P. W.: Cathodoluminescence of Studies of Sediments. In: Scanning Electron Microscopy. Ed.: O. Johari, 409—416. I.I.T. Research Institute, Chicago, Illinois 1971.

Kuwabara, T.: Surface Structure of the Eye Tissue. In: Scanning Electron Microscopy. Ed.: O. Johari, 185—192. I.I.T. Research Institute 1970.

Langmuir, D. B.: Theoretical Limitations of Cathode-Ray Tubes. Proceedings of the Institute of Radio Engineers **25**, 977—991 (1937).

Lewis, E. R., Zeevi, Y. Y., Werblin, F. S.: Scanning Electron Microscopy of Vertebrate Visual Receptors. Brain Research **15**, 559—562 (1969a).

Lewis, E. R., Everhart, T. E., Zeevi, Y. Y.: Studying Neural Organization in Aphysia with the Scanning Electron Microscope. Science **165**, 1140—1143 (1969b).

Lewis, E. R.: Studying Neural Architecture and Organization with the Scanning Electron Microscope. In: Scanning Electron Microscopy. Ed.: O. Johari, 281—288. I.I.T. Research Institute, Chicago, Illinois 1971.

Lim, D. J.: Scanning Electron Microscopic Observation on Non-Mechanically Cryofractured Biological Tissue. In: Scanning Electron Microscopy. Ed.: O. Johari, 257—264. I.I.T. Research Institute, Chicago, Illinois 1971.

LIPKIN, B. L., ROSENFELD, A.: Picture Processing and Psychopictorics, 526 pp. New York: Academic Press Inc. 1970.

MAC DONALD, N. C.: Computer-Controlled Scanning Electron Microscopy. In: Proc. 26th Ann. Meeting Electron Microscope Soc. Amer. Ed.: C. ARCENEAUX, 362. New Orleans: Claitor's Publishing 1968.

MAC DONALD, N. C.: Auger Electron Spectroscopy for Scanning Electron Microscopy. In: Scanning Electron Microscopy. Ed.: O. JOHARI, 89—96. I.I.T. Research Institute, Chicago, Illinois 1971.

MANGER, W. M., BESSIS, M.: Cathodoluminescence Produced in Cells and Proteins by Paraformaldehyde as Seen with the Scanning Electron Microscope. Septieme Congres International de Microscopie Electronique, Grenoble, 483—484. Societe Francaise de microscopie Elecl., Paris 1970.

MANNHEIM, L. A.: Stereoscopic System for Any Size Picture Pairs (Nesch vertical system), Photographic Applications in Sci., Technol. and Med., 38, Nov. (1971).

MANNING, J. S., HAYES, T. L., GLAESER, R. M.: Comparison of Transmission and Scanning Electron Microscope Images of Small Biological Objects. Biophys. J. **8**, A 159 (1968).

MARSZALEK, D. S., SMALL, E. B.: Preparation of Soft Biological Materials for Scanning Electron Microscopy. In: Scanning Electron Microscopy. Ed.: O. JOHARI, 231—239. I.I.T. Research Institute, Chicago, Illinois 1969.

MATAS, B. R., SPENCER, W. H., HAYES, T. L., CHANDLER, R. D.: Morphology of Experimental Vaccinial Superficial Punctate Keratitis — A Scanning and Transmission Electron Microscope Study. Invest. Ophthal. **10**, 348—356 (1971) and Private Communication 1971.

McALEAR, J. H., KREUTZINGER, O., PEASE, R. F. W.: Combining Scanning and Transmission Electron Microscopy in the Study of Heart Muscle Freeze-Etched Replicas. J. Cell Biol. **35**, 89 A (1967).

McCALLISTER, L. P., HADEK, R.: Transmission Electron Microscopy and Stereo Ultrastructure of the T System in Frog Skeletal Muscle. J. Ultrastruct. Res. 33, 360—368 (1970).

McDONALD, L. W., PEASE, R. F. W., HAYES, T. L.: Scanning Electron Microscopy of Sectioned Tissue. Lab. Invest. **16**, 532—538 (1967).

McDONALD, L. W., HAYES, T. L.: The Role of Capillaries in the Pathogenesis of Delayed Radionecrosis of Brain. Amer. J. Path. **50**, 745—764 (1967).

McDONALD, L. W., HAYES, T. L.: Correlation of Scanning Electron Microscope and Light Microscope Images of Individual Cells in Human Blood and Blood Clots. Exp. Molec. Path. **10**, 186—198 (1969).

MICHAELIS, T. W., LANIMER, N. R., METZ, E. N., BALCERZAK, S. P.: Surface Morphology of Human Leukocytes. Blood **37**, 23—30 (1971).

MUIR, M. D., GRANT, P. R., HUBBARD, G., MUNDELL, J.: Cathodoluminescence Spectra. In: Scanning Electron Microscopy. Ed.: O. JOHARI, 401—408. I.I.T. Research Institute, Chicago, Illinois 1971.

NEMANIC, M.: Private communication, 1971.

NIXON, W. C.: Twenty Years of Scanning Electron Microscopy, 1948—1968, in the Engineering Department, Cambridge University, England. In: Scanning Electron Microscopy. Ed.: O. JOHARI, 55—62. I.I.T. Research Institute, Chicago, Illinois 1968.

NOWELL, J. A., PANGBORN, J., TYLER, W. D.: Scanning Electron Microscopy of Avian Lung. In: Scanning Electron Microscopy. Ed.: O. Johari, 249—256. I.I.T. Research Institute, Chicago, Illinois 1970.

OATLEY, C. W., NIXON, W. C., PEASE, R. F. W.: Scanning Electron Microscopy. Adv. in Electronics and Electron Physics, Vol. 21, 181—247 (1965).

PARDUCZ, B.: Ciliary Movement and Coordination in ciliates. Int. Rev. Cytol. **21**, 91—128 (1967).

PAWLEY, J. B., UPHAM, F. T., WINDSOR, A. A., HAYES, T. L.: Color Modulation Display and an Exposure Measuring Circuit for Scanning Electron Microscope Images. Univ. of Calif. Radiation Lab. Report, UCRL 19420 (1969) Fall Semi-Annual Report, Berkeley, 68—76.

PAWLEY, J. B., HAYES, T. L.: A Micromanipulator for the Scanning Electron Microscope. In: Scanning Microscopy. Ed.: O. JOHARI, 105—112. I.I.T. Research Institute, Chicago, Illinois 1971.

PAWLEY, J. B.: Ph. D. Thesis, University of California, Berkeley, in press 1972a.

PAWLEY, J. B.: A Dual Needle Piezoelectric Micromanipulator for the Scanning Electron Microscope. Rev. Sci. Inst. 43 (4), April 1972b.

PAWLEY, J. B.: Charging Artifacts in the Scanning Electron Microscope. In: Scanning Electron Microscopy. Ed.: O. JOHARI. I.I.T. Research Institute, Chicago, Illinois, in press 1972c.

PEASE, D. C.: Histological Technique for Electron Microscopy, 1—381. New York: Academic Press Inc. 1964.

PEASE, R. F. W., NIXON, W. C.: High Resolution Scanning Electron Microscopy. J. Sci. Inst. **42**, 81—85 (1965).

PEASE, R. F. W., HAYES, T. L.: Some Biological Applications of Scanning Electron Microscopy. 6th Int. Congress for Electron Microscopy, Kyoto, Japan 19—20. Tokyo: Maruzen Co., Ltd. 1966a.

PEASE, R. F. W., HAYES, T. L.: Scanning Electron Microscopy of Biological Material. Nature (Lond.) **210**, 1049 (1966b).

PEASE, R. F. W., HAYES, T. L., CAMP, A. S., AMER, N. M.: Electron Microscopy of Living Insects. Science **154**, 1185—1186 (1966).

PEASE, R. F. W., HAYES, T. L.: Scanning Electron Microscopy Using Cathodoluminescence. In: Proc. 25th Ann. Meeting Electron Microscope Soc. Amer. Ed.: C. ARCENEAUX, 122—123. New Orleans: Claitor's Publishing 1967.

PEASE, R. F. W., HAYES, T. L.: Electron Microscopy of Sprouting Seeds. In: Proc. 26th Ann. Meeting Electron Microscope Soc. Amer. Ed.: C. ARCENEAUX, 88—89. New Orleans: Claitor's Publishing 1968a.

PEASE, R. F. W., HAYES, T. L.: Biological Applications of the Scanning Electron Microscope. In: O. JOHARI, Ed., 143—154. I.I.T. Research Institute, Chicago, Illinois 1968b).

PEASE, R. F. W.: Fundamentals of Scanning Electron Microscopy. In: Scanning Electron Microscopy. Ed.: O. JOHARI, 9—16. I.I.T. Research Institute, Chicago, Illinois 1971.

PFEFFERKORN, G.: Preparation Methods and Artifacts in Scanning Electron Microscopy. Proc. 2nd Ann. Stereoscan Colloquium, 81—88. Engis Equipment Co., Morton Grove, Illinois 1969.

POH, T., ALTENHOFF, R. L. J., ABRAHAM, S., HAYES, T. L.: Scanning Electron Microscopy of Myocardial Sections Originally Prepared for Light Microscopy. Exp. Molec. Path. **14**, 404—407 (1971).

RASHEVSKY, N.: Topology and Life: In Search of General Mathematical Principles in Biology and Sociology. Bull. Math. Biophys. **16**, 317—348 (1954).

RASHEVSKY, N.: Mathematical Principles in Biology, 8—114. Springfield, Illinois: Charles C. Thomas 1961.

REMOND, G., KIMOTO, S., OKAZUMI, H.: Use of the SEM in Cathodoluminescence Observations in Natural Samples. In: Scanning Electron Microscopy. Ed.: O. JOHARI, 33—40. I.I.T. Research Institute, Chicago, Illinois 1970.

ROTH, I. L.: Scanning Electron Microscopy of Bacterial Colonies. In: Scanning Electron Microscopy, Ed.: O. JOHARI, 321—328. I.I.T. Research Institute, Chicago, Illinois 1971.

RUSS, J. C.: Progress in the Design and Application of Energy Dispersion X-ray Analyzers for the SEM. In: Scanning Electron Microscopy. Ed.: O. JOHARI, 65—72. I.I.T. Research Institute, Chicago, Illinois 1971.

SALSBURY, A. J., CLARKE, J. A.: New Method for Detecting Changes in the Surface Appearance of Human Red Blood Cells. J. Clin. Path. 20, 603—610 (1967).

SANDERS, J. V.: Colour of Precious Opal. Nature (Lond.) 204, 1151—1153 (1964).

SCHEID, W., TRAUT, H.: Visualization by Scanning Electron Microscopy of Achromatic Lesions ("gaps") Induced by X-Rays in Chromosomes of Vicia faba. Mutation Res. 11, 253—255 (1971).

SCHEIE, P., DAHLEN, H.: Surface Features of Cultured Mammalian Cells. In: Scanning Electron Microscopy. Ed.: O. JOHARI, 127—134. I.I.T. Research Institute, Chicago, Illinois 1969.

SIKORSKI, J., MOSS, J. S., NEWMAN, P. H., BUCKLEY, T.: A New Preparation Technique for Examination of Polymers in the Scanning Electron Microscope. J. Sci. Inst. 1, 29—31 (1968).

SMALL, E. B., MARSZALEK: Scanning Electron Microscopy of Fixed, Frozen and Dried Protozoan. Science 163, 1064—1065 (1969).

SMITH, V., RYAN, J. W., MICHIE, D. D., SMITH, D. S.: Endothelial Projections as Revealed by Scanning Electron Microscopy. Science 173, 925—927 (1971).

SOKOLOFF, A., HAYES, T. L., PEASE, R. F. W., ACKERMAN, M.: Tribolium Castaneum Morphology of "Aureate" Revealed by Scanning Electron Microscopy. Science 157, 443—445 (1967).

SPENCER, W. H., ALVARADO, J., HAYES, T. L.: Scanning Electron Microscopy of Human Ocular Tissues: Trabecular Meshwork. Invest. Ophthal. 7, 651—662 (1968).

SPENCER, W. H., HAYES, T. L.: Scanning and Transmission Electron Microscope Observations of the Topographic Anatomy of Dendritic Lesions in the Rabbit Cornea. Invest. Ophthal. 9, 183—195 (1970).

STACKPOLE, C. W., AOKI, T., BOYSE, E. A., OLD, L. J., LUMLEY-FRANK, J., DE HARREN, E.: Cell Surface Antigens: Serial Sectioning of Single Cells as an Approach to Topographical Analysis. Science 172, 472—474 (1971).

STEERE, R. L.: Freeze-Etching Simplified. Cryobiology 5, 306—323 (1969a).

STEERE, R. L.: Freeze-Etching and Direct Observation of Freezing Damage. Cryobiology 6, 137—150 (1969b).

STEERE, R. L.: High Resolution Stereography of Complementary Freeze-Fracture Replicas Reveals the Presence of Depressions Opposite Membrane Associated Particles of Split Membranes. In: 29th Ann. Proc. Electron Microscopy Soc. Amer., Boston, Mass. Ed.: C. ARCENEAUX, 442—445. New Orleans: Claitor's Publishing 1971a.

STEERE, R. L.: Retention of 3-Dimensional Contours by Replicas of Freeze-Fractured Specimens. In: 29th Ann. Proc. Electron Microscopy Soc. Amer., Boston, Mass. Ed.: C. ARCENEAUX, 242—243. New Orleans: Claitor's Publishing 1971b.

STEWART, A. D. G., BOYDE, A.: Ion Etching of Dental Tissues in a Scanning Electron Microscope. Nature (Lond.) 196, 81—82 (1962).

SUTFIN, L. V., HOLTROP, M. E., OGILVIE, R. E.: Microanalysis of Individual Mitochondrial Granules with Diameters Less than 1,000 Angstroms. Science 174, 947—949 (1971).

SWIFT, J. A., BROWN, A. C.: Transmission Scanning Electron Microscopy of Sectioned Biological Materials. In: Scanning Electron Microscopy. Ed.: O. JOHARI, 113—120. I.I.T. Research Institute, Chicago, Illinois 1970.

THOMAS, R. S.: Microincineration Techniques for Electron-microscopic Localization of Biological Material. In: Advan. in Opt. and Electron Microscopy. Eds.: R. BARER, V. E. COSSLETT, Vol. 3, 99—150. London: Academic Press 1969.

214   T. L. HAYES: Scanning Electron Microscope Techniques in Biology

THORNHILL, J. W., MATTA, R. K., WOOD, W. H.: Examining Three-Dimensional Microstructures with the Scanning Electron Microscope. Grana Palynologica **6**, 3—6 (1965).

THORNTON, P. R.: Scanning Electron Microscopy — Applications to Materials and Device Science, 34—38, 207—212. New York: Barnes & Noble, Inc. 1968.

TOUSIMIS, A. J.: Combined Scanning Electron Microscopy and Electron Probe Microanalysis of Biological Soft Tissue. In: Scanning Electron Microscopy, Ed.: O. JOHARI, 217—230. I.I.T. Research Institute, Chicago, Illinois 1969.

WARFEL, A. H., ELBERG, S. S.: Macrophage Membranes Viewed Through a Scanning Electron Microscope. Science **170**, 446—447 (1970).

WELLS, O. C.: Correction of Errors in Electron Stereomicroscopy. Brit. J. appl. Phys. **11**, 199—201 (1960).

WELLS, O. C.: Bibliography in the Scanning Electron Microscope. In: Record of the Inst. Elect. Electron Engnrs. 9th Symposium on Electron, Ion and Laser Beam Technology, Ed.: R. F. W. PEASE, 412—438. I.E.E.E. 1967.

WELLS, O. C.: Bibliography on the Scanning Electron Microscope. In: Scanning Electron Microscopy. Ed.: O. JOHARI, SEM Symposium 1968, 1969, 1970, 1971. I.I.T. Research Institute, Chicago, Illinois.

WITTRY, D. B., VAN COUVERING, A.: A Stereoscopic Display System for Electron Microprobe Instruments. J. Sci. Inst. **44**, 294—295 (1967).

ZEITLER, E.: Resolution in Electron Microscopy. In: Adv. in Electronics and Electron Physics. Ed.: L. MARTON, **25**, 277—332. New York: Academic Press 1969.

ZEITLER, E.: Scanning Transmission Electron Microscopy. In: Scanning Electron Microscopy. Ed.: O. JOHARI, 25—32. I.I.T. Research Institute, Chicago, Illinois 1971.

Computer Processing of Electron Micrographs

Joachim Frank

A. Introduction

Within the past few years, image processing methods have been introduced into a number of fields where experimental visual data have to be analyzed. Examples in the biological field are radiotherapy (Selzer, 1968) and cytology (Mendelsohn et al., 1968). The implementation in electron microscopy is presently developing very fast. The present work gives a review of some experiences in computer analysis of electron microscopic image data, and tries to show some prospects for the use of this tool in the near future. An attempt has been made to present the material in a way that is ordered according to typical problems of electron microscopy, rather than according to methods of image analysis.

For a number of practical cases, electron micrographs may be considered as images which are obtained from the object by a linear imaging process (i.e., from the projection of a three-dimensional object). In the last decade, a theory of linear system has been developed by generalizing the concepts of the electric network analysis, and this theory has been applied to optical systems (see, for instance, O'Neill, 1963; Goodman, 1968; Papoulis, 1968; Röhler, 1967). It was this development which has put forward a better understanding of image formation and has made what is called "image processing" possible, although some of the newly applied concepts have been known in x-ray crystallography for many years.

The need for an analysis of electron micrographs exceeding visual interpretation is evident for a number of problems such as

1) signal detection in the presence of noise
2) interpretation of phase contrast images
3) three-dimensional reconstruction.

The fact that a human observer is sometimes overtaxed with interpreting visual information is drastically illustrated by examining a focus series at high resolution: each of the pictures appears distinctly different although they all arise from the same object.

An important mathematical tool for image analysis is the Fourier transformation. Some definitions and theorems of the Fourier theory are summarized in section B without proof. Detailed accounts on this subject are to be found in the monographs mentioned above, and in exact treatments of the theory of Fourier transformation. For implementation of image-processing, either computer or coherent optical methods may be used. VANDER LUGT (1968) gives a survey of the theoretical possibilities of optical data processing. (For a critical comparison of processing by computer and analogue optics, see HUANG (1967)). The ability of an optical system to process simultaneously a large amount of data provides certainly an advantage over digital computers. On the other hand, the digital computer has a greater flexibility and can readily be programmed for a particular problem. This flexibility becomes important, for instance, when the complicated modulation transfer functions of electron microscopy have to be taken into account. Nevertheless, rather sophisticated optical filtering techniques have already been applied to electron micrographs (for instance, THON and SIEGEL, 1970).

The computer implementation of the analysis of electron micrographs requires the representation of continuous two dimensional functions in the computer to be fully understood. In section C we will briefly discuss all processes which lie between the electron intensity distribution in the image plane of the electron microscope and the sequence of digital data on the magnetic tape, the input for further processing. This can be done by application of simple Fourier relations.

Straightforward applications of the linear system theory are noise filtering and enhancement of images under assumption of additive noise (section D). Much experience in this field has been gathered in the analysis of pictures obtained from space craft missions, and can be used in electron microscopy, as pointed out by NATHAN (1971). The correlation analysis (section E) will be presented in detail, since the lateral alignment of electron micrographs is an essential part of the image difference method (section G.III) and the restoration from focus series (section F.II, III).

The first specific problem of electron microscopy to be discussed here is the interpretation of high resolution micrographs for which the intensity is governed by the phase contrast mechanism. The formalism of linear transfer theory has been outlined by HANSZEN (1971) for the case of phase contrast imaging. The main effect of the aberrations of the objective lens is to produce a degradation of the image noticeable at high resolution and, in cooperation with mechanical and electrical instabilities of the instrument, to limit the resolution to a value which is far below the wave optical resolution limit at the energies commonly used in electron microscopy. A number of methods have been proposed to compensate the lens aberrations in the instrument: by specific lens design [see review by SEPTIER (1966)] or by use of small filter plates in the back focal plane of the objective lens (HOPPE, 1961; THON and WILLASCH, 1971; UNWIN, 1971; THOMSON and JACOBSEN, 1971). However, these methods require high expenditures and a very delicate adjustment technique. We will show in section F the alternative possibility of using the computer to remove the image degradation and to obtain higher resolution.

Another factor complicates the interpretation of high resolution electron micrographs: the superposition of object and substrate structure. In section G some techniques are discussed which yield a subsequent object/substrate separation from the image.

For the analysis of high resolution micrographs, radiation damage must be considered as a limiting factor since the linear system approach is appropriate only if the number of electrons per unit area is sufficient. Quantitative studies of several authors (see the compilation by GLAESER, 1972) show that many biological objects exposed to the electron beam are disordered before the irradiation dose is attained which is necessary for statistical reasons to attain high resolution. If the dose is minimized, on the other hand, the electron statistics have to be taken into account for optimal interpretation of the image data. Since the noise can no longer be considered additive, the theory of nonlinear systems has to be applied, a field which has yet to be investigated thoroughly and is not covered by this review.

So far, only the relation between the projected object density and the image has been mentioned. The computer methods have very successfully been used for the reconstruction of the three dimensional object structure from the projections (section H). The three dimensional reconstruction using Fourier methods proposed by DE ROSIER and KLUG (1968a) has been developed, by the work of KLUG et al. in Cambridge, into a powerful tool for the study of virus particles and biological macromolecules. The success of this approach was facilitated by the fact that, for medium resolution, there is a simple relation between the projected object density and the electron microscopic image, and that the three dimensional synthesis of the object can be achieved by methods very similar to those proven in x-ray crystallography. Indeed, the electron microscope can be described as a diffraction instrument which, unlike the x-ray diffractometer, is able to record amplitudes *and phases* of the diffracted beams (DE ROSIER and KLUG, 1968; HOPPE, 1971a). Besides the Fourier methods, there are other approaches to three dimensional reconstruction which are still under discussion (section H.II).

B. Linear Systems and Fourier Processing

I. The Concept of Linear Systems

The linear system approach to the description of optical systems will be summarized, following ANDREWS (1970). The response of a linear system is obtained by linear superposition of elementary responses, and can therefore be defined by describing the elementary response. In our case, the input to the system is the object, characterized by a two dimensional function o(r) and the output is the image, p(r). The input function may be represented by a weighted sum of displaced Dirac's delta functions

$$o(r) = \int o(r')\,\delta(r-r')\,dr'. \tag{1}$$

If the linear system is symbolized by an operator H, the output function is obtained from

$$p(r) = \mathsf{H}\{o(r)\} = \mathsf{H}\{\int o(r')\,\delta(r-r')\,dr'\} = \int o(r')\,\mathsf{H}\{\delta(r-r')\}\,dr'. \tag{2}$$

The operator H can be passed through the integral because it is linear, and affects only functions depending on r. The function $\mathsf{H}\{\delta(r-r')\} = h(r-r')$ is called *impulse response or point spread function* of the system. It is the image of an object merely consisting of a δ-shaped peak in $r = r'$. Since any object function can be represented as a superposition of δ-functions, Eq. (1), the function h provides all information necessary to describe the linear system. The system is called isoplanatic if the point spread function depends only on $(r-r')$, and Eq. (2) yields a convolution integral

$$p(r) = \int o(r')\,h(r-r')\,dr'. \tag{3}$$

The analysis of linear systems is facilitated by using orthogonal transformations. Most commonly used is the Fourier transform; for optical systems it can be easily interpreted since it gives the relation between the amplitude distribution in conjugate planes, in Fraunhofer's approximation of Kirchhoff's diffraction integral. There are other useful orthogonal transformations described by ANDREWS (1970a) which also provide complete representations of the image.

II. Fourier Integrals and Theorems

The Fourier transformation in two dimensions is defined by

$$F(k_x, k_y) = \int\int_{-\infty}^{\infty} f(x, y)\exp[2\pi i(k_x x + k_y y)]\,dx\,dy \tag{4}$$

or

$$F(k) = \mathsf{F}\{f(r)\}$$

where $r = (x, y)$ is the vector in real space and $k = (k_x, k_y)$ is the vector in Fourier (or "reciprocal") space. The integral (4) exists if $f(x, y)$ is quadratically integrable. From (4), the function $f(x, y)$ is in turn obtained by the *inverse* Fourier transformation

$$f(x, y) = \int\int_{-\infty}^{\infty} F(k_x, k_y)\exp[-2\pi i(k_x x + k_y y)]\,dk_x\,dk_y \tag{5}$$

or

$$f(r) = \mathsf{F}^{-1}\{F(k)\}.$$

Therefore, $f(r) = F^{-1}\{F\{f(r)\}\}$. Since the Fourier transformation is a linear operation,

$$F\{a\,f_1(r) + b\,f_2(r)\} = a\,F\{f_1(r)\} + b\,F\{f_2(r)\}. \qquad (6)$$

For real $f(r)$,

$$F(k) = F^*(-k) \qquad (7)$$

(*Friedel's Law* in x-ray Crystallography). The *shift theorem* states that

$$F\{f(r + r')\} = F\{f(r)\} \exp[2\pi i\,k\,r']. \qquad (8)$$

In polar coordinates one obtains from (4), (5), by applying the coordinate transformations

$$x = r \cos \varphi \qquad k_x = R \cos \Phi$$
$$y = r \sin \varphi \qquad k_y = R \sin \Phi$$

the new transformations formulas

$$F(R, \Phi) = \int_0^{2\pi}\!\!\int_0^\infty f(r, \varphi) \exp[2\pi i\,r\,R \cos(\Phi - \varphi)]\, r\, dr\, d\varphi \qquad (4')$$

$$f(r, \varphi) = \int_0^{2\pi}\!\!\int_0^\infty F(R, \Phi) \exp[-2\pi i\,r\,R \cos(\Phi - \varphi)]\, R\, dR\, d\Phi \qquad (5')$$

which can be simplified if f is a circular symmetric function $f(r, \varphi) \equiv f(r)$ to yield the *Fourier-Bessel transformations*

$$F(R) = 2\pi \int_0^\infty f(r)\, J_0(2\pi\,r\,R)\, r\, dr \qquad (4'')$$

$$f(r) = 2\pi \int_0^\infty F(R)\, J_0(2\pi\,r\,R)\, R\, dR \qquad (5'')$$

where J_0 is the Bessel function of zero order (see JAHNKE, EMDE and LÖSCH, 1960).

The expression

$$C(r) = \int f_1(r')\, f_2(r' - r)\, dr' \equiv f_1(r) \circ f_2(r) \qquad (9)$$

is called the *convolution product* of the functions f_1 and f_2. The *convolution theorem* states that

$$F\{f_1(r) \circ f_2(r)\} = F\{f_1(r)\}\, F\{f_2(r)\} = F_1(k)\, F_2(k) \qquad (10)$$

and, similarly

$$F^{-1}\{F_1(k) \circ F_2(k)\} = F^{-1}\{F_1(k)\}\, F^{-1}\{F_2(k)\} = f_1(r)\, f_2(r). \qquad (11)$$

Similar Fourier relations hold for the correlation functions which are important for statistical analysis (MARRIAGE and PITTS, 1956; JONES, 1955). The *cross correlation function* is defined as

$$\Phi_{12}(r) = \int f_1(r')\, f_2^*(r' + r)\, dr' \qquad (12)$$

and provides a measure of the mutual statistical dependence of both functions, for each displacement vector r. Especially for $f_1 \equiv f_2$, one obtains the *autocorrelation function*

$$\Phi(r) = \int f(r) f^*(r' + r) \, dr'. \tag{13}$$

It is a measure for the statistical dependence of the values that the function assumes for argument vectors r' which are separated by a vector r. Definitions (12) and (13) are valid for *statistically homogeneous* functions, i.e. functions whose statistical properties do not change when translating the argument by an arbitrary vector. The Fourier theorems for the correlation functions are

$$\mathsf{F}\{\Phi(r)\} = |\mathsf{F}\{f(r)\}|^2 \equiv \mathsf{W}(k) \tag{14}$$

$$\mathsf{F}\{\Phi_{12}(r)\} = \mathsf{F}\{f_1(r)\} \, \mathsf{F}^*\{f_2(r)\} \equiv \mathsf{W}_{12}(k). \tag{15}$$

The relation (14) is known as the *Wiener-Khinchin theorem*. The autocorrelation function and the *power spectrum* ("Wiener spectrum") $\mathsf{W}(k)$ are two equivalent statistical descriptions of two-dimensional stochastic functions. The Wiener spectrum gives the contribution of the Fourier coefficients at each spatial frequency k to the mean square deviation σ^2 of the function representing the image, independently of their phase. This is a consequence of *Parseval's theorem*

$$\Phi(r = o) = \int f(r') f^*(r') \, dr' = \int \mathsf{W}(k) \, dk. \tag{16}$$

Similar to the Wiener-Khinchin theorem, the relation (15) states that the Fourier transform of the cross correlation function is the *cross power spectrum* $\mathsf{W}_{12}(k)$, which is the conjugate product of the Fourier transforms of two images.

The convolution theorem (10) is very useful for describing the connexion between input and output function of the linear system. Instead of the convolution integral in (3), we have

$$\mathsf{F}\{p\} = \mathsf{F}\{o\} \, \mathsf{F}\{h\}. \tag{17}$$

The advantage of "working in Fourier space" both for theoretical analysis and for computation becomes more obvious when a chain of linear systems with point spread functions $h_1, h_2 \ldots h_n$ has to be examined. While the notation in real space leads to an n-fold interlocked convolution integral, the Fourier notation is simply

$$\mathsf{F}\{p\} = \mathsf{F}\{o\} \, \mathsf{F}\{h_1\} \, \mathsf{F}\{h_2\} \ldots \mathsf{F}\{h_n\}. \tag{18}$$

As an example, the optical imaging (h_{opt}) and the photographic recoding (h_{ph}) may be considered a chain of linear systems which has the resultant transfer function

$$\mathsf{F}\{h\} = \mathsf{F}\{h_{opt}\} \, \mathsf{F}\{h_{ph}\}. \tag{19}$$

III. Implementation on the Computer

In the last paragraph, it has been shown that both the convolution and correlation integrals can be obtained either directly from their defining equations or by using the equivalent Fourier relations. The direct calculation corresponds to a matrix multiplication and requires, in the one dimensional case, $N(N-1)$ operations if the array contains N elements. The advantage of the Fourier method is a result of the fact that there exists a fast transformation algorithm which requires only $N \log_2 N$ operations. The calculation of a convolution expression can be done by executing three Fourier transformations and one scalar multiplication, yielding $3 N \log_2 N + N$ operations. The resulting speed factor for computation by direct versus Fourier methods is $(N-1)/(3 \log_2 N + 1)$ which increases quickly when N is increased.

In practice, this factor is usually smaller since the range of the point spread functions is limited. Therefore, also direct computations of the convolution product are used. NATHAN (1971) and SELZER (1968) do their analyses in real space with point spread matrices small compared with the size of the picture. The speed of their computations is further increased by using a multiplier-processor, a hardware device especially designed for convolution operations.

In the computer, the Fourier integral has to be replaced by a sum. Let us assume that a two dimensional function $f(x, y)$ is given by the values it assumes on $M \times N$ equidistant points of a grid $f_{jk} = f(x_j, y_k) = f(j \Delta x, k \Delta y)$. The discrete Fourier transform on the reciprocal grid $(k_{xl}, k_{ym}) = (l/\Delta x, m/\Delta y)$ is then given by (for inst. HILDEBRAND, 1956; compare with Eq. (1))

$$F(k_{xl}, k_{ym}) = \sum_{j=0}^{M-1} \sum_{k=0}^{N-1} f_{jk} \exp \left[2\pi i \left(\frac{x_j k_{xl}}{M} + \frac{y_k k_{ym}}{N} \right) \right]. \tag{20}$$

This representation implies that the given function f extended over a finite area is replaced by an infinite periodic function obtained by translating f in vertical and horizontal directions.

The computation of (20) is made easy when the *Cooley-Tukey algorithm* is used (COOLEY and TUKEY, 1965). This algorithm takes advantage of a decomposition of the transform matrix which becomes possible if N is a power of 2. (For details see, for instance, ANDREWS, 1970a.) Subroutines using this principle are found in most user program libraries.

C. Digitizing of Electron Micrographs
I. Photographic Recording

The effect of the photographic plate can be described by a point spread function $h_{ph}(r-r')$ or, alternatively, by the corresponding transfer function $H_{ph}(k) = F\{h_{ph}\}$. The ideal point spread function $h_{ph}(r-r') = \delta(r-r')$

cannot be realized for two reasons: 1) the photographic grain size is finite, and 2) the elementary event (the impact of a single electron) produces a cluster of grains with finite lateral extension. The finite width of the photographic point spread function is expressed, in terms of the transfer function, as a decrease of the magnitude of the Fourier coefficients with increasing radius vector in the Fourier plane, $|k|$. This resolution limiting effect can be minimized by choosing an emulsion with suitable scattering properties and grain size. An additional parameter is the electron optical magnification. An increase in magnification by a factor ξ corresponds to a scale transformation in Fourier space

$$k \to \frac{1}{\xi} k.$$

The effect is that the spatial frequency region essential for transferring the electron microscopical information is contracted with respect to the Fourier coordinate system belonging to the real space coordinate system of the photographic plate. If the factor ξ is chosen appropriately, this frequency region will now cover a central area of the Fourier plane not affected by the decrease of the photographic transfer function.

In order to facilitate a formal description of the densitometric measurement in the next paragraphs, it is necessary to assume a continuous density distribution on the electron micrograph. It is clear from the nature of the imaging process that this assumption is an idealization. In fact, we have on the plate a statistical distribution of grain clusters each of which documents the impact of a single electron. The mean number of clusters per unit area is a measure of the local electron intensity. The size of the cluster and the number of grains per cluster is dependent on the electron energy and the scattering properties of the emulsion. (For a detailed account of this subject see VALENTINE and WRIGLEY (1964) and BURGE, GARARD and BROWNE (1968)). From this complicated discrete distribution the continuous density distribution is generated only by the action of the densitometer which integrates the transparency over the measuring aperture. Since the integration extends only over a small area, the result of the measurement is a superposition of an ideal continuous function representing the image, and a noise function describing the statistical fluctuations of the number of electrons and the number of grains. In the following paragraphs of this section, we will only regard the continuous function, leaving the discussion of the noise contribution for section D.

II. The Densitometer

The densitometer is used for decomposing and digitizing the transparency. The values representing the transparency on the sampling grid points are recorded on a magnetic tape and are subsequently converted to density values by taking the logarithm. Any scanning device may be used if its spatial resolution is sufficient. There are mechanical scanners (see ARNDT et al., 1969) and flying spot systems (see ARNDT, CROWTHER and MALLETT, 1968). The main disadvantage of the mechanical scanners, the low scanning speed, has recently been overcome by the development of rotating drum instruments. The scanner is often incorporated in an image processing system, and the special way of data acquisition is dependent on the particular installation. Some problems of data acquisition in connexion with densitometry are discussed by BILLINGSLEY (1971); MENDELSOHN et al. (1968); ARNDT, CROWTHER and MALLETT (1968) and BENDER and ROWE (1971).

Image intensifier systems permit the recording of the electron intensity directly, without the photographic plate. The analogue video output can then be digitized and used as direct input to the computer. Installations of this type are under development in a number of laboratories (see, for instance, GLAESER, KUO and BUDINGER, 1971).

The densitometer modifies the input image. We will distinguish between artifacts which are produced by defects of the scanner and those inherent to the principle of the measurement.

Distortions of the sampling grid belong to the first group. They are due to mechanical inaccuracies or, for flying spot densitometers, to non-linearities of the deflection, and will not be discussed since they can be compensated by appropriate electronic devices or calibration methods.

Artifacts inherent to the measurement are caused by

1) the discontinuous sampling
2) the scanning aperture
3) the effect of the image boundary.

They will be discussed in the next paragraphs. Similar accounts of this subject are given by BILLINGSLEY (1971) and by DE ROSIER and MOORE (1970).

III. Sampling

How can the error be characterized which arises when a continuous function is represented by its values at discrete arguments? Let us assume, for a moment, that we have a point-like aperture. It is evident that details smaller than the sampling distance are not represented by the discrete function. By applying the Fourier theory, we obtain a quantitative answer to our question (SAYRE, 1951). Supposed the sampling grid $g(r)$ is an infinite quadratic point lattice that is periodic with the translation constant a. The output function $d'(r)$ representing the discrete values on the grid may be considered as being generated by multiplying the input density function $d(r)$ with the grid function

$$d'(r) = d(r) g(r) . \tag{21}$$

With the notation $D(k) = F\{d(r)\}$, $D'(k) = F\{d'(r)\}$, $G(k) = F\{g(r)\}$ and the convolution theorem, we obtain by Fourier transforming (21)

$$D'(k) = D(k) \circ G(k) \qquad (22)$$

where $G(k)$ is again a quadratic point lattice with the constant $1/a$ (GOOD-MAN, 1968b). Therefore, the Fourier transform of the output function is

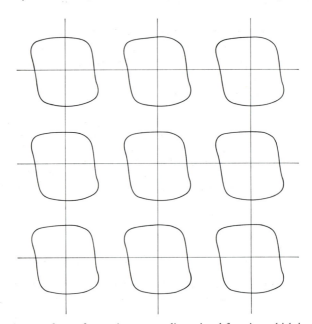

Fig. 1. Fourier transform of a continuous twodimensional function which is sampled on a square lattice. (After SAYRE, 1951)

obtained by placing the Fourier transform of the continuous input function onto each point of the reciprocal lattice $G(k)$, see Fig. 1. In general, $D(k)$ is extended infinitely, and so the contributions from all reciprocal points are overlapping each other. Since, however, the input image has originated from a physical imaging process, the spatial frequency range where $D(k)$ is appreciably different from zero is limited ("band limit"). Let us assume that the band limitation is given by $|k| \leq k_0$. Apparently, the overlapping of adjacent Fourier transforms can be avoided as long as the period of the reciprocal lattice $G(k)$ is

$$\frac{1}{a} \geq k_0. \qquad (23)$$

This criterion defines the period a of the sampling grid. The band limit may be estimated by examining the Fourier intensity of the input picture in the light optical diffractometer (KLUG and BERGER, 1964; THON, 1965). In the case of coherent bright field micrographs of phase objects the band limit is indicated by the frequency region where the characteristic contrast zones (THON, 1965) disappear. For the determination of the band limit, it is helpful to produce an additional modulation of that part of the Fourier intensity which arises from the object, by using a technique developed by FRANK et al. (1970a; see also HOPPE et al., 1970).

In order to keep the number of sampling points (and thus the computer time and storage) as small as possible it is reasonable to work just at the limit of condition (23). There are, however, particular cases where it is desirable to have additional sampling points, for instance to avoid interpolation if different pictures have to be aligned by correlation methods (section E).

In the context of the theory of stationary stochastic functions, the densitometric measurement is considered a discontinuous sampling process which yields an estimate of the input function (cf. DAVENPORT and ROOT, 1958). It has been proved for one-dimensional functions that decreasing the sampling distance a impairs the signal to noise ratio as soon as a becomes smaller than the correlation distance (= halfwidth of the auto-correlation peak) of the noise (PORCHET and GÜNTHARD, 1970). If we apply this result to the two-dimensional case, we find a lower limit for the sampling distance which depends on the properties of the photographic material.

IV. The Effect of the Scanning Aperture

So far, the discussion has assumed a point-like sampling aperture. In fact, the densitometer integrates the transparency over a finite aperture area. The transparency measured at r_i, $\tilde{t}(r_i)$ can be expressed as a convolution product of the transparency $t(r)$ and the aperture function $b(r)$

$$\tilde{t}(r_i) = [t(r) \circ b(r)]_{r=r_i} = [\exp(-d(r)) \circ b(r)]_{r=r_i}. \tag{24}$$

Taking the logarithm yields the optical density

$$\tilde{d}(r_i) = \log[\exp(-d(r)) \circ b(r)]_{r=r_i} \tag{25}$$

which can approximately be replaced by a convolution product of the continuous density function with the aperture function, provided that the density does not change too much within the aperture.

$$\tilde{d}(r) = d(r) \circ b(r). \tag{26}$$

Therefore, the transform $D(k)$ in (22) has to be replaced by

$$\tilde{D}(k) = D(k) B(k). \tag{27}$$

The Fourier transform of a circular aperture (radius r) is

$$B(R) = \frac{J_1(2\pi r R)}{2\pi r R} \tag{28}$$

where J_1 is the Bessel function of the first order (see JAHNKE, EMDE and LÖSCH, 1960). In the flying-spot densitometer, the light spot on the emulsion corresponds to the aperture, and the aperture function is now given by the intensity distribution over this spot. According to BILLINGSLEY (1971), this distribution can be approximated by a gaussian or a $\cos^2$-function. In any case, the aperture produces an attenuation of the high order Fourier coefficients and reduces the resolution. This effect is sometimes desired for too wide sampling grids because it reduces the artifacts due to overlapping of adjacent transforms mentioned in the last paragraph. If, on the other hand, the sampling grid is narrow enough to fulfilll condition (23), then one could think of making the aperture as small as possible so that it does not impair the resolution. The last conclusion is false, however, since it does not take statistical considerations into account which favour large apertures: The number of electrons per unit area, n, is limited by the radiation tolerance of the object. As the number of electrons falling into the aperture area F is $N = nF$, the standard deviation of the electron noise is $\sigma_n = (nF)^{1/2}$. This noise fluctuation has to be compared with the local fluctuation due to the signal, $\sigma_s = nF(\Delta I/I_0)$ where ΔI is the intensity variation in the bright field image, and I_0 is the background intensity. Therefore, the signal-to-noise ratio is $\sigma_s/\sigma_n = (nF)^{1/2}\,\Delta I/I_0$, which is proportional to the linear dimensions of the aperture. If the aperture is too small when compared with the sampling distance, an appreciable amount of information is lost by the scanning procedure.

We obtain as a result of our discussion in the paragraphs III and IV that the sampling distance and the aperture radius are constants which have to be chosen carefully, with regard to both band limit of the object spectrum and the noise statistics.

V. The Effect of the Image Boundary

The densitometer cuts out a finite area from the image, generally with a rectangular boundary (dimensions x_0, y_0). The cutting can be thought as having been generated by a multiplication with a function of r which is one within the rectangle and zero outside. In the Fourier representation this is equivalent to a convolution of the Fourier transform of the infinitely extended image with the transform of this cutting function

$$C(k_x, k_y) = \frac{\sin \pi x_0 k_x}{\pi x_0 k_x} \frac{\sin \pi y_0 k_y}{\pi y_0 k_y} \tag{29}$$

and has mainly the effect of producing strong spikes along the axes of the Fourier plane. The spikes are not an artifact but are necessary to represent the steep density transitions on the boundary. As mentioned in section B.II the representation of the Fourier integral by a finite Fourier sum implies that, instead of the individual image, a two-dimensional periodic array is transformed which is obtained by repeating the image. In this repeating "image", upper and lower, and left and right edges of the image are thus adjoining each other, respectively. In many cases the mean densities on adjacent edges are not very different, therefore the spiking is less pronounced. The effect is strong, on the other hand, if the dimensions of the Fourier transformed array are greater than those of the scanned image, and if the missing data are replaced by zeros. A similar situation arises when the reconstruction problem requires a particular area of the image to be separated for further analysis (cf. DE ROSIER and MOORE, 1970). Setting the unwanted data to zero manifests itself, as in the case of the rectangular boundary, in convoluting the image transform with the "shape transform" associated with the particular boundary. The spikes of the shape transform can disturb the analysis of the object transform, for instance the indexing of the helical diffraction pattern. DE ROSIER and MOORE (1970) therefore use a constant instead of zero, which is defined in such a way that the density step on the boundary is minimized.

D. Noise Filtering

The image is disturbed by random processes from various sources most commonly described as "noise". There are three idealizations used in the literature to facilitate the discussion of the image interpretation in the presence of noise: that the noise be white, additive and gaussian. "White" means that the energy of the noise function is evenly distributed over the spatial spectrum, or in other terms: that the Wiener spectrum of the noise $W(k) \approx$ const. "Additive" means that the disturbed image $p'(r)$ can be thought as being generated by simple superposition of an ideal image $p(r)$ and a noise function $n(r)$. Finally, "gaussian" implies that the probability to encounter a given noise amplitude in a given image element follows a gaussian distribution.

The interpretation of an image disturbed by noise can be considered as a decision between different hypotheses which have different a priori probabilities according to the information available. A decision theory for the particular case of binary decision process has been formulated by HARRIS (1964) under the assumption of gaussian noise.

I. Noise Sources

The predominant noise sources in electron microscopy are the electron noise ("shot noise") and the noise of the photographic plate. (The supporting film can also be thought of as a noise source lying on the object side of the imaging system; see section G). We already talked about the photographic noise in connexion with the representation problems, section C.I. As a description of the photographic noise, it is a fair approximation to assume an additive function which is independent of the electron intensity. In their theoretical treatment of the photographic granularity JONES (1955) and MARRIAGE and PITTS (1956) pointed out that either the Wiener spectrum or the autocorrelation function provide an adequate description for the photographic noise.

Obviously it is not possible to describe the electron noise as additive and as independent from the signal. The number n of electrons hitting the plate per unit time and area is given by the Poisson statistics which has the standard deviation $n^{1/2}$. Therefore, the electron fluctuation is signal-dependent. RÖHLER (1967) has shown, however, that the additivity is a good approximation provided two conditions are both fulfilled:

1) the signal and the noise amplitude are both small when compared with the carrier of the signal;

2) the signal and the noise amplitude are in the same order of magnitude. The first condition is fulfilled for bright field phase contrast imaging of weakly scattering objects: the intensity modulations produced by the phase contrast are small compared with the background intensity. The same is true for the intensity fluctuations due the electron statistics under normal radiation conditions (500 ... 1000 el/$Å^2$). The second condition holds in those cases where the noise is particularly bothersome and the analysis is most desirable.

On the other hand, the additive approach fails if the background intensity is very low or if the signal-to-noise ratio is large. These restrictions should always be kept in mind when the noise is considered additive in the following sections.

II. Noise Filtering in the Case of Periodic Objects

For periodic objects, the signal/noise ratio can be improved considerably by translating the image by integer multiples of the period and superposing all images created by this procedure. The same principle is the basis of the photographic superposition technique of MARKHAM, FREY and HILLS (1963). The signal/noise ratio improves with the square root of the number of superpositions. Nathan applied the averaging method to an electron micrograph of bovine liver catalase (1970, 1971), see Fig. 2.

It should be mentioned, however, that an equivalent noise suppression can be accomplished more efficiently in Fourier space. It is only necessary to mask off, by an appropriate filter function, all Fourier coefficients except those lying on the points k_i of the reciprocal lattice, similarly to the optical filter technique used by KLUG and DE ROSIER (1966). Since the contribution of a given spatial frequency k to the mean square deviation σ^2 is

Fig. 2. Noise suppression by averaging over a number of periods of the image of a crystalline object. (Density display by CRT output). a) Original electron micrograph of bovine liver catalase. Horizontal periodicity 68 Å; b) Image computed by averaging over a number of periods; c) Contour display of b) by removing high order digital bits. (From NATHAN, 1970; by courtesy of Claitor's Publishing Division)

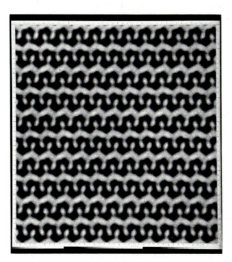

Fig. 2b

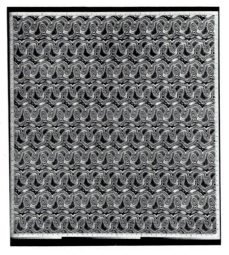

Fig. 2c

given by the value of the Wiener spectrum at k, $W(k)$ (see section B.II), and since the Fourier contributions of the signal are concentrated at the reciprocal points, the signal/noise ratio is improved due to the action of the masking filter by the factor

$$\varkappa = \frac{\sigma_n}{\left[\sum_{i\,(k_1,b)} \int W_n(k)\,dk \right]^{1/2}} \tag{30}$$

where $W_n(k)$ is the Wiener spectrum of the noise, σ_n is the standard deviation of the noise, and (k_1, b) denote the small areas of integration around the reciprocal points the size of which is a property of the particular filter function chosen. The factor $\varkappa$ can become very large because these areas are small when compared to the total Fourier area contributing to σ_n.

The averaging method and the Fourier masking method, respectively, have an interesting aspect in connexion with the problem of overcoming the radiation damage. In many cases of biological interest the radiation dose permitted per object resolution area is not sufficient to form an image of the object structure with adequate contrast (see, for instance, GLAESER, 1971). GLAESER, KUO and BUDINGER (1971) proposed, as a way to overcome this difficulty, to gather the information from several periods of the image of a periodic object, thus reducing the radiation dose per resolution area by using the spatial redundancy. A model calculation by these authors shows how the vectors of the reciprocal lattice can be determined from the Wiener spectrum of a statistical image, and demonstrates the gain in the signal/noise ratio obtained by superposing the image on itself, translated by multiples of the period (Fig. 3). In order to exhaust the statistical information of the image it is then necessary to replace the photo plate by an efficient image intensifier system capable of registering single electron events (GLAESER, KUO, BUDINGER, 1971).

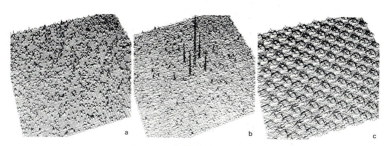

Fig. 3. Model calculation to illustrate the efficacy of the superposition method for imaging at levels of electron irradiation. a) Statistically noisy image of a checkerboard object; b) Wiener spectrum of a); c) Image constructed by spatial superposition. (From GLAESER, KUO and BUDINGER, 1971; by courtesy of Claitor's Publishing Division)

III. Noise Filtering in the Case of Aperiodic Objects

For aperiodic objects, the Fourier transform is not concentrated at discrete spots in the reciprocal plane but is spread out over a certain domain of the reciprocal plane where it is superimposed by the noise transform. Although it is impossible in this situation to obtain an effective signal/noise separation, the Fourier filtering can still be very helpful for the interpretation of the image, because those Fourier coefficients of the signal can be sacrificed which are deemed not very significant for the analysis. For instance, low frequent density fluctuations visible in the electron micrograph of a membrane do arise from the signal but may hinder rather than help the detection of small low-contrast details. Therefore, one may decide to suppress the low frequency region of the image transform in order to eliminate both the insignificant part of the object transform and a part of the noise transform which contributes a considerable portion to the noise rms σ_n. The Fourier filtering consists of multiplying the image transform $P(k)$ with a filter function $H(k)$

$$P'(k) = P(k) H(k) \tag{31}$$

which is equivalent, according to the convolution theorem, to a convolution of the image with the corresponding point spread function $h(r) = F^{-1}\{H(k)\}$. The design of specific spatial filters is closely connected with the restoration and separation problems to be discussed in detail later. At present, only a class of unspecific filters will be mentioned: low- and high-pass filters. The low-pass filter attenuates the Fourier coefficients at high spatial frequencies but passes those at low spatial frequencies. The high-pass filter, on the other hand, attenuates the Fourier coefficients at low spatial frequencies, leaving the high spatial frequencies unchanged (Fig. 4). For the

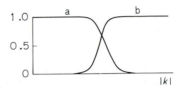

Fig. 4. Spatial filter functions in Fourier space. a) Low-pass filter; b) High-pass filter

construction of these filters it is necessary to avoid discontinuities as these produce strong artifacts in the inverse transform ("overshot effect"). To give an example for the possible use of the low-pass filter in electron microscopy, one might wish to eliminate the high-frequency focus-dependent granularity of the supporting grid, if only object details larger than about 10 Å are relevant for the investigation. A more detailed discussion of high- and low-pass filters may be found in the paper by SELZER (1968) who has successfully used these filter functions for analyzing medical x-ray pictures with the computer.

E. The Cross Correlation Function and its Use for Image Alignment

If two or more electron micrographs of the same objects are to be evaluated, their position with respect to the common origin has to be found. As a coarse reference, distinct object details may be used for adjusting the films or plates in the densitometer. This is insufficient, however, to yield the accuracy needed in high resolution analysis since the boundaries of such details are not well defined. For refining the translational position, the cross correlation function (CCF), which has been defined in section B II, can be used successfully. The CCF of two statistically dependent images has a peak whose displacement from the origin gives the mutual displacement of the images. Two particular cases shall be considered here which are important for some practical applications to be discussed later: a) the CCF of carbon film micrographs with equal defocus, taken before and after specimen preparation and b) the CCF of micrographs with different defocus. The applications are the image difference method (section G.III) and the complex object restoration (section F.III), respectively.

I. Two Electron Micrographs with Identical Defocus Value

The electron micrographs taken before and after preparation are

$$p_1(r) = n_s(r) \circ h(r) + n_1(r), \qquad p_2(r) = (o(r) + n_s(r + r')) \circ h(r) + n_2(r) \quad (32)$$

$n_s(r)$ the "noise" due to the support film structure
$o(r)$ the object structure added by the preparation
$h(r)$ the point spread function belonging to a linear imaging process such as the phase contrast imaging
r' the unknown displacement vector
$n_1(r), n_2(r)$ the noise contributions on the side of the image (i.e. photographic noise).

For calculating the CCF, one has first to compute

$$W_{12}(k) = F\{p_1(r)\} F^*\{p_2(r)\} \qquad \text{(see section B.I)}$$

from the calculated Fourier transforms of the micrographs and then obtains

$$\Phi_{12}(r) = F^{-1}\{W_{12}(k)\}.$$

From Eq. (32) and from the shift theorem Eq. (8) it follows that

$$W_{12}(k) = (N_s(k) H(k) + N_1(k)) (O(k) H(k) + \underline{N_s(k) H(k) \exp[2\pi i r' k]} + N_2(k))^* \tag{33}$$

where capital letters denote the Fourier transform of the corresponding real space function (for instance $N_s = F\{n_s\}$).

The essential Fourier contribution to the CCF $\Phi_{12}(r)$ comes from the multiplication of the terms underlined in (33):

$$\Phi(r) = F^{-1}\{|N_s(k)|^2 |H(k)|^2 \exp[-2\pi i k r']\} \tag{34}$$

$$= (\Phi_{n_s}(r) \circ \Phi_h(r)) \circ \delta(r - r')$$

This is a convolution product of the autocorrelation function (ACF) of the film structure, Φ_{n_s} with the ACF of the point spread function, Φ_h. Therefore the sharpness of the peak and thus the accuracy of the alignment varies strongly with the shape of the point spread function, which in turn is determined by the defocus value and other electron optical parameters. Fig. 5 shows as an example the CCF of two carbon film micrographs taken at the same defocus ($\Delta z \approx 3000$ Å, from LANGER et al. (1970a)). The peak allows the position to be determined with 0.5 Å accuracy or better.

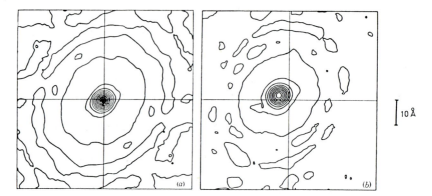

Fig. 5. Crosscorrelation function of two electron micrographs of a carbon film at the same defocus value ($\Delta z \approx 3000$ Å). The peak is displaced by (3.5 Å, 1.7 Å) from the central position, indicating that the second image has to be shifted by (-3.5 Å, -1.7 Å) in order to bring it into register with the first image. (From LANGER et al., 1970a; by courtesy of Berichte der Bunsengesellschaft für Physikalische Chemie)

II. Two Electron Micrographs with Different Defocus Values

Some problems require the mutual positions of micrographs of one focus series to be known. Let us consider the images of the supporting film for different defocus values

$$p_1(r) = n_s(r) \circ h_1(r) + n_1(r)$$

$$p_2(r) = n_s(r) \circ h_2(r) + n_2(r) \tag{35}$$

where h_1, h_2 are the point spread functions for different Δz. Now we obtain as the essential contribution to the CCF instead of (34)

$$\Phi(r) = \mathbf{F}^{-1}\{|N_s|^2 H_1 H_2^* \exp[-2\pi i k r']\}$$

$$= (\Phi_{ns} \circ \Phi_{h_1 h_2}) \circ \delta(r - r'). \tag{36}$$

This is a displaced CCF which is a convolution product of the ACF of the film structure with the CCF of the point spread functions, $\Phi_{h_1 h_2}$. The shape of the CCF peak is strongly dependent on the Δz-values. There are unfavourable Δz-combinations where the peak is so broad that the displacement vector r' cannot be determined with sufficient accuracy.

This can be understood by examining the term $H_1 H_2^* = \sin \gamma_1 \sin \gamma_2$ in (36). $\sin \gamma_1$ and $\sin \gamma_2$ are the phase contrast transfer functions for phase objects, with γ denoting the phase shift produced by the objective lens, see section F. I. These functions have different sign on subsequent zones in the Fourier plane. The location of these zones will change for different values of Δz. Therefore the product $\sin \gamma_1 \sin \gamma_2$ has normally positive and negative zones which are counteracting each other as for their contributions to the CCF peak.

III. A Technical Note

The displacement vector obtained from the CCF usually does not lie on a point of the scanning grid, but in some intermediate position. Therefore, the position of the image to be aligned with the reference image has to be calculated by interpolating from the values on the grid. The interpolation can be avoided, however, if the scanning distance is much smaller than the limiting value of the resolution.

Finally, it should be mentioned that a procedure for refining the mutual orientation of digitized pictures has been suggested by HOPPE et al. (1969) but has not been used so far. The orientation is usually adjusted with the aid of distinct details of the supporting film that are clearly visible in both electron micrographs under consideration. This is conveniently done when the micrographs are mounted on the densitometer.

F. Two-Dimensional Restoration

The image differs from the object in two respects:
1) It is distorted due to the aberrations of the imaging system.
2) It contains noise resulting from random errors of the imaging and the recording system.

The problem of restoring the object function from the image has been examined in detail for the case of linear systems with additive noise (ELIAS, GREY, and ROBINSON, 1952; HARRIS, 1966; LEVI, 1970).

I. Restoration of Phase Objects from a Single Phase Contrast Image

In electron microscopy, there is a particular case where the relation between image and object can approximately be described by a convolution integral (compare section B, Eq. (3))

$$p(r) = o(r) \circ h(r) \tag{37}$$

This is for the phase contrast imaging of weak phase objects[1] for coherent bright field illumination (HANSZEN and MORGENSTERN, 1965).

This theory takes only the elastically scattered electrons into account. Recently it has been pointed out by PARSONS (1970) and by CRICK and MISELL (1971) that the contribution of the inelastically scattered electrons to the image has been underestimated in the literature. However, it is reasonable to assume that the images obtained from the inelastic part are blurred due to the cooperation of chromatic aberration and finite energy width, and that therefore only the low order Fourier coefficients of the image are affected. This intuitive conclusion has recently been confirmed by MISELL and CRICK (1971).

---

1 Theoretically, the same is true for pure amplitude objects (HANSZEN and MORGENSTERN, 1965), but they do not exist in electron microscopy.

One has to identify

1) $p(r) \equiv \Delta I(r)$

where ΔI is the intensity variation in the bright field image;

2) $o(r) \equiv V(r)$

where V is the real two-dimensional potential function derived from the three-dimensional potential distribution of the object by projecting along the optical axis. The Fourier transform of the projected potential is the $k_z = 0$ central section of the structure factor. In our notation, with $r_j = (x_j, y_j)$ as the location of the jth atom in the projection plane, and $k = (k_x, k_y)$, the Fourier coefficients on this section are

$$F(k) = \sum_j f_j(k) \exp[2\pi i k r_j] \tag{38}$$

where f_j are the atomic scattering amplitudes. Since we are in this section dealing with two dimensional restoration, we will call (38) simply the "structure factor", without danger of causing confusion.

3) $h(r) \equiv F^{-1}\{\sin \gamma(k) A(k)\}$ \hfill (39)

where $A(k)$ is the aperture function and $\gamma(k)$ describes the phase shifts in the back focal plane resulting from defocusing and the lens aberrations of the objective lens (SCHERZER, 1949).

$$\gamma(k) \equiv \gamma(\vartheta, \varphi) = -\frac{\pi}{\lambda}\left(\Delta z + \frac{\Delta z_A}{2} \sin 2(\varphi - \varphi_0)\right)\vartheta^2 + \frac{\pi}{2\lambda} C_s \vartheta^4 \tag{40}$$

λ wavelength
Δz defocus value
Δz_A focus difference of axial astigmatism
C_s spherical aberration constant
φ_0 reference azimuth of axial astigmatism.

An additional transfer term F_{ph} has to be inserted in the operand in (39) in order to allow for the degradation produced by a detector system. This function will be omitted, however, since it has no effect for high magnification, for the reasons given in section C.I. With the identifications made above, Eq. (37) takes now the form

$$\Delta I(r) = V(r) \circ F^{-1}\{\sin \gamma(k) A(k)\} \tag{41}$$

which yields after Fourier transformation

$$J(k) = F\{\Delta I(r)\} = F(k) \sin \gamma(k) A(k). \tag{42}$$

The *phase contrast transfer function* $\sin \gamma$ usually has several zeros in the aperture region where it changes its sign (see Fig. 6 as an example). It can be shown, by calculating the corresponding point spread functions (HOPPE and LANGER, 1965), that the resolution is impaired very much by these

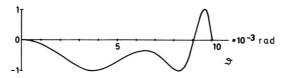

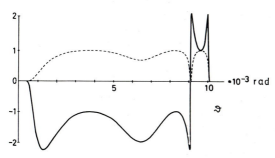

Fig. 6. Restoration of the image of a pure phase object: a) phase contrast transfer function
($U = 80$ KV, $C_s = 3.9$ mm, $\Delta z = 1600$ Å); b) weighting function (dotted line) and
filter function (solid line). (From FRANK et al., 1970b; by courtesy of Berichte der Bunsengesellschaft für Physikalische Chemie)

phase reversals when compared with the resolution of a purely aperture-
limited lens.

In order to outline the restoration procedure, we start again from the
general notation

$$p(r) = o(r) \circ h(r) + n(r) \tag{43}$$

where we have included an additive noise function on the side of the image.
By Fourier transforming, we obtain

$$P(k) = O(k) H(k) + N(k) \tag{44}$$

with capital letters denoting the Fourier transform of the corresponding
real space function. We call restoration any procedure which results in a
good approximation of the object function $o(r)$[2].

The quality of the approximation can be stated in a quantitative way
and can be used for constructing an optimal filter, as will be shown later.

---

2 i. e., within the limits set by the aperture limit. It should be mentioned, however,
that it is possible theoretically to recover Fourier coefficients lying outside of the aperture, by using the analytical properties of the Fourier transform of a bounded object
(see HARRIS, 1964b; BARNES, 1966).

In the restoration scheme used by HARRIS (1966), Eq. (44) is multiplied by a filter function

$$H'(k) = H_W(k) H^{-1}(k) \tag{45}$$

to yield

$$P'(k) = H'(k) P(k) = O(k) H_W(k) + N(k) H_W(k) H^{-1}(k). \tag{46}$$

$H_W(k)$ is a weighting function which has to be designed in such a way that the noise amplification in (46) is limited. Before application of the filter the signal-to-noise ratio is

$$\tau_1 = \frac{\int\limits_{(ap)} W_0(k) |H(k)|^2 \, dk}{\int\limits_{(ap)} W_n(k) \, dk} \tag{47}$$

and after

$$\tau_2 = \frac{\int\limits_{(ap)} W_0(k) |H_W(k)|^2 \, dk}{\int\limits_{(ap)} W_n(k) |H(k)|^{-2} |H_W(k)|^2 \, dk} \tag{48}$$

(see, for instance, HARRIS, 1964a), where $W_0(k)$ and $W_n(k)$ are the Wiener spectra of the object and the noise, respectively, and (ap) symbolizes integration over the aperture range. From (48) it is clear that the signal-to-noise ratio vanishes if $H(k)$ contains zeros within the aperture and $H_W(k)$ is set to unity. Therefore it is necessary to adjust H_W carefully to the transfer function $H(k)$. Reverse transformation of (46) results in

$$p'(r) = o(r) \circ h_W(r) \circ a(r) + n(r) \circ h(r) \circ h_W(r) \circ a(r). \tag{49}$$

Here, the notations

$$h_W = \mathbf{F}^{-1}\{H_W\}, \qquad h = \mathbf{F}^{-1}\{H^{-1}\},$$

$$a = \mathbf{F}^{-1}\{A\} = \int\limits_{(ap)} \exp[2\pi i k r] \, dk$$

have been used. The first term on the right hand side in (49) has a reduced resolution when compared with the object function. This is due to the effect of aperture limitation, expressed by convolution with $a(r)$, and to the point spread function of the weighting function, $h_W(r)$. Obviously, restoration without loss of resolution is only possible when noise is absent. HOPPE (1970a, 1970b) has discussed the application of this restoration scheme in phase contrast electron microscopy. First computations were shown by FRANK et al. (1970a, 1970b). As weighting function,

$$H_W(k) = \exp\left[\alpha\left(1 - \frac{1}{|\sin \gamma|}\right)\right] \qquad (\alpha \text{ is a parameter})$$

was used (Fig. 6). Because of insufficient output facilities (optical density was obtained by printing symbols upon each other) it was difficult to

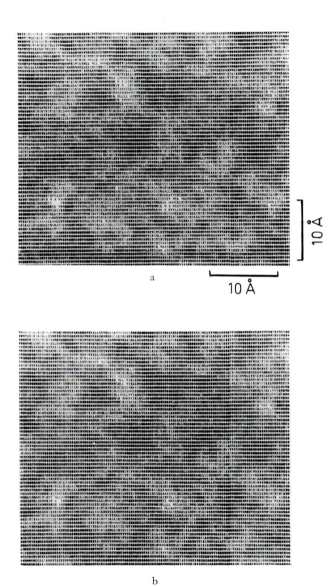

Fig. 7. Restoration formalism applied to a small portion of an electron micrograph of the murein sacculus of *Spirillum serpens*. (Density display by printing symbols upon each other). a) Input image; b) Restored image. (From FRANK et al., 1970b; by courtesy of Berichte der Bunsengesellschaft für Physikalische Chemie)

assess the improvement in image quality and resolution obtained by filtering (Fig. 7).

One of the critical problems of the restoration of phase contrast micrographs is the determination of $\sin \gamma$ for the particular exposure. While spherical aberration, wavelength and magnification are constants of the instrument, the other parameters occurring in (40), namely Δz, Δz_A, φ_0 vary from one experiment to the other. FRANK et al. (1970a, 1970b) developed a least squares method for determining these missing parameters from the Fourier transform of the carbon film image near the object. This method uses the fact that the Fourier transform of a carbon film micrograph under coherent bright field conditions is given by

$$J(k) = \sin \gamma(k) F(k) = \sin \gamma(k) f_{carbon}(k) \sum_j \exp[2\pi i k r_j] \qquad (50)$$

The sum in (50) is, for an amorphous structure, a function with random amplitudes and random phases. Therefore, in the optical diffraction pattern there appears a white, strongly fluctuating intensity distribution modulated by $\sin^2 \gamma$, providing the basis for Thon's observations (1965). $|J(k)|^2$ can be calculated, as well, and shows the characteristic Fourier zones (Fig. 8). In the least squares approach, the parameters Δz, Δz_A and φ_0 are determined in such a way that $\sin \gamma$ describes the location and shape of these zones optimally. (For a detailed description of this method, see BUSSLER, to be published (1972)).

Since (50) is valid only for coherent illumination, the error of the phase determination increases with increasing ϑ. It should be possible to take the partial incoherence of the illumination into account for the determination of the electron optical parameters, as well.

Another way for determining these parameters is the evaluation of the dark field/bright field image displacement effect (HEINEMANN, 1972). Recently, BUDINGER (1971) showed that the data obtained from the displacement experiment agree fairly well with those obtained from diffractograms.

WELTON (1971) recently proposed to apply Wiener's optimal filter (1949) to the phase object restoration problem. This filter is optimal in the sense of the least squares criterion

$$\int [o(r) - p'(r)]^2 \, dr = \min \qquad (51)$$

where $p'(r)$ is the filtered image, and allows for noise on the side of the image ("detector noise"). The Wiener filter is

$$H'(k) = \frac{H^*(k)}{|H(k)|^2 + W_n(k)/W_0(k)} \qquad (52)$$

with the same notation as introduced earlier in this paragraph. A short derivation of this formula for two-dimensional problems is presented, for instance, in the review article by LEVI (1970). WELTON (1971) shows in a model calculation that the image quality and the resolution is substantially

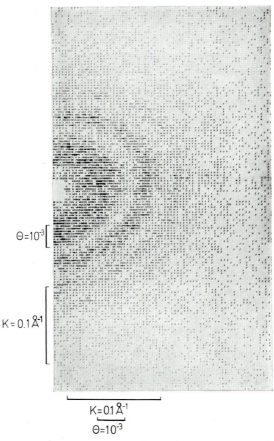

Fig. 8. Fourier transform of an electron micrograph of a carbon film. Displayed is the logarithm of the absolute value of the Fourier coefficients, by printing symbols upon each other. (U = 100 KV, C_S = 3.9 mm, Δz = 3000 Å, Δz_A = 1200 Å). (From FRANK et al., 1970b; by courtesy of Berichte der Bunsengesellschaft für Physikalische Chemie)

enhanced by the Wiener filter. The model object is a double stranded DNA molecule with thallium atoms attached to the phosphate groups, on a carbon film (Fig. 9). The imaging conditions assumed in this calculation are not realistic, however: the aperture opening corresponds to 1 Å resolution, and the radiation dose is assumed to be 3000 el./Å². Although the Wiener filter seems to be the optimal solution of the restoration problem, it has an essential disadvantage: It requires the Wiener spectrum of the object (that is, of the class of possible objects) to be known, an amount of a priori information not available in many situations.

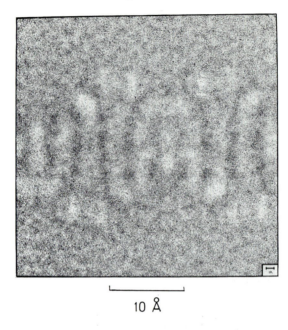

10 Å

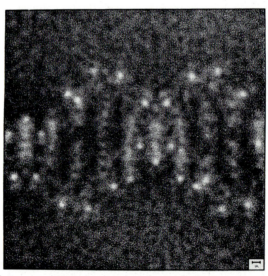

Fig. 9. Model calculation using the Wiener filter function: Restoration of the image of double stranded DNA, with thallium atoms attached to the phosphate groups (amplitude contrast neglected, highly coherent illumination assumed; $U = 150$ KV, $C_S = 1$ mm, aperture 4×10^{-2} rad). a) Simulated phase contrast image; b) Restored image. (From WELTON, 1971 (computation of (b) was in error and has been corrected); courtesy by Claitor's Publishing Division)

II. Restoration of Phase Objects from a Focus Series

Because of the presence of noise, even optimal filtering leaves an information gap near the zeros of the phase contrast transfer function. These gaps can be filled by evaluation of several electron micrographs of one focus series, as proposed by HANSZEN (1968). At the same time, however, the amplitude portion of the object can be obtained, as will be shown in the next paragraph.

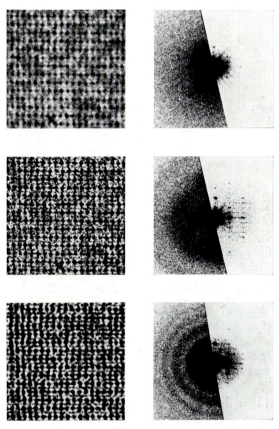

Fig. 10. Restoration of a crystalline object (negatively stained catalase crystal) from a bright field focus series. a) Electron micrographs and the corresponding optical diffraction patterns at $\Delta z = 0$ Å, 5400 Å, 14500 Å (from top to bottom, respectively). The apparent lattice spacings in the micrograph are 65.5 Å in the horizontal and 89 Å in the vertical direction. Left hand side of diffraction patterns overexposed to show phase contrast zones from amorphous background. b) Noise filtered, averaged images reconstructed from the discrete catalase transforms at different defocus. (From ERICKSON and KLUG, 1970; by courtesy of Berichte der Bunsengesellschaft für Physikalische Chemie)

ERICKSON and KLUG (1970, 1971) use a focus series in order to reconstruct a periodic object, catalase crystals (Fig. 10a). The calculated Fourier transforms of the micrographs of the series shows discrete diffraction maxima on the reciprocal lattice which are changing both in amplitude and

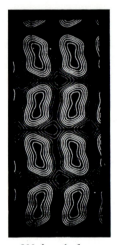 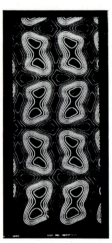 

800 Å underfocus   5,400 Å underfocus   14,500 Å underfocus

Fig. 10b

phase when the focus is changed. The Δz dependence of the diffraction spots could be described by

$$J(\vartheta, \varphi) = F(\vartheta, \varphi) [\sin \gamma(\vartheta) + Q(\vartheta) \cos \gamma(\vartheta)]. \tag{53}$$

$\gamma(\vartheta)$ is the phase shift defined in Eq. (40); it is in (53) not dependent on φ because the astigmatism was compensated in this experiment. It becomes visible in this equation that this object cannot be considered as a pure phase object, but an additional "amplitude term" has to be introduced which will be discussed in more detail in the next paragraph. ERICKSON and KLUG were able to calculate the "true" transform $F(\vartheta, \varphi)$ from the transforms of the series. They thereby obtained, by inverse transformation, several repeats of the restored object. Since only the discrete diffraction spots are included in the transformation, the aperiodic noise is automatically eliminated (Fig. 10b).

III. Restoration of the Complex Object

So far only the restoration of phase objects has been dealt with. In the last paragraph an experiment has been outlined which shows that the phase contrast equation (42) has to be modified in order to allow for amplitude effects when heavy atom stains are present in the object. This is because the imaginary part of the elastic scattering cannot be neglected for atoms with high atomic number (GLAUBER and SCHOMAKER, 1953). Image formation for the case of complex scattering has been discussed by ZEITLER (1968), REIMER (1969) and HOPPE (1970a).

With complex atomic scattering amplitudes

$$f_j(\vartheta) = f_j'(\vartheta) + i f_j''(\vartheta), \tag{54}$$

the structure factor now has the form

$$\begin{aligned} F(k) &= \sum_j f_j'(\vartheta) \exp[2\pi i k r_j] + i \sum_j f_j''(\vartheta) \exp[2\pi i k r_j] \\ &= F_r(k) + i F_i(k). \end{aligned} \tag{55}$$

This decomposition of the structure factor corresponds to the decomposition of the object function into a "phase" and an "amplitude" part (HANSZEN and MORGENSTERN, 1965). The Fourier analysis of the bright field intensity distribution gives, save a constant factor and a quadratic term,

$$\begin{aligned} J(k) &= \delta(k) + i F(k) \exp[-i\gamma(k)] - i F^*(-k) \exp[i\gamma(k)] \\ &= \delta(k) + 2F_r(k) \sin\gamma(k) - 2F_i(k) \cos\gamma(k) \end{aligned} \tag{56}$$

Therefore the Fourier contributions from both real and the imaginary scattering are mixed up in the transform of the image[3]. It is desirable to separate both contributions since in this way additional information about the object becomes available. An interesting application is the separation of heavy and light elements of the object on the basis of the Z-dependence of the imaginary scattering (HOPPE, 1970a). The structure factor can be computed from (56) if at least two "measurements" $J_n(k)$ are made with different phase functions $\gamma_n(k)$ that are known from an independent measurement. A convenient way to change $\gamma(k)$ is to change the defocus Δz. Other methods for obtaining the structure factor are suggested by HOPPE, LANGER, and THON (1970) and by HOPPE (1971b). If the focus series contains $N > 2$ pictures, a least squares method can be used, as proposed by SCHISKE (1968). The normal equations give the solution

$$F(k) = \frac{-i \sum_{m=1}^{N} J_m(k) \exp[i\gamma_m(k)] \left\{ N - \sum_{n=0}^{N} \exp[2i(\gamma_n(k) - \gamma_m(k))] \right\}}{N^2 - \left| \sum_{m=1}^{N} \exp[-2i\gamma_m(k)] \right|^2}. \tag{57}$$

Inverse transformation of $F(k)$ yields a complex image since $F(k) \neq F^*(-k)$, which follows from (55) for $f_j'' \neq 0$. This image is a complete restoration

---

3 ZEITLER and OLSEN have emphasized the importance of the cosine term for the image (1967). Quantitative estimations of the contrast of single atoms and small groups of atoms, taking the imaginary scattering into account, have been carried out by ZEITLER (1966) and REIMER (1969).

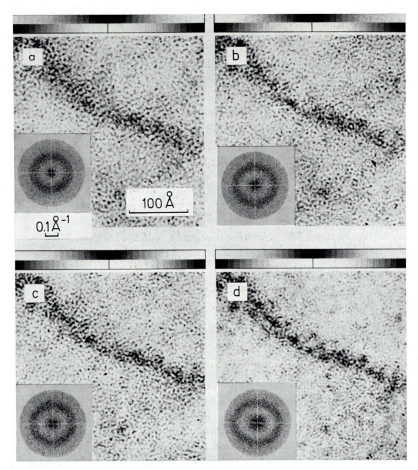

Fig. 11. Restoration of the image from the imaginary part of the atomic scattering amplitudes. a—d Defocus series of uranyl stained DNA on a thin carbon film, together with calculated Fourier transform (U = 100 KV, $\Delta z = -80, +40, +160, +280$ Å, $\Delta z_A \cong 0$ Å, $C_S = 2.9$ mm); e "Imaginary image", shows enhanced contrast for the uranium atoms. (From FRANK, 1972; by courtesy of Biophysical Journal)

of the object projection as it is imaged by the elastically scattered electrons. The imaginary part represents the image arising from the imaginary scattering and can be expected to have a high contrast for heavy atoms. FRANK (1972) used this selective contrast enhancement for the study of a focus series of DNA stained with uranyl acetate on a carbon film.[4]

---

4 Earlier studies have been done on Ferredoxin by BUSSLER and HOPPE but have not yet been published. Preliminary results were shown in the lecture by FRANK et al. (1970a).

The phase functions γ_n were obtained from a portion of each micrograph which contained only the carbon film, by using the least squares approach mentioned previously in section F.I. The micrographs had to be brought into register by calculating the mutual cross correlation functions (see section E). Fig. 11 shows the original series and the imaginary part

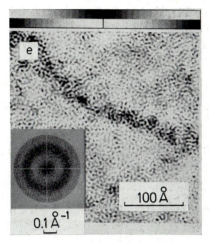

Fig. 11e

of the image obtained from the structure factor which was calculated from (57). The latter has indeed an enhanced contrast for the DNA molecule but the enhancement is less than expected. Further experiments will show whether the method described above can be developed into a tool for the study of macromolecular structures. The practical resolution is limited by two effects: First, the error in the γ-determination which increases with increasing scattering angle and second, the curvature of Ewald's sphere which produces $F(k) \neq F^*(-k)$ even for objects with pure real scattering (HOPPE, 1970a; ZEITLER, 1966).

IV. Restoration from Dark Field Images

Finally, a completely different restoration method will be mentioned which has been proposed by COWLEY (1953) and is being pursued, in modified form, by NATHAN (1971). By tilting the illumination with respect to the optical axis, different parts of the diffraction spectrum are brought into the center of the aperture plane where the lens aberrations are small. In the method described by NATHAN (1971), a small aperture is displaced in the rear focal plane of the objective lens so that only a small number of diffraction maxima pass through. In each position a dark field image is taken. It can be shown that, by analyzing these dark field images, one is able to calculate the relative phases of all diffraction maxima. From the phases and the amplitudes the object can be reconstructed, with a resolution comparable to the resolution of the electron diffraction pattern. Since the diffraction patterns of some objects show diffraction maxima corresponding to 1 Å resolution or less, the synthesis from dark field images may well result in atomic resolution provided that the object is stable enough to permit a dark field series to be taken. NATHAN (1971) has demonstrated the usefulness of his method by simulating the experiment on the optical bench.

G. Object / Support Separation

The effectiveness of the electron microscopic imaging relies on the properties of the supporting film, if the tolerance dose per unit area of the object is limited. For, the contrast deemed to be sufficient for detecting an object detail (see GLAESER, 1971) is affected by the magnitude of the fluctuations due to the image of the film structure. The supporting film may be considered to be a noise contribution ("substrate noise") on the object side of the system, as opposed to the electron noise and the photographic noise on the detector side. The noise function representing this contribution is *additive* for bright field conditions, which will exclusively be assumed in this section. To some extent, it is possible to minimize the effect of the substrate noise by using careful preparation techniques and by controlling the focus. HARRIS discusses these methods in detail (1970).

For *subsequent* separation of object and substrate noise in the computer, some a priori information about the object function $o(r)$ and/or the noise function $n_s(r)$ must be known. We may classify the separation methods according to the type of information available.

I. Optimal Filtering

Suppose we know the statistical description of both functions, $o(r)$ and $n_s(r)$ in terms of their Wiener spectra $W_0(k)$ and $W_s(k)$, respectively. Then the separation problem is solved by the optimal filter already mentioned in section 5a. However, since the noise is on the *object* side of the imaging process, the filter function now has a different form (LEVI, 1970).

$$H'(k) = \frac{1}{1 + W_s(k)/W_0(k)} , \tag{58}$$

independent of the transfer function. The filter has no effect at spatial frequencies where $W_s(k)$ is small compared to $W_0(k)$ but comes into action where $W_s(k)$ is large. The Wiener spectrum of the supporting film may be determined from an electron diffraction pattern whereas $W_0(k)$ may be obtained by calculation from the class of possible objects (WELTON, 1971). But, as the possible objects are in general significantly different for high resolution details, the usefulness of the optimal Wiener filter is limited.

II. Matched Filtering

Now let us assume that the structure of the object (and thus the transform $O(k)$) is already known and the problem is to find out whether it is contributing to our image or not. This task is essentially a pattern recognition problem, and can be solved by applying the matched filter (see, for instance, ANDREWS, 1970b; LEVI, 1970). Under the hypothesis that the particular object is anticipated at r, the Fourier transform of the image is

$$P(k) = H(k) O(k) \exp[2\pi i k r] + H(k) N_s(k) \tag{59}$$

where $N_s(k) = F\{n_s(r)\}$.

The matched filter for the image of $o(r)$ is then, for white noise,

$$H'(k) = H^*(k) O^*(k) \exp[-2\pi i k r']. \tag{60}$$

For, multiplication of (59) with this filter function and subsequent inverse transformation yields

$$F^{-1}\{|H(k)|^2 |O(k)|^2\} \circ \delta(r-r') + F^{-1}\{N(k) O^*(k) |H(k)|^2\}. \tag{61}$$

The first term is, according to section B, the autocorrelation function of the image undisturbed by noise, and it has a sharp peak at $r' = r$. The second term is the cross correlation function (CCF) of the noise image and the object image. Because of the statistical independence of object and noise functions, the CCF has no such pronounced maximum. Therefore the appearance of a peak in (61) can be used as a criterion of whether the object specified by the matched filter is present or not. Although this technique seems to be promising in all cases where only a small number of distinct models are under consideration, it has not yet been used in electron microscopy.

A detailed examination shows that the matched filter is optimal in a sense that it maximizes the signal-to-noise ratio in $r = r'$ (ANDREWS, 1970b). It should also be mentioned that the matched filter can be designed for non-white noise functions as well (VANDER LUGT, 1964).

III. Separation Based on Knowledge of the Film Structure

The separation of the object from the support becomes easy when the supporting film is crystalline (HOPPE et al., 1969). The contributions of a periodic structure are confined to δ-like peaks on the reciprocal lattice. These diffraction maxima can be masked off either in the computer or in a coherent optical device. Inverse transformation then results in an undisturbed image, save a small loss of information in the masked portions of the reciprocal plane. RIDDLE and SIEGEL (1971) have recently succeeded in preparing very thin graphite films with monocrystalline areas. It is likely therefore that the method described above will be applied in the near future.

With the image difference method proposed by HOPPE et al. (1969) the structure of an amorphous film can be eliminated as well. One electron micrograph of the film is taken before specimen preparation, and one after the object under investigation has been deposited on the film. In our linear approximation of the bright field intensity (see section F.I), the image of the object ought to be the difference between the micrograph of film plus object and the micrograph of the film. A similar method has been used successfully by SELZER (1968) in the computer analysis of x-ray pictures, in order to detect long term changes in tissues, for instance those resulting from tumor growth. The mutual position of the pictures is easily determined by calculating the cross correlation function (see section E). It has been mentioned already in Fig. 5, section E that the mutual position of two carbon foil pictures can be determined with an accuracy of better than 0.5 Å. LANGER et al. (1970a, 1970b) developed a computer program which performs the cross correlation, positions the images and subsequently takes the difference. For a model experiment, two electron micrographs of a plain carbon film were taken within a short time. It was found that the changes of the film, due to contamination and radiation damage, were already sufficient to produce a pronounced difference image. Therefore the difference image method cannot be used successfully until an efficient ultra-high-vacuum-microscope becomes available.

IV. Separation Based on the Z-dependence of the Imaginary Scattering

If the object and the support consist of atoms of different species, the strong Z-dependence of the imaginary (or "anomalous") electron scattering can be used for the object/support separation. For this purpose, the complex object function has to be restored. In the imaginary part of this function the contrast of heavy atoms lying in a light atom environment is enhanced considerably. An application of this principle has already been discussed in section F.IV.

H. Three-Dimensional Reconstruction

Because of the large depth of focus of the electron microscope, the micrographs are essentially images of an object projected onto a plane normal to the incident beam. In order to recover the three-dimensional structure it is necessary to have a number of projections of the object at different angles. They may be produced by tilting the specimen, or, alternatively by selecting a few particles from an array of particles randomly oriented on the supporting grid. From the projections two different computational methods can be used to obtain the three dimensional structure:

1) DE ROSIER and KLUG use the projection theorem of the Fourier theory ("the transform of a two-dimensional projection of a three-dimensional structure is identical with the corresponding central section of the three-dimensional transform") to construct the object transform plane by plane from the transforms of the projections. Subsequently, the object is regained by inverse transformation.

2) In the "real space" methods the algebraic equation system is solved which represents the projection relations (CROWTHER, DE ROSIER and KLUG, 1970). For all practical purposes the matrices of this equation system are so large that a direct solution seems to be impossible, even when fast computers with a large memory are used. It was only recently that an iterative method was proposed for finding solutions which are consistent with the projection equations (GORDON, BENDER and HERMAN, 1970). It is possible that this approach will make the direct method a practicable alternative to the Fourier method.

The two principal ways of reconstruction are, of course, equivalent since any operation on functions in real space has its counterpart in an operation on their Fourier representations. The formal equivalence of the reconstruction methods has been emphasized by LAKE (1971b), see the comparison of the reconstruction schemes in Fig. 12.

I. The Fourier Method

In order to present the theory of the three dimensional reconstruction by the Fourier method we will closely follow the account given by CROWTHER, DE ROSIER and KLUG (1970) on this subject.

1. Principle of the Three Dimensional Fourier Reconstruction

The projection theorem for projecting along one of the Cartesian axes is easily obtained from the definition of the three-dimensional Fourier transform. The transform of the object density distribution $\varrho(x, y, z)$ is

$$F(k_x, k_y, k_z) = \iiint \varrho(x, y, z) \exp[2\pi i(x k_x + y k_y + z k_z)] \, dx \, dy \, dz. \quad (62)$$

For $k_z = 0$ we obtain a central section of F:

$$\begin{aligned} F(k_x, k_y, 0) &= \iint \{ \int \varrho(x, y, z) \, dz \} \exp[2\pi i(x k_x + y k_y)] \, dx \, dy \\ &\equiv \iint \sigma(x, y) \exp[2\pi i(x k_x + y k_y)] \, dx \, dy. \end{aligned} \quad (63)$$

It is obviously identical with the two-dimensional transform of the object projection along the z-axis. An equivalent result is obtained for an arbitrary

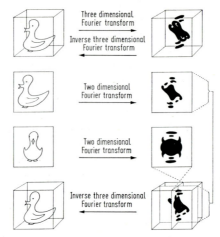

Fig. 12. Equivalence of the two principal threedimensional reconstruction methods: a) Fourier space reconstruction method. Real space structures are shown on the left and their respective Fourier transforms are shown on the right. (From LAKE, 1971a; courtesy by Claitor's Publishing Division); b) Real space reconstruction method. (From J. A. LAKE, "Biological Structures" in "Optical Transforms". Ed.: H. LIPSON, Academic Press Inc. (London) Ltd.; courtesy by Academic Press)

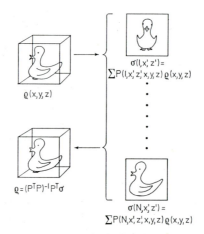

Fig. 12b

direction of projection. Let us assume that the predominant contrast in the image arises from the loss of electrons scattered elastically in wide angles and stopped by the objective aperture. Then the transmitted intensity I decreases with increasing mass thickness following an exponential law (HALL, 1953)

$$I(x, y) = I_0 \exp[-\varkappa \, \sigma(x, y)] \tag{64}$$

where I_0 is the intensity of the primary beam, $\varkappa$ is the mass absorption coefficient and $\sigma(x, y)$ the projection of the object mass density defined in (63). Since the optical density on the photographic plate is proportional to the electron intensity in a wide density range (see VALENTINE and WRIGLEY, 1964), the projected object mass density is proportional to the logarithm of the measured optical density.

The exponential law assumed above is a fairly good approximation for describing the amplitude contrast of negatively stained biological objects at medium and low resolution (for a discussion of the validity of (64) see ZEITLER and BAHR, 1965). Indeed, only electron micrographs taken under such conditions have been used so far for three dimensional reconstruction. At high resolution the inhomogeneous distribution of the stain becomes visible, and in addition the phase contrast can no longer be neglected. Three-dimensional reconstruction at high resolution requires therefore 1) that no negative staining is used for specimen preparation, and 2) that the distorted object projections are restored previous to the three-dimensional synthesis by methods discussed in section F.

The main effort of the reconstruction is to fill the Fourier space with data up to the resolution required, either by evaluating as many independent projections as possible, or by using symmetry relations. As opposed to the situation in x-ray crystallography, the amplitudes *and phases* of the Fourier coefficients are known. Inverse transformation of $F(k)$ yields then the three dimensional object density

$$\varrho(r) = \int F(k) \exp[-2\pi i k r] \, dk \qquad (65)$$

which can be represented in the computer by a summation

$$\varrho'(r) = \sum_l F(k_l) \exp[-2\pi i k_l r] . \qquad (66)$$

This representation holds only if the sample points k_l are regularly spaced. For, $\varrho'(r)$ in (66) may be considered as being a convolution of $\varrho(r)$ with the inverse transform of the sample function. Regularly spaced sample functions have inverse transforms which consist of sharp spikes on a lattice reciprocal to the sample lattice. Therefore, (66) represents an infinite repeat of the object on a three dimensional lattice. By appropriately choosing the sampling lattice, an overlapping of the object in the inverse transform can be avoided. If the object is bounded with diameter 2a, a sampling distance $1/2a$ in the Fourier space is sufficient to prevent overlapping (SHANNON, 1949).

Now the essential problem of the 3-d reconstruction is to determine all Fourier coefficients on a portion of the lattice whose size is determined by the required resolution. For, "blank" regions in the Fourier transform have the effect that the inverse transform is so strongly distorted that it has only little similarity with the desired object.

The coordinate systems practically being used are the Cartesian and the cylindrical systems. The Fourier transformation becomes a Fourier-Bessel inversion in the cylindrical coordinate system. For both systems the Fourier data obtained from the experiment usually do not come to lie on the sampling grid. This means that the Fourier coefficients on the grid have to be derived from the available coefficients lying in intermediate positions (*interpolation problem*). Furthermore, there are large domains of Fourier space where no data are available for experimental reasons. Even the best tilting devices have an angular limit of ± 45 degrees, which leaves a double cone-shaped region unexplored. We may call this the *extrapolation problem* but there is no difference between this and the interpolation problem from a systematic point of view, as will become clear in the next paragraph.

2. The Interpolation Problem

The general solution of the interpolation problem has been presented by CROWTHER, DE ROSIER and KLUG (1970). Their derivation will be outlined here since it shows both the possibilities and the limitations of the three dimensional Fourier reconstruction. The Fourier coefficients "measured" at arbitrary points, $F_{k'}$ and the Fourier coefficients lying on the regular sampling grid F_k are linked together by the "observational equations"

$$\sum B_{kk'} F_k = F_{k'} \tag{67}$$

where

$$B_{kk'} = \int_U \exp[2\pi i (k - k') r] dr \tag{68}$$

U is a volume large enough to include the object. These equations follow from (65) and (66) when a valid representation $\varrho'(r) \equiv \varrho(r)$ is required. The equation system can be solved as usual by matrix inversion. Since the individual measurements of the $F_{k'}$ are due to errors, and since the observation points may be unevenly distributed in Fourier space, it is necessary to have more observation equations than unknowns and to use a least squares procedure. The corresponding normal equations are, in matrix notation

$$B^H B F = B^H F' \tag{69}$$

with B^H denoting the Hermitian transpose of the matrix B. Provided that the normal matrix $B^H B$ is non-singular, the solution of (69) is

$$F = (B^H B)^{-1} B^H F'. \tag{70}$$

For Cartesian coordinate systems we may choose a cube of 2a length on a side as the "including Volume", U, in (68) and a sampling of 1/2a. Then the matrix element (68) turns out to be the well-known interpolation function of WHITTAKER (1915) and SHANNON (1949)

$$B_{kk'} = \frac{\sin \pi(k_x' - k_x)}{\pi(k_x' - k_x)} \frac{\sin \pi(k_y' - k_y)}{\pi(k_y' - k_y)} \frac{\sin \pi(k_z' - k_z)}{\pi(k_z' - k_z)}. \tag{71}$$

The importance of this function for three-dimensional reconstruction has been emphasized by HOPPE (1969). It is an apparent advantage of (71) that the three dimensional interpolation can be split up into independent sets of two and one dimensional interpolations, thus making the computation easier.

For the cylindrical coordinate system (that is, the polar coordinate systems in real and Fourier space as introduced in section B.II, plus a cylinder axis z and Z, respectively) the interpolation procedure comes out in a natural way, if the cylindrical expansion of KLUG, CRICK and WYCKOFF (1958) is used:

$$\varrho\,(r,\,\varphi,\,z) = \sum_n \int g_n\,(r,\,Z)\,\exp[i\,n\,\varphi]\,\exp[2\,\pi\,i\,z\,Z]\,dZ\,. \tag{72}$$

The Fourier transform in cylindrical coordinates has the form

$$F\,(R,\,\varPhi,\,Z) = \sum_n G_n\,(R,\,Z)\,\exp\left[i n\left(\varPhi + \frac{\pi}{2}\right)\right] \tag{73}$$

where $G_n(R,\,Z)$ and $g_n(r,\,Z)$ of (72) are interrelated by the Fourier-Bessel transformation

$$g_n\,(r,\,Z) = \int G_n\,(R,\,Z)\,J_n\,(2\,\pi\,R\,r)\,2\,\pi\,R\,dR \tag{74}$$

with J_n as the Bessel function of the nth order. The Eq. (73) may as well be considered as an observation equation on an annulus with radius R in the Fourier space, which states, for a given $\varPhi_j$, the relation between the observed Fourier coefficients $F\,(R,\,\varPhi_j,\,Z)$ and the set of $G_n(R,\,Z)$. The maximum number n, for which effectively $G_n \neq 0$, is limited for a given radius R since the radius of the object is limited. If more observations are made than this number, the additional information can again be used by a least squares method. With the abbreviation $B_{jn} = \exp\{i\,n\,(\varPhi_j + \pi/2)\}$, the observation equations are in matrix notation

$$F' = BG \tag{75}$$

and the normal equations of the least squares problem

$$B^{\mathrm{H}} F' = B^{\mathrm{H}} BG\,. \tag{76}$$

Assuming that $B^{\mathrm{H}} B$ is non-singular, the solution is

$$G = (B^{\mathrm{H}} B)^{-1} B^{\mathrm{H}} F'\,. \tag{77}$$

We see that for both coordinate systems considered, all Fourier coefficients lying on the grid within the resolution sphere or box are in principle available from the least squares interpolation of the observed data, as long as the normal matrices $B^{\mathrm{H}} B$ are non-singular, that is as long as their eigenvalues are nonzero. In practice, a lower limit has to be set for the eigenvalues in order to prevent an excessive noise amplification when the solution (77) is calculated. CROWTHER, DE ROSIER and KLUG (1970) specify the condition that the mean inverse eigenvalue is less than unity. This condition does not restrict the use of the interpolation for calculating

Fourier coefficients on sampling points lying close to the observation points, but it practically excludes the possibility of obtaining the coefficients lying in the whole Fourier region not accessible by tilting.

Hoppe (1971a) shows that there is a possible way to overcome this difficulty but his method requires the electron density peaks of atoms to be resolved, a situation not yet achieved in electron microscopy.

3. The Use of Symmetries

The amount of data to be determined can be greatly reduced when the object has known symmetries. The symmetries can be used either implicitly by choosing an appropriate coordinate system in which the least squares problem takes a simpler form, or by explicitly using the symmetry relations in Fourier space for filling out the unknown regions. All three-dimensional reconstructions done so far rest on the presence of symmetries, which are commonly detected by examining the optical diffraction pattern of the micrographs (Klug and Berger, 1964; Moody, 1971). DeRosier and Klug (1968a) illustrate the advantage of symmetry. They estimate the number m of independent views required for three dimensional reconstruction of structures of linear dimension $D = 250 \text{ Å}$ with a resolution of $d = 30 \text{ Å}$. The numbers obtained are $m = 1$ for helical (T4 phage tail), $m = 2$ for icosahedral (Tomato bushy stunt virus) and $m = 30$ for no symmetry (ribosome). A general estimation for m ($m \approx \pi D/d$) has been given (for the case of no symmetry) by Crowther, DeRosier and Klug (1970). In the derivation of this formula it was assumed, however, that the m views are equally spaced in the angular range of the cylindrical coordinate system.

For helical symmetry one projection already provides enough information to fill the Fourier space with data to a certain resolution limit. This is the reason why most of the reconstruction work has been done with helical structures.

The particular role of helical symmetry comes out when the mathematical formulation of the diffraction theory of helical structures is considered. This has been worked out by Klug, Crick and Wyckoff (1958) and by Cochran, Crick and Vand (1952). Supposed the axis of helical symmetry lies on the axis of the cylindrical coordinate system, then the Fourier transform is [cf. (73)]

$$F(R, \Phi, l/c) = \sum_n G_{n,1}(R) \exp \left[i n \left(\Phi + \frac{\pi}{2} \right) \right] \qquad (78)$$

where l is the layer line number and c is the axial repeat of the structure: $\varrho(r, \varphi, z) = \varrho(r, \varphi, z + c)$. $F(R, \Phi, Z)$ is zero everywhere except at $Z = l/c$. The integers l and n are connected by the "selection rule"

$$l = tn + um \qquad (79)$$

t   number of turns of the helix per axial repeat
u   number of subunits per repeat
m   any integers satisfying (79).

The reconstruction consists of determining the coefficients $G_{n,1}$ from the measured Fourier coefficients F, and subsequently calculating $\varrho\,(r,\ \varphi,\ z)$ by Fourier-Bessel inversion

$$\varrho\,(r,\ \varphi,\ z) = \sum_1 \sum_n g_{n,1}(r) \exp\,(-\,2\pi ilz/c) \tag{80}$$

and

$$g_{n,1}(r) = \int G_{n,1}(R) J_n\,(2\pi R\,r)\,2\pi R\,dR \tag{81}$$

with J_n as the Bessel function of the nth order.

DeRosier and Moor (1970) have specified the condition under which a reconstruction is possible from a single view only (see also DeRosier, 1970; DeRosier and Klug, 1968a). This is possible, if in the Fourier region specified by the resolution, there is on any layer plane 1 only one of the G_n contributing to the sum (78). The coefficients G_n have, as functions of R, the property that they are effectively zero for $R < (n-2)/2\pi a$ where a is the largest radial dimension of the particle. In the region where only one n contributes we have, instead of (78), the equation

$$F\,(R,\ \Phi,\ l/c) = G_{n,1}(R) \exp\left[i\,n\!\left(\Phi + \frac{\pi}{2}\right)\right]. \tag{82}$$

4. Implementation

The practical way of performing the three dimensional reconstruction of a helical object has been described in detail by DeRosier and Moore (1970). We will give a brief outline of this paper since it characterizes the procedure in a number of studies to be mentioned later. The main effort of the reconstruction consists of utilizing Eq. (82) for reconstruction of the full three dimensional transform from the two dimensional transform of the digitized micrograph.

Fig. 13 is a schematic presentation of the flow of data in the three-dimensional reconstruction. The images are examined by optical diffraction. From the optical diffraction patterns of single particles masked out from the image, the axial repeat distance can be estimated, and an indexing can be proposed on the basis of helical diffraction theory. Furthermore, the quality of the diffraction patterns is the basis for selecting a limited number of images for further analysis. These images are then scanned and converted to arrays of optical densities. From these arrays the Fourier transforms are computed, using the Cooley-Tukey algorithm. Before the Fourier transformation is carried out, two procedures are performed, in Fig. 13 called "box" and floating". The box algorithm is used to mask off all density values but those belonging to the particle. In the "floating" process the effect of the steep density step at the boundary of the particle is reduced (see discussion in section C.V). The next step, after Fourier transformation, is the interpolation of layer line data from the computed transform. This is necessary if the layer lines of the helical diffraction pattern do not coincide with the rows of the transform. As we know from the general discussion of the interpolation problem in section H.I.2, a Fourier coefficient lying between sampling points is obtained by summing over the contributions of

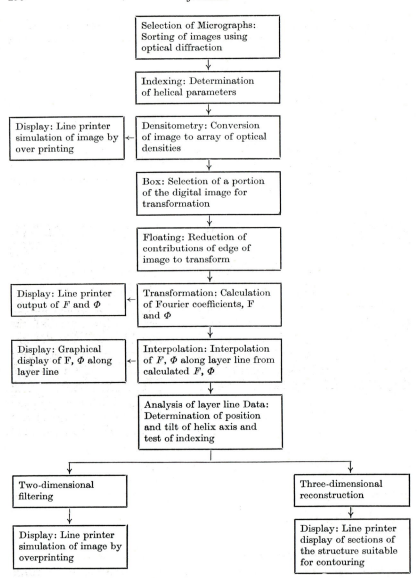

Fig. 13. Data flow diagram for the three dimensional reconstruction of helical objects. (From DeRosier and Moore, 1970; by courtesy of Journal of Molecular Biology)

all sampling functions centered in the sampling points (cf. (71) in three dimensions). However, this complicated procedure can be replaced by a bilinear interpolation, as DeRosier and Moore pointed out, if 1) the sampling distance in reciprocal space is fine relative to the inverse size of the object and 2) the position of the origin of the coordinate system is known with respect to the digitized image. Both the position of the origin and the tilt angle of the helical axis can be obtained from the image transform, by making use of the symmetry relation on the layer lines. In this step, denoted by the box "Analysis of Layer Line Data", the indexing can also be confirmed. Once the indices n and the coefficients $F(R, \Phi, l/c)$ are known with proper phases, the reconstruction is a straightforward evaluation of the Eqs. (73), (81), (80) in this order. The result is a three dimensional density function $\varrho(r, \varphi, z)$ representing the object. For display, a number of two-dimensional sections $r = const$, $z = const$ or $\varphi = const$ are produced, which can then be used for constructing a three dimensional model.

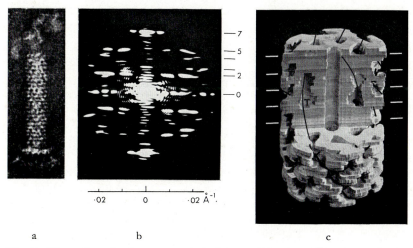

a   b   c

Fig. 14. Three dimensional reconstruction of the extended T4 phage tail at 15 Å resolution. a) Electron micrograph of a single tail; b) Optical diffraction pattern; c) Model showing the three dimensional reconstruction. (From DeRosier and Klug, 1968a; by courtesy of Nature)

The first three dimensional reconstruction by DeRosier and Klug (1968a) resulted in a three dimensional model of the extended T4 phage tail, at a resolution of 15 Å (Fig. 14). It is the goal of their work to study the contraction mechanism by comparing the three dimensional structures of extended and contracted phage tail.

Moore, Huxley and DeRosier (1970) examined thin filaments of F-Actin, decorated and undecorated with myosin (radial resolution 23 Å, axial resolution 45 Å). Particularly interesting in the context of this work is a method developed by Moore and DeRosier (1970) to isolate single

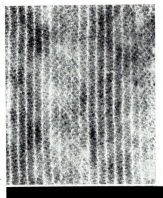
a

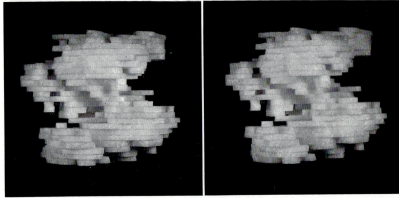
b

Fig. 15. Helical arrangement of the chromatoid bodies of *Entamoeba invadens*. a) Electron micrograph of positively contrasted chromatoid bodies sectioned along the helical axis; b) Stereopair of the threedimensional reconstruction. The model shows six-seventeenth of a repeat. The diameter of the chain is 600 Å. (From LAKE and SLAYTER, 1970b; by courtesy of Nature)

strands of F-Actin, from the paracrystallire array, for further analysis. The authors make use of the fact that the point lattice is known which describes the location of each particle so that an undisturbed particle can be obtained by deconvolution. This is done most conveniently by using the convolution theorem.

Another helical object is available in the thin, stained sections of the chromatoid bodies of *Entamoeba invadens*. These were studied by MORGAN (1968) and LAKE and SLAYTER (1970a, 1970b). The chromatoid bodies consist of ribonucleoprotein particles, which are arranged in closely packed helices and are thought to be ribosomes. Morgan starts from the intensities in the optical diffraction pattern of an image area which contains a number of helical chains, and uses for the three dimensional synthesis the phases corresponding to a single hypothetical model helix with point-like ribosomes. De Rosier and Klug (1968b) have raised objections against the principle of this approach and have questioned the result. Indeed, a rigorous treatment of the same object by Lake and Slayter (1970a, 1970b) — isolation of a single chain in the digitized image and using the phases of the computed transform-resulted in a ribosome model different from that obtained by Morgan (Fig. 15). The working resolution was 40 Å, mainly limited by the inhomogeneity of the stain. A particular problem arose due to the fact that the sections in this study were thinner than the diameter of a single helix. Therefore the electron micrograph shows projections of sectioned helices, and the application of the reconstruction method of De Rosier and Klug leads to errors (Lake and Slayter, 1970b). Lake showed how the helical reconstruction method can be modified to account for the sectioning (1971a, 1971b).

To complete the list of helical reconstructions, De Rosier's recent work on the tubes of catalase should be mentioned (1970, 1971).

An object with cylindrical symmetry was examined by Finch and Klug (1971): the stacked-disk rod of tobacco mosaic virus protein. The rods are built from rings made of 17 protein subunits. The rings are associated in pairs ("discs") in the direction of the longitudinal axis of the particle. The three dimensional synthesis was expected to answer the question of whether the rings in these disks face each other in the same or in the opposite direction. The outcome of the reconstruction (at a resolution of 20 Å) favoured the first hypothesis, a result bearing important implications for the understanding of the assembly of the TMV particle from protein and RNA (Durham, Finch and Klug, 1971; Durham and Klug, 1971; Butler and Klug, 1971).

The interpolation scheme of Crowther, De Rosier and Klug (1970) has first been used for studying virus particles with icosahedral symmetry by Crowther et al. (1970; see also Crowther, 1971). It is known from the work of Crowther, De Rosier and Klug cited above that for this symmetry class one general view suffices to construct 59 other views related by symmetry. The coefficients lying on the corresponding sections of the Fourier space can now be used to calculate Fourier data lying equidistantly on the annuli by means of the interpolation method. The eigenvalue criterion mentioned in paragraph 2 of this section necessitated two views to be included for the reconstruction at 60 Å (human wart virus) and more than two for the reconstruction at 28 Å (tomato bushy stunt virus). The working resolution was determined by the resolution at which icosahedral symmetry was still preserved. The mutual orientation of the different particles taken for the analysis, i.e. the orientation of the particular view to the symmetry axes, is accomplished by comparing "common lines". These are lines in Fourier space related by symmetry and passing through the origin, which ought to carry identical Fourier coefficients. The degree of agreement of the common lines gives, at the same time, an estimation of how well the symmetry of the particle is preserved in the different preparations.

5. A Two-dimensional Fourier Reconstruction Scheme

Another Fourier reconstruction method has been proposed by RAMA-CHANDRAN and LAKSHMINARAYANAN (1971). The analysis is simplified if the object is tilted about one axis only. Then the projection theorem establishes a simple relation between the object density on the plane normal to the tilting axis and the optical density on the micrograph along the line upon which this plane is projected. Thus the three dimensional problem is reduced to the reconstruction of a plane from different one-dimensional projections. The authors show that the plane can be reconstructed by calculating a one-dimensional convolution product and a one-dimensional Fourier summation. Subsequently, the independently obtained planes are stacked together to yield the three-dimensional object. This method has advantages for computation but is essentially equivalent to a two dimensional Fourier reconstruction method, which in turn is a specialized version of the method of DE ROSIER and KLUG. Therefore the reconstruction of an object plane without symmetry is governed by the same rules as have been formulated by CROWTHER, DE ROSIER and KLUG (1970) for the case of three dimensions. This means in particular that there is no way to make up for missing data outside of the Fourier region covered by the tilting stage.

II. Real Space Methods

The direct methods for three dimensional reconstruction make use of the algebraic relations between the three dimensional object density and the projected densities.

1. Exact Solution

CROWTHER, DE ROSIER and KLUG present the principle of the real space reconstruction in general form (1970; see also KLUG, 1971). Suppose that the object consists of point-like atoms which have been well resolved in an ideal electron microscope. Then the spatial position of each atom can be obtained by a simple back-triangulation (Fig. 16a). However, this procedure already fails to have an unequivocal solution, if the atoms are thought to be extended spheres (Fig. 16b). In the realistic case of continuous density distributions the object density has to be represented by its values on a three-dimensional grid. If a Cartesian coordinate system is used then it is necessary to introduce suitable weighting functions P_{kl} such that in the

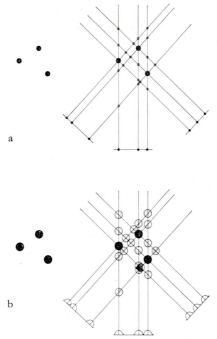

Fig. 16. Real space reconstruction of a simple structure by backtriangulation. a) point-like atoms : unequivocal solution; b) disk-like atoms : ambiguity. (From KLUG, 1971; by courtesy of the Royal Society)

discrete representation of the projection relation all experimental data contribute with equal weight. Then the projected densities σ_k measured on a regular grid and the object densities ϱ_l are related by

$$\sigma_k = \sum_l P_{kl}\, \varrho_l. \qquad (83)$$

Suppose we have so many independent projections that the number of "observations" σ_k is greater than the number of unknown object densities ϱ_l. Then (83) can be solved by the least squares method. The solution of the normal equation is, in matrix notation

$$\varrho = (P^T P)^{-1} P^T \sigma \qquad (84)$$

for a non-singular normal matrix $(P^T P)$. In all practical cases, however, the matrices in (84) are so large that they cannot be handled in reasonable time even by the fastest computers available.

By tilting about one axis the equation system (83) becomes separated into a set of independent equation systems for the object densities on the planes normal to the tilting axis, and the computation is simplified. This fact has been used by VAINSHTEIN (1970) and by GORDON, BENDER and HERMAN (1970). The approach of these authors will be outlined in the following paragraphs.

2. Superposition Method

VAINSHTEIN (1970) shows that a reconstruction of an object plane can be achieved by superposing functions which are simply obtained by translating the one-dimensional function representing the projected density, along the direction of projections (Fig. 17). The intersecting bands produce

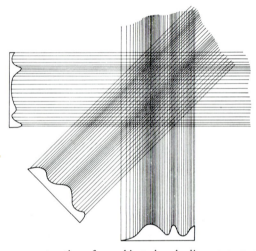

Fig. 17. Real space reconstruction of one object plane by linear superposition. The superposing functions are derived from the density distributions along corresponding projection lines of the micrographs of the tilting series. (After VAINSHTEIN, 1970)

a "star". In the polygon-shaped center of the "star" a two dimensional function is generated which approximates the desired section of the object provided a sufficient number of independent views have been taken. The detailed analysis of VAINSHTEIN shows that for the reconstruction to be valid at the boundary of the particle, an additional term has to be included. The same reconstruction procedure has been proposed independently by GORDON, BENDER and HERMAN (1970, Appendix C) as the "additive moire method" or "fast arithmetic method".

As VAINSHTEIN points out, the superposition method is equivalent with the two dimensional Fourier reconstruction method discussed earlier in section H.I.3. The limitations of the method can therefore be discussed in Fourier terminology. If the number of views taken for the synthesis is small, or if the views are concentrated in a narrow angular range, then the

Fourier plane is not sufficiently covered with data. From the theoretical treatment by CROWTHER, DE ROSIER and KLUG (1970) it is known that gaps in the Fourier transform lead to severe artifacts in the reconstructed object (see paragraph a) of this section). Such artifacts are apparent, indeed, in the "additive moire" reconstructions of a test pattern by GORDON, BENDER and HERMAN (1970) although they are difficult to explain without using Fourier equivalence relations.

Symmetries of the object reduce the number of views required for the real space method, as in the case of the three-dimensional Fourier reconstruction. VAINSHTEIN, BARYNIN and GURSKAYA (1968) have applied the superposition method to reconstructing the hexagonal structure of bovine liver catalase, and obtained satisfactory agreement with the results of x-ray analysis.

3. Iterative Approximation

A new concept has been introduced by GORDON, BENDER and HERMAN (1970; also GORDON and BENDER, 1971). The authors use a random process to obtain solutions which are optimally compatible with the projection equations. Their term "Algebraic Reconstruction Technique" (ART) is somewhat misleading since it fails to indicate the characteristic feature of the technique, namely the participation of a random process, as opposed to other algebraic reconstruction methods mentioned previously. GORDON, BENDER and HERMAN (1970) start from the relations between the density values ϱ_{ij} on an object plane and the observed values on the corresponding one-dimensional projection, σ_k, without allowing for weighting functions. When tilting is about one axis, for each tilting angle there are equations of the form

$$\sigma_k = \sum_{(i,j) \varepsilon S_k} \varrho_{ij} \tag{85}$$

where the summation runs over the subset S_k of those grid points which are cut out by two adjacent rays of the projecting beam of parallel rays, see Fig. 18. (The distance between two rays is taken to be close to the object grid distance.)

Now let us assume that q electron micrographs are taken with different tilting angles. Then for each object plane there are about np equations for the n^2 unknowns ϱ_{ij}. For an object of reasonable size, and for a reasonable resolution, $n \gg q$; i.e. we have far less equations than unknowns. It is known from the theory of linear equations that there exists a $(n-q)$-fold infinite number of solutions compatible with this underdetermined equation system. However, from simple permutational considerations it is evident that a large number of these solutions are not *essentially* different from the true solution. GORDON, BENDER and HERMAN (1970) show how *one* solution can be obtained by use of suitable algorithms. The computation starts with a blank array ϱ_{ij} and fills it iteratively. Each step secures that the equations (85) are fulfilled for *one* particular view. After a number of steps for different views an array ϱ'_{ij} is obtained which is consistent with all q sets of equations (85) in an optimal way. Model calculations, starting from projections of a test pattern, have shown that the solutions obtained by

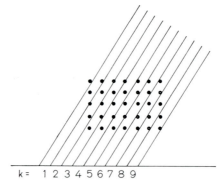

Fig. 18. Definition of the projection sum (85) of one object plane, used in the "Algebraic Reconstruction Technique" by GORDON et al. (1970)

means of this random process are similar to the original. Averaging over a number of solutions was found to increase the similarity. As a quantitative criterion for "similarity", the generalized Euclidean distance between original ϱ_{ij} and trial structure ϱ_{ij}' was used

$$\left[\frac{1}{n^2} \sum_{i}^{n} \sum_{j}^{n} (\varrho_{ij} - \varrho_{i'j}')^2 \right]^{1/2} \tag{86}$$

which assumes its minimum for identical arrays ϱ_{ij}, $\varrho_{i'j}'$.

Also, the significance of the entropy function for the assessment of the reconstruction has been discussed by GORDON and HERMAN (1971). But as yet the information theoretical aspects of the ART approach to the reconstruction are not fully understood.

A first application of the "Algebraic Reconstruction Technique" was reported by BENDER, BELLMANN and GORDON (1970). Electron micrographs of negatively stained ribosomes taken under six different tilting angles ($-36°$ to $+45°$) were digitized and evaluated by the method described above. The resolution of the three dimensional synthesis (obtained by stacking the reconstructed object planes ϱ_{ij} together) is 20 Å (Fig. 19). The reconstruction of three different particles results in structures resembling each other only in their overall shape. This discrepancy may be ascribed to the effects of preparation and to the radiation damage. It could as well result from the ambiguity in determining a common origin for different tilted views. The definition of the origin with respect to each projection is a problem no less critical than in the hypothetical case of the Fourier reconstruction of a general object. It can be solved at moderate resolution

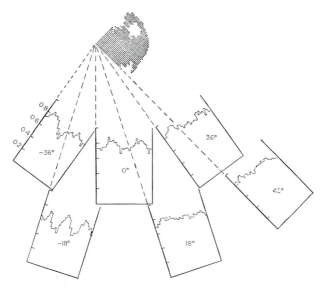

Fig. 19. Three dimensional reconstruction of a ribosome particle by the "Algebraic Reconstruction Technique". a) A reconstructed section of the ribosome and the corresponding one-dimensional projections in proper orientation; b) Different views of the three-dimensional model. Rectangular boundary visible on the upper left view is due to data truncation. (From BENDER, BELLMANN and GORDON, 1970; by courtesy of Journal of Theoretical Biology)

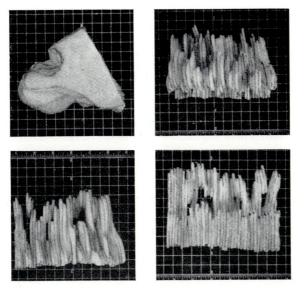

Fig 19b

with the aid of discrete reference particles by back-triangulation, as has been demonstrated by HART (1968). In the ribosome experiment, however, no markers were used, and the authors had to rely on the position of an annular distribution of stain surrounding the particle as reference.

The advantages and disadvantages of the "Algebraic Reconstruction Technique", when compared to the Fourier method, are the subject of a controversy between the ART group (GORDON, BENDER and HERMAN, 1970; BELLMANN et al., 1971) and CROWTHER and KLUG (1971). Computer time and storage requirements are mentioned by the ART group but their relative importance is hard to estimate for different installations. The main question at issue is whether the essential restriction of the Fourier method imposed by the limitation of the tilting range can be overcome by the algebraic construction technique. While the ART group claims that the angular range accessible with normal tilting devices ($+30°$) suffices to reconstruct asymmetric objects, CROWTHER and KLUG argue that the angular limitation produces an anisotropic resolution as in the case of the Fourier method, which merely does not become evident when the Fourier notation is not used (however, cf. Appendix in BELLMANN et al., 1971). The same authors criticize the fact that the projection equations do not include the weights which are, according to CROWTHER, DE ROSIER and KLUG (1970), necessary to guarantee equal sampling frequency on all projections (see paragraph a) of this section). BELLMANN et al. point out, however, that this effect is important only in the immediate vicinity of an object raster point, and that it can be minimized by appropriately choosing the sampling distance.

A clear advantage of the Fourier method over all real space methods comes out as a consequence of the orthogonality of the eigen functions of the Fourier expansions: The reconstruction can be refined to higher resolution, simply by including new terms in the synthesis, without the need to start the computation from the beginning (CROWTHER, DE ROSIER and KLUG, 1970).

It should be added that, even while facing essentially the same restrictions as the Fourier method, the reconstruction method of GORDON, BENDER and HERMAN remains a remarkable approach which will probably have an increasing number of applications in the near future. It may be expected that current accounts on this matter (HERMAN and ROWLAND, 1971; FRIEDER and HERMAN, 1971) will clarify a number of open questions. A vivid account of the controversy outlined above has been presented by a correspondent of Nature New Biology (1971).

Acknowledgements

This work was carried out as part of the Harkness Fellowship program of the Commonwealth Fund, New York. I would like to thank Dr. ROBERT GLAESER for a detailed discussion, Mrs. ROWENA PRELOCK for her help with the preparation of the manuscript.

References

ANDREWS, H. C.: Computer Techniques in Image Processing. New York—London: Academic Press 1970 Ch. 2; 1970a Ch. 6; 1970b Ch. 4.

ARNDT, U. W., CROWTHER, R. A., MALLET, J. F. W.: A Computer Linked Cathode Ray Tube Microdensitometer for X-Ray Crystallography. J. sci. Instrum. Ser 2, **1**, 510—516 (1968).

ARNDT, U. W., BARRINGTON, L. J., MALLET, J. F. W., TWINN, K. E.: A Mechanical Microdensitometer. J. sci. Instrum. Ser 2, **2**, 385—387 (1969).

BARNES, C. W.: Object Restoration in a Diffraction-Limited Imaging System. J. Opt. Soc. Amer. **56**, 575—578 (1966).

BELLMANN, S. H., BENDER, R., GORDON, R., ROWE, J. E. (Jr.): ART is Science Being a Defense of Algebraic Reconstruction Techniques for Three Dimensional Electron Microscopy. J. theor. Biol. **32**, 205—216 (1971).

BENDER, R., BELLMANN, S. H., GORDON, R.: ART and the Ribosome: A Preliminary Report on the Three-Dimensional Structure of Individual Ribosomes Determined by an Algebraic Reconstruction Technique. J. theor. Biol. **29**, 483—487 (1970).

BENDER, R., ROWE, J. E. (Jr.): Microdensitometers and Data Processing for Electron Microscopy. 29th Ann. Proc. Electron Microscopy Soc. Amer., 98—99 (1971).

BILLINGSLEY, F. C.: Image Processing for Electron Microscopy II. A Digital System. Advances in Optical and Electron Microscopy Vol. 4. Ed.: BARER and COSSLETT, 127—159. London-New York: Academic Press 1971.

BUDINGER, T. F.: Tranfers Function Theory in Image Evaluation in Biology-Applications in Electron Microscopy and Nuclear Medicine. Ph. D. Thesis, University of California, Berkeley 1971.

BURGE, R. E., GARRARD, D. F., BROWNE, M. T.: The Response of Photographic Emulsions to Electrons in the Energy Range 7—60 KeV. J. sci. Instrum. Ser. 2, **2**, 707—714 (1968).

BUSSLER, P.: To be published (1972).

BUTLER, P. J. G., KLUG, A.: Assembly of the Particle of Tobacco Mosaic Virus from RNA and Disks of Protein. Nature (Lond.) New Biol. **229**, 47—50 (1971).

COCHRAN, W., CRICK, F. H. C., VAND, V.: Structure of Synthetic Polypeptides I. The Transform of Atoms on a Helix. Acta Cryst. **5**, 581—586 (1952).

COOLEY, J. W., TUKEY, J. W.: An Algorithm for the Machine Calculation of Complex Fourier Series. Math. Comput. **19**, 297—301 (1965).

Correspondent: The ART of the Possible. Nature (Lond.) New Biol. **232**, 131 (1971).

COWLEY, J. M.: A New Microscope Principle. Phys. Soc. Proc. Sec. B. **66**, 1096—1100 (1953).

CRICK, R. A., MISELL, D. L.: A Theoretical Consideration of Some Defects in Electron Optical Images. A Formulation of the Problem for the Incoherent Case. J. appl. Physics. **4**, 1—20 (1971).

CROWTHER, R. A.: Procedures for Three-Dimensional Reconstruction of Spherical Viruses by Fourier Synthesis from Electron Micrographs. Phil. Trans. **B 261**, 221—230 (1971).

CROWTHER, R. A., AMOS, L. A., FINCH, J. T., DEROSIER, D. J., KLUG, A.: The Reconstruction of a Three-Dimensional Structure from Projections and Its Application to Electron Microscopy. Nature (Lond.) **226**, 421—425 (1970).

CROWTHER, R. A., DEROSIER, D. J., KLUG, A.: The Reconstruction of a Three-Dimensional Structure from Projections and Its Application to Electron Microscopy. Proc. roy. Soc. A, **317**, 319—340 (1970).

CROWTHER, R. A., KLUG, A.: ART and Science or Conditions for Three-Dimensional Reconstruction from Electron Microscope Images. J. theor. Biol. **32**, 199—203 (1971).

DAVENPORT, W. B., ROOT, W.: Random Signals and Noise, 76. New York: McGraw-Hill Book Co. 1958.

DeROSIER, D. J.: Three-Dimensional Image Reconstruction of Helical Structures. Ber. Bunsenges. Phys. Chem. **74**, 1127—1128 (1970).

DeROSIER, D. J.: Three-Dimensional Image Reconstruction of Helical Structures. Phil. Trans. **B 261**, 209—210 (1971).

DeROSIER, D. J., KLUG, A.: Reconstruction of Three Dimensional Structures from Electron Micrographs. Nature (Lond.) **217**, 130—134 (1968a).

DeROSIER, D. J., KLUG, A.: Positions of Ribosomal Subunits. Science **163**, 1470 (1968b).

DeROSIER, D. J., MOORE, P. B.: Reconstruction of Three-Dimensional Images from Electron Micrographs of Structures with Helical Symmetry. J. molec. Biol. **52**, 355 to 369 (1970).

DURHAM, A. C. H., FINCH, J. T., KLUG, A.: States of Aggregation of TMV Virus Protein. Nature (Lond.) New Biol. **229**, 37—42 (1971).

DURHAM, A. C. H., KLUG, A.: Polymerisation of TMV Protein and Its Control. Nature (Lond.) New Biol. **229**, 42—46 (1971).

ELIAS, P., GREY, D. S., ROBINSON, D. Z.: Fourier Treatment of Optical Processes. J. Opt. Soc. Amer. **42**, 127—134 (1952).

ERICKSON, H. P., KLUG, A.: The Fourier Transform of an Electron Micrograph: Effects of Defocusing and Aberrations, and Implications for the Use of Underfocus Contrast Enhancement. Ber. Bunsenges. Phys. Chem. **74**, 1129—1137 (1970).

ERICKSON, H. P., KLUG, A.: Measurement and Compensation of Defocusing and Aberrations by Fourier Processing of Electron Micrographs. Phil. Trans. **B 261**, 105—118 (1971).

FINCH, J. T., KLUG, A.: Three-Dimensional Reconstruction of the Stacked-Disk Aggregate of Tobacco Mosaic Virus Protein From Electron Micrographs. Phil. Trans. **B 261**, 211—219 (1971).

FRIEDER, G., HERMAN, G. T.: Resolution in Reconstructing Objects from Electron Micrographs. J. theor. Biol. **33**, 189—211 (1971).

FRANK, J.: A Study on Heavy/Light Atom Discrimination in Bright Field Electron Microscopy Using the Computer. Biophys. J. **12**, 484—511 (1972).

FRANK, J., BUSSLER, P. H., LANGER, R., HOPPE, W.: A Computer Program System for Image Reconstruction and Its Application to Electron Micrographs of Biological Objects. VII Intern. Congr. El. Micr., Vol. I, 17—18, Grenoble 1970a.

FRANK, J., BUSSLER, P. H., LANGER, R., HOPPE, W.: Einige Erfahrungen mit der rechnerischen Analyse und Synthese von elektronenmikroskopischen Aufnahmen hoher Auflösung. Ber. Bunsenges. Phys. Chem. **74**, 1105—1115 (1970b).

GLAESER, R. M., Limitations to Significant Information in Biological Electron Microscopy as a Result of Radiation Damage. J. Ultrastruct. Res. **36**, 466—482 (1971).

GLAESER, R. M.: Representative Electron Exposures for Damaging Effect. Microstructures., in press (1972).

GLAESER, R. M., KUO, I., BUDINGER, T. F.: Method for Processing of Periodic Images at Reduced Levels of Electron Irradiation. 29th Ann. Proc. Electron Microscopy Soc. Amer. 466—467 (1971).

GLAUBER, R., SCHOMAKER, V.: The Theory of Electron Diffraction. Phys. Ther. Rev. **89**, 667—671 (1953).

GOODMAN, J. W.: Introduction to Fourier Optics, p. 4. New York: McGraw-Hill 1968.

GORDON, R., BENDER, R., HERMAN, G. T.: Algebraic Reconstruction Techniques (ART) for Three-Dimensional Electron Microscopy and X-Ray Photography. J. theor. Biol. **29**, 471—481 (1970).

GORDON, R., BENDER, R.: New Three-Dimensional Algebraic Reconstruction Techniques (ART). 29th Ann. Proc. Electron Microscopy Soc. Amer. 82—83 (1971).

GORDON, R , HERMAN, G. T., Reconstruction of Pictures from Their Projections. Quarterly Bulletin of the Center of Theoretical Biology, State University of New York at Buffalo 4 (1), 71—151 (1971).

HALL, C. E.: Introduction to Electron Microscopy, Ch. 9. New York: McGraw-Hill 1953.

HANSZEN, K.-J.: Lichtoptische Anordnungen mit LASER-Lichtquellen als Hilfsmittel für die Elektronenmikroskopie. IV Eur. Reg. Conf. El. Micr. Vol. I, Rome, 153—154 (1968).

HANSZEN, K.-J.: The Optical Transfer Theory of the Electron Microscope: Fundamental Principles and Applications. Advances in Optical and Electron Microscopy Vol. 4. Ed.: BARER and COSSLETT, 1—84. London-New York: Academic Press 1971.

HANSZEN, K.-J., MORGENSTERN, B.: Die Phasenkontrast- und Amplitudenkontrastübertragung des elektronenmikroskopischen Objektivs. Z. angew. Phys. 19, 215—227 (1965).

HARRIS, J. L.: Resolving Power and Decision Theory. J. Opt. Soc. Amer. 54, 606—611 (1964a).

HARRIS, J. L.: Diffraction and Resolving Power. J. Opt. Soc. Amer. 54, 931—936 (1964b).

HARRIS, J. L.: Image Evaluation and Restoration. J. Opt. Soc. Amer. 56, 569—574 (1966).

HARRIS, W. W.: Reducing the Effect of Substrate Noise in Electron Images of Biological Objects. Some Biological Techniques in Electron Microscopy. Ed.: PARSONS, 147 to 174. New York-London: Academic Press 1970.

HART, R. G.: Electron Microscopy of Unstained Biological Material: The Polytropic Montage. Science 159, 1464—1467 (1968).

HEINEMANN, K.: In-situ Measurement of Objective Lens Data of a High-Resolution Electron Microscope. Optik, in press (1972).

HERMAN, G. T., ROWLAND, S.: Resolution in ART: An Experimental Investigation of the Resolving Power of An Algebraic Reconstruction Technique. J. theor. Biol. 33, 213—223 (1971).

HILDEBRAND, F. B.: Introduction to Numerical Analysis, 258. New York: McGraw-Hill Book Co., Inc. 1956.

HOPPE, W.: Ein neuer Weg zur Erhöhung des Auflösungsvermögens des Elektronenmikroskops. Naturwissenschaften 48, 736—737 (1961).

HOPPE, W.: Das Endlichkeitspostulat und das Interpolationstheorem der dreidimensionalen elektronenmikroskopischen Analyse aperiodischer Strukturen. Optik 29, 617 bis 621 (1969).

HOPPE, W.: Principles of Electron Structure Research at Atomic Resolution Using Conventional Electron Microscopes for the Measurements of Amplitudes and Phases. Acta Cryst. A 26, 414—426 (1970a).

HOPPE, W.: Principles of Structure Analysis at High Resolution Using Conventional Electron Microscopes and Computers. Ber. Bunsenges. Phys. Chem. 74, 1090—1100 (1970b).

HOPPE, W.: The Use of Zone Correction Plates and Other Techniques for Structure Determination of Aperiodic Objects at Atomic Resolution Using a Conventional Electron Microscope. Phil. Trans. B 261, 71—74 (1971a).

HOPPE, W.: Zur „Abbildung" komplexer Bildfunktionen in der Elektronenmikroskopie. Z. Naturforsch. 26a, 1155—1168 (1971b).

HOPPE, W., LANGER, R.: Numerical Calculations of the Images of Single Atoms in Electron Microscopes. Intern. Conf. Electr. Diffr. and Crystal Defects., Melbourne, 10—5 (1965).

HOPPE, W., LANGER, R., FRANK, J., FELTYNOWSKI, A.: Bilddifferenzverfahren in der Elektronenmikroskopie. Naturwissenschaften 56, 267—272 (1969).

HOPPE, W., LANGER, R., HIRT, A., FRANK, J.: An Equipment for Structure Research at High Resolution Using an Electron Microscope as a Tool, VII Int. Conf. E.. Micr. Vol. II, Grenoble, 5—6 (1970).

HOPPE, W., LANGER, R., THON, F.: Verfahren zur Rekonstruktion komplexer Bildfunktionen in der Elektronenmikroskopie. Optik 30, 538—545 (1970).

HUANG, T. S.: Combined Use of Digital Computers and Coherent Optics in Image Processing. SPIE Computerized Imaging Techniques Symp., Washington, D. C. (1967).

JAHNKE, E., EMDE, F., LÖSCH, F.: Tables of Higher Functions. New York: McGraw-Hill 1960.

JONES, R. C.: New Methods of Describing and Measuring the Granularity of Photographic Materials. J. Opt. Soc. Amer. 45, 799—808 (1955).

KLUG, A.: Optical Diffraction and Filtering and Three-Dimensional Reconstructions from Electron Micrographs. Phil. Trans. B261, 173—179 (1971).

KLUG, A., CRICK, F. H. C., WYCKOFF, H. W.: Diffraction by Helical Structures. Acta Cryst. 11, 199—213 (1958).

KLUG, A., BERGER, J. E.: An Optical Method for the Analysis of Periodicities in Electron Micrographs and Some Observations on the Mechanics of Negative Staining. J. molec. Biol. 10, 565—569 (1964).

KLUG, A., DeROSIER, D. J.: Optical Filtering of Electron Micrographs. Reconstruction of One-Sided Images. Nature (Lond.) 212, 29—32 (1966).

LAKE, J. A.: Reconstruction of Three Dimensional Structures from Electron Micrographs of Sectioned Helices. Proc. First Europ. Biophysics Congress, Baden 6, 453 to 457 (1971a).

LAKE, J. A.: Reconstruction of Three-Dimensional Structures from Electron Micrographs: The Equivalence of Two Methods. 29th Ann. Proc. Electron Microscopy Soc. Amer. 90—91 (1971b).

LAKE, J. A., SLAYTER, H. S.: Three Dimensional Fourier Analysis of the Ribonuclein Particle (Ribosome) Helix of Entamoeba invadens. 28th Ann. Proc. Electron Microscopy Soc. Amer. 266—267 (1970a).

LAKE, J. A., SLAYTER, H. S.: Three-Dimensional Structure of the Chromatoid Body of Entamoeba invadens. Nature 227, 1032—1037 (1970b).

LANGER, R., FRANK, J., FELTYNOWSKI, A., HOPPE, W.: Anwendung des Bilddifferenzverfahrens auf die Untersuchung von Strukturänderungen dünner Kohlefolien bei Elektronenbestrahlung. Ber. Bunsenges. Phys. Chem. 74, 1120—1126 (1970a).

LANGER, R., FRANK, J., FELTYNOWSKI, A., HOPPE, W.: Application of the Difference Image Method to the Study of Structural Changes in Carbon Foils. VII Intern. Cong. El. Micr. Vol. I, Grenoble, 19—20 (1970b).

LEVI, L.: On Image Evaluation and Enhancement. Optica Acta 17, 59—76 (1970).

MARKHAM, R., FREY, S., HILLS, G. J.: Methods for Enhancement of Image Detail and Accentuation of Structure in Electron Microscopy. Virology 20, 88—102 (1963).

MARRIAGE, A., PITTS, E.: Relation Between Granularity and Autocorrelation. J. Opt. Soc. Amer. 46, 1019—1027 (1956).

MENDELSOHN, L., MAYALL, B. H., PREWITT, J. M. S., BOSTROM, R. C., HOLCOMB, W. G.: Digital Transformation and Computer Analysis of Microscopic Images. Advances in Optical and Electron Microscopy Vol. 2. Ed.: BARER and COSSLETT, 77. London-New York: Academic Press 1968.

MISELL, D. L., CRICK, R. A.: An Estimate of the Effect of Chromatic Aberration in Electron Microscopy. J. appl. Physiol. 4, 1668—1674 (1971).

MOODY, M. F.: Application of Optical Diffraction to Helical Structures in the Bacteriophage Tail. Phil. Trans B261, 181—195 (1971).

MOORE, P. B., DeROSIER, D. J.: Deconvolution. J. molec. Biol. 50, 293—295 (1970).

MOORE, P. B., HUXLEY, H. E., DeROSIER, D. J.: Three-Dimensional Reconstruction of F-Actin, Thin Filaments and Decorated Thin Filaments. J. molec. Biol. **50**, 279—292 (1970).

MORGAN, R. S.: Structure of Ribosomes of Chromatoid Bodies: Three-Dimensional Fourier Synthesis at Low Resolution. Science **162**, 670—671 (1968).

NATHAN, R.: Computer Enhancement of Electron Micrographs. 28th Ann. Proc. Electron Microscopy Soc. Amer., 28—29 (1970).

NATHAN, R.: Image Processing: Enhancement Procedures. Advances in Optical and Electron Microscopy Vol. 4. Ed.: BARER and COSSLET, 85—125. London-New York: Academic Press 1971.

O'NEILL, E. L.: Introduction to Statistical Optics, Reading, Mass.: Addison-Wesley 1963.

PAPOULIS, A.: Systems and Transforms with Applications in Optics. New York: McGraw-Hill 1968.

PARSONS, D. F.: Problems in High Resolution Electron Microscopy of Biological Materials in Their Natural State. Some Biological Techniques in Electron Microscopy. Ed.: PARSONS, 1—68. New York-London: Academic Press 1970.

PORCHET, J. R., GÜNTHARD, H. H.: Optimum Sampling and Smoothing Conditions for Digitally Recorded Spectra. J. sci. Instr. **3**, 261—264 (1970).

RAMACHANDRAN, G. N., LAKSHMINARAYANAN, A. V.: Three-Dimensional Reconstruction from Radiographs and Electron Micrographs: Application of Convolutions instead of Fourier Transforms. Proc. Nat. Acad. Sci. (U.S.A.) **68**, 2236—2240 (1971).

REIMER, L.: Elektronenoptischer Phasenkontrast. II Berechnung mit komplexen Atomstreuamplituden für Atome und Atomgruppen. Z. Naturforsch. **24**a, 377—389 (1969).

RIDDLE, H. N., SIEGEL, B. M.: Thin Pyrolytic Graphite Films for Electron Microscope Substrates. 29th Ann. Proc. Electron Microscopy Soc. Amer. 226—227 (1971).

RÖHLER, R.: Informationstheorie in der Optik. Optik und Feinmechanik in Einzeldarstellungen Bd. **6**. Ed.: Günther, Wiss. Verlagsgesellschaft 1967, p. 171.

SAYRE, D.: The Calculation of Structure Factors by Fourier Summation. Acta Cryst. **4**, 362—367 (1951).

SCHERZER, O.: The Theoretical Resolution Limit of the Electron Microscope. J. appl. Physics. **20**, 20—29 (1949).

SCHISKE, P.: Zur Frage der Bildrekonstruktion durch Fokusreihen. IV Eur. Reg. Conf. El. Micr. Vol. I, Rome, 145—146 (1968).

SELZER, R.: The Use of Computers to Improve Biomedical Image Quality. Amer.Fed. of Information Processing Societies (AFIPS) Fall Joint Computer Conference. 817 to 834 (1968).

SEPTIER, A.: The Struggle to Overcome Spherical Aberration in Electron Optics. Advances in Optical and Electron Microscopy. Vol. 1. Ed.: BARER and COSSLET, 204—274. London-New York: Academic Press 1966.

SHANNON, C. E.: Communications in the Presence of Noise. Proc. I.R.E.N.Y. **37**, 10—21 (1949).

THOMSON, M. G. R., JACOBSEN, E. H.: Quadrupole-Octopole and Foil Lens Corrector Systems. 29th Ann. Proc. Electron Microscopy Soc. Amer. 16—17 (1971).

THON, F.: Elektronenmikroskopische Untersuchungen an dünnen Kohlefolien. Z. Naturforsch. **20**a, 154—155 (1965).

THON, F., SIEGEL, B. M.: Experiments with Optical Image Reconstruction of High Resolution Electron Micrographs. Ber. Bunsenges. Phys. Chem. **74**, 1116—1120 (1970).

THON, F., WILLASCH, D.: High Resolution Electron Microscopy Using Phase Plates. 29th Ann. Proc. Electron Microscopy Soc. Amer. 38—39 (1971).

UNWIN, P. N. T.: Phase Contrast and Interference Microscopy with the Electron Microscope. Phil. Trans. **B261**, 95—104 (1971).

VAINSHTEIN, B. K.: Finding the Structure of Objects from Projections. Soviet Physics-Crystallography **15**, 781—787 (1971). Transl. from Kristallografiya **15**, 894—902 (1970)

VAINSHTEIN, B. K., BARYNIN, V. V., GURSKAYA, G. V.: The Hexagonal Crystalline Structure of Catalase and Its Molecular Structure. Soviet Physics-Doklady **13**, 838 to 841 (1969). Transl. from Doklady Akademii Nauk SSSR **182**, 569—572 (1968).

VALENTINE, R. G., WRIGLEY, N. G.: Graininess in the Photographic Recording of Electron Microscope Images. Nature (Lond.) **203**, 713—715 (1964).

VANDER LUGT, A.: Signal Detection by Complex Spatial Filtering. IEEE Trans. Inform. Theory IT-10, 139—145 (1964).

VANDER LUGT, A.: A Review of Optical Data-Processing Techniques. Optica Acta **15**, 1—33 (1968).

WELTON, T. A.: Computational Correction of Aberrations in Electron Microscopy. 29th Ann. Proc. Electron Microscopy So. Amer. 94—95 (1971).

WIENER, N.: The Extrapolation, Interpolation and Smoothing of Stationary Time Series, 175. New York: J. Wiley & Sons 1949.

WHITTAKER, E. T.: On the Functions Which Are Represented by the Expansions of the Interpolation Theory. Proc. roy. Soc. Edinb. A **35**, 181 (1915).

ZEITLER, E.: Contrast of Single Atoms in an Aberration-Free Electron Microscope. VI Int. Conf. El. Micr., Kyoto, 43—44 (1966).

ZEITLER, E.: Resolution in Electron Microscopy. Advances in Electronics and Electron Physics Vol. 25. Ed.: L. MARTON, 277—332. New York-London: Academic Press 1968.

ZEITLER, E., BAHR, G. F.: Contrast and Mass Thickness. Lab. Invest. **14**, 946—954 (1965).

ZEITLER, E., OLSEN, H.: Complex Scattering Amplitudes in Elastic Electron Scattering. Phys. Rev. **162**, 1439—1447 (1967).

High Voltage Electron Microscopy

KIYOSHI HAMA

A. Introduction

The high energy electron microscope has been used mainly in the field of material science and of minerology. Especially in minerology, the advantage of the higher penetrating power of electrons at higher accelarating voltage which enables the observation of material in bulk is enormous and the high voltage device becomes indispensable.

However, the initial impetus towards high voltage operation came from biologists desiring to examine whole cells at a time when thin sectioning techniques had not been developed. Further efforts on the project were stopped by the development of microtomy which enabled the preparation of specimens thin enough for ordinary transmission microscopy.

Recently, the remarkable merits of high voltage operation in the field of biology have been recognized by several investigators (COSSLETT, 1971a, b; DUPOUY, 1968; DUPOUY, PERRIER and DURRIEU, 1960; HAMA and NAGATA, 1970a, b; HAMA and PORTER, 1969; NAGATA and FUKAI, 1971; NAGATA and HAMA, 1970; PARSONS, 1972; RICE, MOSES and WRAY, 1971; RIS, 1969; SZIRMAE, FISHER and MOSES, 1971). It is expected that the high voltage electron microscope will be more common in biology laboratories in the near future.

B. Merits of the High Voltage Electron Microscope

The following advantages can be expected using the high voltage electron microscope in the observation of biological materials.

1) Thicker specimens can be observed at higher accelerating voltage.

2) The practical resolution obtainable from a specimen of given thickness is expected to improve mainly because of less chromatic aberration.

3) The damage to the specimen caused by the electron beam is expected to be smaller at higher accelerating voltages.

The theoretical and experimental basis for these points will be discussed shortly.

I. Specimen Penetration

The maximum thickness of a specimen observable in an electron microscope may be defined both in terms of the fraction of the incident beam collected in the objective aperture (thus determining the overall image brightness) and the energy lost by the beam in transversing the specimen, which limits the resolution obtained in the image owing to chromatic aberration. If one considers only image brightness, the maximum thickness

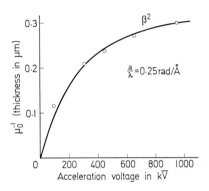

Fig. 1. This figure shows the relation between the accelerating voltage and penetrating power of electrons using glass fibers as a test specimen. μ_0 is the absorption coefficient

observable may be very large even at relatively low voltage using high current density and a larger objective aperture. However in this case, beam damage on the specimen and chromatic aberration cause serious problems. Consequently there is a practical limitation to the maximum specimen thickness observable at a given accelerating voltage.

Theoretically, it is known that the penetrating power of electrons is proportional to β^2 where β equals the velocity of electrons at a given voltage divided by the velocity of light, when α/λ is kept constant ($\alpha =$ semi angle of objective aperture, $\lambda =$ wave length of electrons) (Fig. 1) Dupouy, 1968; Dupouy and Perrier, 1962; Fujita et al., 1967; Hashimoto, 1964; Nagata and Hama, 1970). On the other hand, the actual increase of maximum usable thickness with rise in accelerating voltage has been postulated to be more remarkable than was expected from the β^2 relationship mentioned above (Dupouy, 1968; Ueda and Nonoyama, 1967). This fits well with experiences had with biological specimens (Hama and Porter, 1969; Szirmae et al., 1971).

According to our own experiments, 1 μ, 1.5 μ and 2 μ thick biological sections could be observed at 500, 800 and 2000 KV respectively with a resolution high enough to resolve the "unit membrane" structure clearly (HAMA and NAGATA, 1970a; HAMA and PORTER, 1969). We tried the observation of very thick biological specimens using very high accelerating voltage (2000 KV) and it was found that in the case of observations on specimens thicker than 3 μ, the image was bright enough but thermal effects on the specimen, drift of the specimen and sublimation of stain substances, became important. A good image quality could not be expected

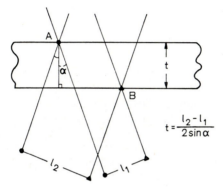

Fig. 2. This figure shows a method of estimating the section thickness from a stereo pair. Both l_1 and l_2 are distances between the two points, A and B, projected on each of the stereo pair. α is a tilting angle from the original position

under these conditions. Thus the limitation on examing very thick specimens appears to be set by temperature rise and radiation damage rather than by inadequate transmission or resolution caused by chromatic aberration. At present, it seems that about 3 μ would be the limit of the observable specimen thickness even at very high accelerating voltage when reasonable image quality is required.

There is another serious problem with thick specimen observations besides damage to the specimens. When the specimen is thick images from the whole depth of the specimen are superimposed, consequently electron stereoscopy is required to unscramble the confusion. Electron stereoscopy will be described in the section on biological applications of the high voltage electron microscope.

With regards to stereo scopy, resolution in the z axis, in other words, resolution in the direction of incident beam, should be mentioned. In this connection, a method for estimating the thickness of the specimen will be briefly discussed.

The thickness of the sectioned specimen has been estimated by means of the microtome scale, interference color, interferometer etc. The thickness judged from the microtome scale is hardly accurate. The accuracy of interference color or interferometer is rather high for thin specimens but not for thick specimens, over 1 μ. In our experiments, the thickness of the specimens was estimated in the following way (Fig. 2). Two photographs were taken by tilting the stage in the opposite azimuth at a given angle $\pm \alpha$. Two image points A and B on opposite surfaces of the specimen were selected. The projected distances between the two points, l_1 and l_2 were measured on each of the two photographs. The thickness of the section (t) was estimated from the formula.

$$t = \frac{l_2 - l_1}{2 \sin \alpha} \tag{1}$$

The accuracy of this method depends mainly on the accuracy of the microscope magnification.

II. Resolving Power

Electrons in the incident beam lose energy by inelastic collisions with atoms in the specimen resulting in chromatic aberration. The radius of the disk (δc) of least confusion in the paraxial image plane is given by the formula,

$$\delta c = \alpha_0 \, C_c \, (\Delta V / V_0) \tag{2}$$

where α_0 is the angular aperture of the objective lens, C_c is the chromatic aberration coefficient of the objective lens, ΔV is a measure of the energy loss in a given thickness of the specimen and V_0 is initial voltage.

When the specimen is rather thin, it is not easy to decide what value to assign to ΔV, as the imaging beam still contains quite a proportion of electrons which have suffered no energy loss and the problem of the chromatic aberration is not serious (COSSLETT, 1971a; NAGATA and HAMA, 1971). However, in the case of very thick specimens, the problem of energy loss becomes significant and the chromatic aberration causes serious difficulties. If one uses a high accelerating voltage with thick specimens, the amount of energy lost in a specimen of given thickness decreases, so that the ratio $\Delta V / V_0$ falls rapidly. This effect can be thought of as corresponding to an increase in the thickness of a specimen that will result in a given value of resolution. The image resolution of 11 nm for 1 μ of collodion at 200 KV and 2.7 nm at 500 KV was estimated with $\alpha_0 = 5 \times 10^{-3}$, $C_c = 4$ mm and using the measured ΔV in Eq. (2) mentioned above (COSSLETT, 1971a). Actually, as already mentioned, better than 5 nm resolution could be achieved in a 1.5 μ biological section at 800 KV. This value of resolution fits quite well with that of the above mentioned theoretical one.

The resolution limit for spherical aberration is expressed by the equation

$$\delta \varrho = C \times \alpha_0^5 \qquad (3)$$

where C is the spherical aberration coefficience of the objective lens, α_0 is the objective aperture angle expressed in radians. At a higher accelerating voltage, the scatter angle of electrons becomes smaller, and as a consequence the effective aperture angle becomes smaller. Thus smaller spherical aberration can be expected at higher accelerating voltage. This is especially true with dark field observation.

Therefore, a higher resolution can be expected under conditions of high voltage operation because of the reduction both in chromatic and spherical aberration. The smaller beam effect on the specimen at higher accelerating voltage (see below) may well also contribute to the better image quality. Actually, in the case of observation of thin specimens, it was found that the image resolution at 800 KV was also better as compare with that at 75 KV (see biological applications).

III. Beam Damage

The specimen damage caused by the electron beam is mainly of two types, ionization and displacement of atoms. The first effect decreases as beam voltage is raised due to the reduction in energy loss. Thus the use of a higher accelerating voltage in electron microscopy is regarded as desirable if beam damage is to be reduced to a minimum. Though hardly adequate as a measure of relative beam effects, the following observation is of some significance. In microscopy of relatively thick (over 100 nm thick) sections with 75 KV or 100 KV, the section frequently breaks, and thermal drift ensues. Moreover, the sublimation of staining materials is frequently observed. These same phenomena, however, were not noticed with the 500 KV and 800 KV beam even when the beam was brought to cross over. This is taken to indicate that far less energy is dissipated from the high voltage beam and less beam damage results.

It has been postulated that the damage to the specimen is inversely proportional to the energy of the incident electrons (GLAESER, BUDINGER and AEBERSOLD, 1970; KOBAYASHI and SAKAOKA, 1964). The following experiment shows the voltage dependent beam effect some what quantitatively. Lead acetate was selected as a test specimen. After the diffraction pattern was observed the beam intensity was increased to a level sufficient for a five second exposure photographic record. The records were made every 15 seconds. The beam intensity at the level of the specimen was measured by using a Faraday cup after the records were finished. The begining of the change of diffraction pattern was chosen as an indication of the beam effect on the specimen. The amount of exposure, $I \times T$, where I is beam intensity at the level of specimen, T is exposure time, needed for the change of diffraction pattern, each at 75 KV and 800 KV were measured and compared to each other. Table 1 shows the type of experiment (upper table) and the results obtained (lower table). As shown in the table, the amount of exposure, $I \times T$, needed for the change of diffraction pattern at 800 KV is about three times as large as that at 75 KV. As the beam effect on the specimen can be expressed as $1/I \times T$, the beam effect on the specimen at 800 KV is about one third of that at 75 KV. But

Table 1

Acc. volt	beam current	cross over size	maximum current density on the specimen	aperture size (intermediate lens)
75 kv	20 μA	3 μΦ	0.2 A/cm	50 μΦ
800 kv	12 μA	2 μΦ	1 A/cm	50 μΦ

Acc. volt	I	$I \times T$ (mA. sec.)	number of data
75 kv	0.7—1.4 mA	80—130 (mA. sec.)	3 cases
800 kv	4—9 mA	340—360 (mA. sec.)	3 cases

since very thick specimens can now be examined, the total amount of energy absorbed in the specimen will increase. The result is a rise in temperature and ultimately destruction of the specimen. This seems to set a practical limit of 2—3 μ for the thickness of specimens that can be examined even at very high accelerating voltage.

In addition to this, the displacement of atoms becomes a more serious effect at higher accelerating voltages. The threshold for the displacement of atoms in carbon is about 120 KV, so, no atoms are removed in conventional electron microscopy. However, above this voltage the cross section for the process increases rapidly, so at higher accelerating voltage, there is a high probability of disruption of molecules during the examination at high magnification which needs high beam density at the specimen (COSSLETT, 1971 b).

The problem of radiation damage is quite serious especially in the case of biological applications of the high voltage electron microscope. The problem is also intimately associated with the sensitivity of the recording system. If high sensitivity recording systems become more readily available, then the current intensity at the specimen level may be reduced below the threshold. Otherwise, the observation of living matter using the high voltage electron microscope which has long been an attractive aim for many biologists seems quite difficult and impractical.

C. Biological Applications

Some examples of biological applications of the high voltage electron microscope will be described which correlate with some of the merits mentioned above.

I. Specimen Preparation

Most of the observed specimens were fixed by osmium tetroxide alone or by glutaraldehyde followed by osmium tetroxide. The specimens were stained en-block with a 4% aqueous solution of uranyl acetate for 45 minutes in ice before dehydration. Blocks were embedded in Epon 812 (LUFT, 1961) after ethanolic dehydration. Sections ranging from 60 nm to 3 μ were cut with Porter Blum MT-1 or MT-2 ultra microtomes using glass knives. Sections were mounted on collodion covered copper grids and double stained with 1.5% uranyl acetate in 70% ethanol for 20 minutes and with lead citrate for 20 minutes.

As indicated, the whole procedure is simillar to that of ordinary transmission electron microscopy of biological thin sections. The problem of inadequate penetration of the stain through the thick specimen was solved by employment of an effective en-block stain and prolonged staining duration. The problem of low contrast on the viewing screen was solved by replacing the metalic screen base with a thin plastic plate, so that the back scatter of electrons from the screen base was reduced. The contrast on the viewing screen was then adequate for focussing and photographic recording even during the observation of thin specimens.

Cellular elements such as mitochondria, isolated and negatively stained, or whole cells such as bacterial spores, unfixed and unstained, are also favorably observed (Fig. 3, Fig. 4).

II. High Resolution Observation

As mentioned before, better resolution can be expected at a higher accelerating voltage. The following examples will show this using ordinary thin sections. Fig. 5 is a cross section of glycerinized insect flight muscle. The thickness of the section is 60 nm. The picture was taken at 800 KV. The images of thin and thick filaments are surprisingly clear in this high voltage electron micrograph in view of the fact that the high voltage

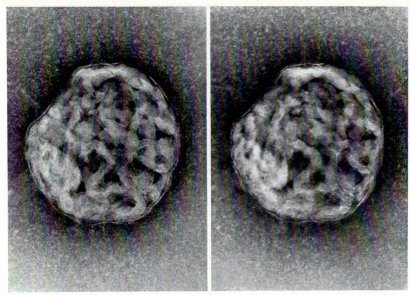

Fig. 3. A stereo pair showing a mitochondrion isolated and negatively stained. The photographs were taken at 800 KV. Tubular cristae are seen meandering in this three dimensional figure. 41,000 ×

Fig. 4. This stereo pair shows unfixed bacterial spores (*Bacillus megaterium*). The diameter of the spore is about 1.5 μ, so the penetrating power of electrons at 800 KV is high enough to enable the whole cell observation. 15,000 ×

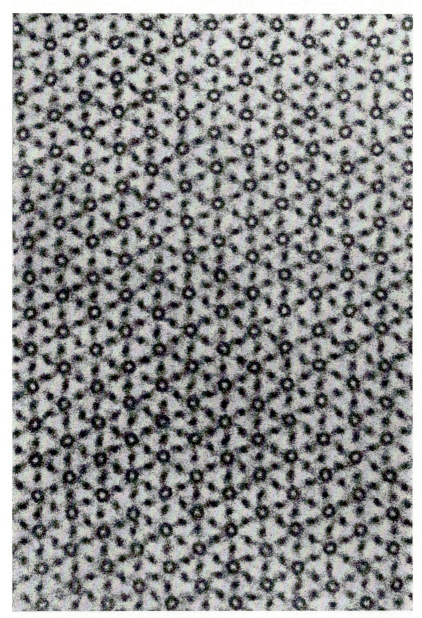

Fig. 5. High voltage electron micrograph showing a cross section of glycerinized insect flight muscle. Regular arrangement of thin and thick filaments are shown. 224,000 ×

electron microscope has not been refined mechanically and electrically to the extent that the conventional electron microscope has.

The fine structure of the gap junction between the two supporting cells in the saccular macula of the gold fish was observed at 75 KV (Fig. 6) and at 500 KV (Fig. 7). Similar thin sections, about 60 nm thick, prepared from the same block were used to compare the image quality. It is clearly

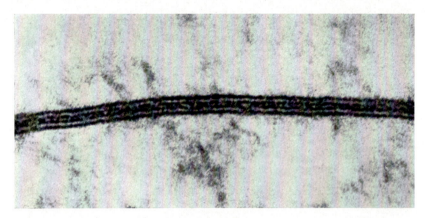

Fig. 6. A high power electron micrograph showing a gap junction between the two supporting cells in the saccular macula of the gold fish taken at 75 KV. The apposing membranes are seen to be separated by a gap of 2 nm. The thickness of the section is 60 nm. 315,000 ×

shown both in Fig. 6 and in Fig. 7 that the adjacent two outer leaflets of the apposing plasma membranes are separated by a gap of about 2 nm. In the 500 KV picture (Fig. 7) it is seen that the adjacent two outer leaflets are connected by electron dense bridges repeated at a regular periodicity across the 2 nm gap. These details are not so clear in the 75 KV picture (Fig. 6).

Fig. 8 and Fig. 9 show "unit membrane" structure covering the microvilli in the intracellular secretory canaliculus of a parietal cell of mouse stomach taken at 75 KV and 800 KV respectively using similar thin sections cut from the same block. The two electron dense layers are more clearly separated in the 800 KV picture than in the 75 KV picture. Densitometry traces across the unit membrane structures were done on the negative plates (Fig. 10). Needless to say, the peak to peak distances between the two electron dense layers are equal in both pictures. The height of the peak density in the 75 KV picture is higher than that of the 800 KV picture. This means that the contrast of the 75 KV picture is higher than that of the 800 KV picture. The density of the middle, electron-lucent layer

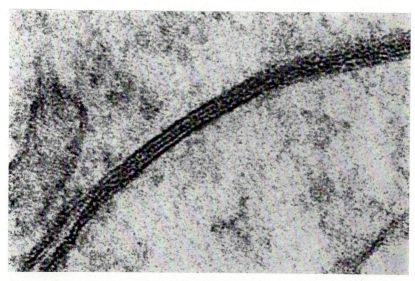

Fig. 7. A gap junction found in a specimen similar to the previous figure. The photograph is taken at 500 KV. Two apposing membranes are seen to be separated more clearly than in the previous 75 KV picture. Details of the 2 nm gap between the two membranes are clearly seen. 315,000 ×

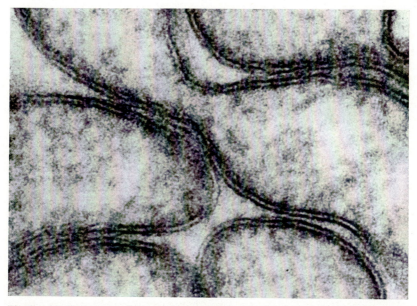

Fig. 8 and 9. High power electron micrographs of microvilli in the intra-cellular secretory canaliculi of mouse stomach. Each photograph was taken at 75 KV (Fig. 8) and 800 KV

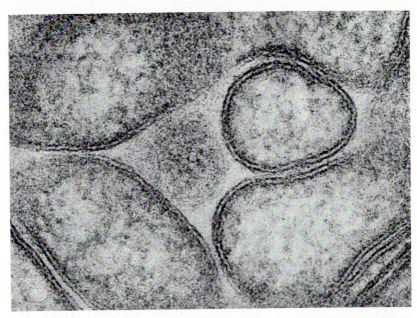

(Fig. 9) using the similar thin section prepared from the same block. Three layered "unit membrane" structure is more clearly resolved in the 800 KV picture (Fig. 9) than in the 75 KV picture (Fig. 8). 163,000 ×

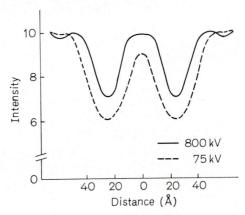

Fig. 10. Densitometry traces across the "unit membrane" structures shown in the previous 75 KV picture (dotted line) and 800 KV picture (solid line). The explanation is shown in the text

returns to the background level in the 800 KV picture, whereas in that of the 75 KV picture it goes to 25% of the peak density. The width of the peak density at half height is about 3 nm in the 75 KV picture and about 2 nm in the 800 KV picture. Thus the two peak densities are more clearly separated in the 800 KV picture than in the 75 KV picture.

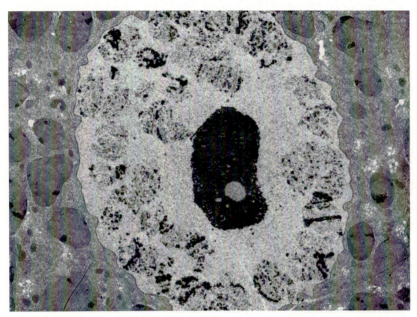

Fig. 11. A nucleus of a drosophila salivary gland cell seen in a thin section (80 nm thick). The photograph was taken at 75 KV. The familiar band patterns of giant chromosomes can not be seen in this thin section picture. 5,000 ×

These results show that the image quality obtained at a higher accelerating voltage is better than that at lower voltage which is in accordance with the theoretical consideration mentioned earlier. The better image quality at higher accelerating voltage is probably due to less background fog as well as to smaller chromatic and spherical aberrations (NAGATA and HAMA, 1971).

III. Observation of Thick Specimens

As mentioned in the preceding paragraph, the employment of the high voltage electron microscope has enabled the direct observation of specimens thicker than one micron. The merit of thick specimen observation in biology will be shown in the following two figures. Fig. 11 shows a usual electron micrograph of a salivary gland cell of a Drosophila larva. This cell is known to have giant chromosomes which show beautiful band patterns in the light microscope. However, in this micrograph, the images of giant chromosomes appear as dissociated electron dense masses and do not show

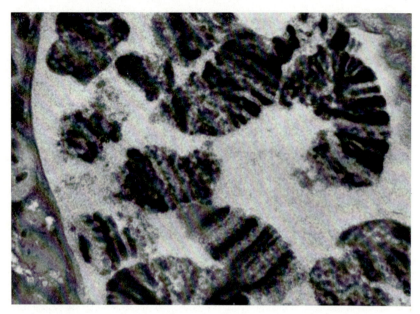

Fig. 12. Giant chromosomes are observed at 800 KV using a thick section (1 μ thick) cut from the same block as the previous picture. Band patterns of the giant chromosomes are clearly seen in this high voltage electron micrograph. 10,000 ×

the characteristic band patterns because the specimen is to thin for integration of the overlapped band pattern to occur. The thickness of the specimen is about 80 nm. Fig. 12 shows the giant chromosomes of a similar specimen observed at 800 KV using a 1 μ thick section. The section was cut from the same block as the one shown in Fig. 11. The band pattern of the giant chromosome is clearly seen at much better resolution than that of ultraviolet or light microscopy.

As shown here, the advantage of thick specimen observation is apparent in the field of biology, however, at the same time, there is a problem of overlapping of images through the depth of the specimen. Methods of unscrambling the confusion of the images and utilization of the information from the third dimension should be developed.

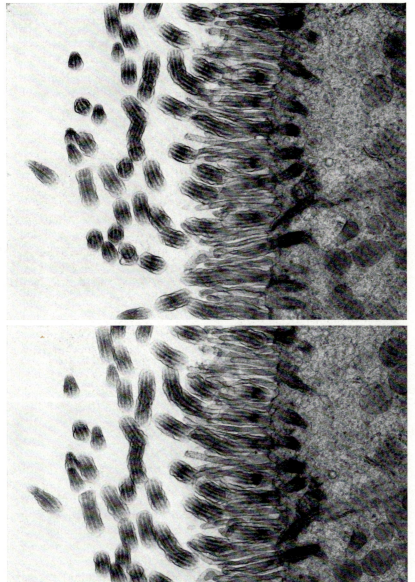

Fig. 14. A stereo pair showing the surface of the mouse trachea taken at 800 KV. The specimen thickness is about 1 μ. Three dimensional organization of basal bodies, kino cilia and microvilli are seen in this stereo picture. 22,000 ×

bacterial spores (Bacillus megaterium) which are unfixed and undehydrated previously but probably dehydrated during the observation in the micro-scope. The fine structure of the spore itself is obscured because of the high electron density, however some details are shown inside the capsule with the aid of stereoscopy.

In addition to these examples, stereo observations can effectively be applied to biological specimens in combination with various cytochemical or enzymatic reactions. For example, the demonstration of Golgi apparatus using TPPase and ACPase reactions (NOVIKOFF et al., 1971) can be done more easily with high voltage device.

V. Observation of Undehydrated Specimens

Taking advantage of the higher penetrating power and reduction in beam damage at a higher accelerating voltage, the observation of bio-materials in the hydrated state have been attempted using various techniques.

1. Ultracryotome Method

Thin sections were cut from briefly fixed and quickly frozen specimens using a cryo-ultramicrotome (RICE et al., 1971). The sections were observed in the frozen state using a cold specimen stage. A problem in this case was how to prevent the specimen from drying during the observation in the microscope.

2. Wet Cell Method

Two types of wet cell have been described. One is a dynamic wet cell design based on a differential pumping system so that specimens are in contact with a 24 torr pressure of water vapor (PARSONS, 1972). The other is a closed wet cell using films of SiO (NAGATA and FUKAI, 1971) or metal (DUPOUY, 1968) (100—200 nm thick) and is open to the exterior through fine tubing so that water can be introduced to the cell externally (Fig. 15).

The survival rate of wet bacterial spores (Bacillus megaterium) after electron bombardment was examined to determine the radiation damage on living matter. A photograph of the bacterial spores was taken at low electron intensity prior to incubation. A second photograph of the same field as the previous picture was taken after incubation. At 800 KV, it was found that about 100% budding was observed when very low current density was used for the first recording (NAGATA and FUKAI, 1971). Since the examined specimens were not living cells in the usual sense, but rather spores, this result can not simply be generalized for other living cells.

Also, the magnification used was very low and the details of cellular fine structure could not be observed. More over, there have been many disputes concerning this type of work. For example, KOBAYASHI (1971a, b) claims that since water molecules have a very large cross-section for inelastic scattering, high resolution can not be expected using a wet atmosphere and that Brownian motion of the particles in the cell or in the suspending liquid may disturb high resolution observations. He concluded that since resolution better than that of light microscopy can not be expected the observation of biological materials in a humidified condition may be

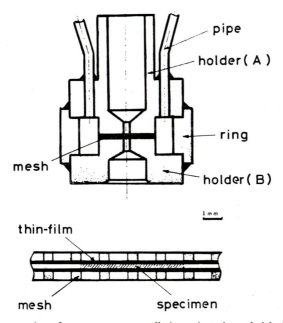

Fig. 15. A cross section of an open type wet cell. A specimen is sandwiched between the two meshes which are covered with thin SiO film. The specimen space is connected to the atmosphere through pipes

useless. Radiation damage is considered to be another limiting factor especially in attempting to observe living cells in a wet atmosphere. It seems that present studies are insufficient to make any definite conclusion on this matter. Looking at the positive aspects it can be said that either the cryoultratome technique or the wet cell method combined with the high voltage electron microscope open a possibility of observing biomaterials under more native conditions than conventional methods which have been employed thus far.

D. Design and Construction of High Voltage Electron Microscopes

The microscope column itself is simillar to that of a considerably scaled up conventional microscope. The high voltage generator is in principle the same as in a 100 KV microscope: Cockcroft-Walton voltage-multiplying rectifier, symmetrically designed and with negative feed back for stabilization of the out put. It is air-insulated in the Cavendish, Toulouse, and U.S. Steel Laboratories or is in a pressuer-tank as in all commercial models (AEI, Hitachi, JEM and Shimazu).

The stabilized output is fed to a conventional electron gun which injects electrons into a multistage accelerator column, containing 20 to 100 sets of electrodes to which an appropriate voltage is applied via a series of dividing resistors. The accelerator and generator are contained in the same pressure tank (Hitachi and JEM) or in a separate tank (AEI and Shimazu). The lens and recording systems are similar to those of a 100 KV electron microscope.

For an understanding of the construction and specifications of the high voltage microscope, a survey of the 3000 KV electron microscope at Osaka University will be described.

The high voltage unit is of the Cockcroft-Walton type. The accelerating tube consists of 100 steps with 25—30 KV for each step. In addition to the employment of a multistep type accelerating tube, both high voltage unit and accelerating tube are enclosed in a high pressure tank filled with 2.5 to 3 atmospheres pressure of freon gas to prevent small discharges which may affect stability. The final stability of the electron beam which contributes to image formation is of the order of 1^{10-5} at the maximum voltage. The lens system consists of double condenser, double objective, intermediate and double projector lenses. Because of the high energy of the electrons, very strong magnetic fields are required. The maximum magnetic field of the objective lens is estimated to be 30,000 gausses (about 30,000 ampere turns). Consequently the coil and pole piece are very large, about 80 cm and 20 cm in diameter respectively. Although the leakage of X rays are minimized to 0.5 mr/hr, all operations can be done with remote control if it is desired. The optical allignment including that of condenser and objective apertures, beam intensity control, focusing and recording can be done using the remote control system. As the image is received by a vidicon target, all these adjustments are done while watching the television screen. The image recording is done with a photo-plate or with video tape.

The height of the microscope is about 12 meters and the weight is about 70 tons. Attention was given to the prevention of vibrations. The whole system is suspended with rubber cushions and air buffers at the level

Fig. 16. A photograph showing outer view of 3000 KV electron microscope. The upper part of the picture shows pressure tank which consists of three parts. The lower part of the picture shows a part of the column and the operation desk

of the junction between the accelerating tube and the electron microscope column in order to reduce the vibration over 3 to 5 Hz. The resolution achieved at 2500 KV is 2 Å. Fig. 16 shows the outer view of the 3000 KV electron microscope.

The application of the high voltage electron microscope in the field of biology has just begun. Many merits and disadvantages have been claimed by various investigators and problems of radiation damage, recording systems and the better preservation and preparation of specimens are to be resolved in the future through patient study. The present status and some fundamental problems in this field have been discussed in this short paper.

References

Cosslett, V. E.: High voltage electron microscopy and its application in biology. Phil. Trans. B. **261**, 35—44 (1971a).

Cosslett, V. E.: The scope and limitations of high voltage electron microscopy. J. Ultrastruct. Res. **37**, 255—257 (1971b).

Dupouy, G.: Electron Microscopy at Very High Voltages, Advances in Optical and Electron Microscopy, Vol. **2**, 168—250 (1968). Academic Press. Ed.: Barer and Cosslett.

Dupouy, G., Perrier, F., Durriru, L.: Microscopie électronique. L'observation de la matière vivante au moyen d' un microscope électronique fonctionnant sous très haute tension. C. R. Acad. Sc. **251**, 2836—2841 (1960).

Dupouy, G., Perrier, F.: Principaux résultats obtenus aves un microscope électronique fonctionnant sous un million de volts. Fifth International Congress for Electron Microscopy, Philadelphia, Breese Jr. S. S., ed. **1**, A-3 (1962).

Fujita, H., Kawasaki, Y., Fujibayashi, E., Kajiwara, S., Taoka, T.: Metallurgical investigation with 500 Kv electron microscope, Japan. J. appl. Physics **6**, 214—230 (1967).

Glaeser, R. M., Budinger, T. F., Aebersold, P. M., Thomas, G. T.: Radiation damage in biological specimens. Microscopie électronique Vol. **1**, 463—464 (1970).

Hama, K., Nagata, F.: High resolution observation of biological section with a high voltage electron microscope. J. Electron Microscopy **19**, 170—175 (1970a).

Hama, K., Nagata, F.: A stereoscopic observation of tracheal epithelium of mouse by means of the high voltage electron microscope. J. Cell Biol. **45**, 654—659 (1970b).

Hama, K., Porter, K. R.: An application of high voltage electron microscopy to the study of biological materials. J. de Microscopie **8**, 149—158 (1969).

Hashimoto, H.: Energy dependence of extinction distance and transmissive power for electron waves in crystals. J. appl. Physics **35**, 277—290 (1964).

Huxley, H. E.: The double array of filaments in cross striated muscle. J. biophys. bio-chem. Cytol. **3**, 631—647 (1957).

Kobayashi, K.: Fundamentals of HVEM for organic materials, reported at U.S.A.-Japan Seminar on "New trends in high voltage electron microscopy" (1971a). Personal Communication.

Kobayashi, K.: On the electron microscopy of colloidal particles in liquids, Ibd. (1971b). Personal Communication.

Kobayashi, K., Sakaoka, K.: The changes of polymer crystals due to irradiation with electrons accelerated at various voltages. Bull. Institute Chem. Res. (Kyoto University) **42**, 473—493 (1964).

Luft, J. H.: Improvements in epoxy resin embedding methods. J. Cell Biol. **9**, 409—414 (1961).

Nagata, F., Fukai, K.: Effects of electron irradiation on growth of spore, reported at U.S.A.-Japan Seminar on "New trends in high voltage electron microscopy" (1971). Personal Communication.

NAGATA, F., HAMA, K.: Observation of biological specimens with a high voltage electron microscope. 27th Annual Proceedings EMSA, (1970).

NAGATA, F., HAMA, K.: Chromatic aberration on electron microscope image of biological sectioned specimen. J. Electron Microscopy **20**, 6—10 (1971).

NOVIKOFF, P. M., NOVIKOFF, A. B., QUINTANA, N., HAUW, J. J.: Golgi apparatus, gerl, and lysosomes of neurons in rat dorsal root ganglia, studied by thick section and thin section cytochemistry. J. Cell Biol. **50**, 859—886 (1971).

PARSONS, D. F.: Beam efficiency, plasmon inelestic scatter and radiation damage in the high voltage microscope, J. appl. Physics **43**, (1972) in press.

RICE, R. V., MOSES, J., WRAY, G.: High voltage electron microscopy of unstained frozen striated muscle, reported at U.S.A.-Japan Seminar on "New trends in high voltage electron Microscopy" (1971). Personal Communication.

RIS, H.: Use of the high voltage electron microscope for the study of thick biological specimens. J. de Microscopie **8**, 761—766 (1969).

SZIRMAE, A., FESHER, R. M., MOSES, J., RICE, R. V.: Contrast and resolution of thick sections of muscle in the HVEM, reported at U.S.A.-Japan Seminar on "New trends in high voltage electron Microscopy" (1971). Personal Communication.

UEDA, R., NONOYAMA, M.: The observation of thick specimens by high voltage electron microscopy. Experiment with molybdenite films at 50—500 Kv. Japan. J. appl. Physics **6**, 557—566 (1967).

Subject Index

Aberration, chromatic 158, 276, 278, 279, 287
— coefficient 159
— spherical 158, 279, 287
Acetylcholinesterase sites 127
Acrolein 49, 60
Agents, cross-linking 15
— cryoprotective 36, 56, 61
Amoeba proteus 19
Amplitude contrast 253
Antimitotic properties 21
Antistatic agent 177
Aperture function 225
Aquon 10
Area correction 130, 138
Artifact 48, 164, 199, 223, 226, 232, 265
— charging 176, 185, 198
— drying 172
— ice crystal 49, 61
— shrinkage 57, 58
Autocorrelation function 220
Autoradiography, extended sources 119

Bacteria, fine structure 126
— spores 292
Band limitation 224, 225
Bleach, hypochlorite 78
Block staining 48
Bonding, ester 9
— hydrogen 61
— hydrophobic 36, 62, 101
Brain tissue 59
Butyl methacrylate 2, 4

Carbon, platinum source 77
— rods 77
Carcinogenicity 22, 25
Catalysts, nonoxidizing 15
Cathode, lanthinum hexaboride 160
Cathodoluminescence 156, 197
Cathodoluminescent mode 203
Celloidin 1, 2
Cells, blood 166
— tissue-cultured 166

Central nervous system 167
Charging artifact 176, 185
— effects 179
Chondrocyte 128
Chondroitin sulfate 41
Chromatic aberration, coefficient 276, 278, 287
Chromatin 46
Chromatoid bodies, *Entamoeba invadens* 260
Chromic acid 78
Chromosomes, giant 287, 288
— isolated, stained 167
Cleaning solution 78
Clearing agent 18
Coagulants, protein 35
Coding information 154
Coefficient, secondary electron 175
Collection diameter 159, 160
Complementary replica technique 169
Complete mixing 18
Conduction, improvement in 177
Contamination 67, 80, 84, 90, 99, 102, 103, 173, 183, 202, 250
— particulate 104, 105
— specific 103, 106
— water vapour 105
Contrast 246
— DNA 247, 250
— tissue 14
Convolution 221
— theorum 219, 224, 231
Cooley-Tukey algorithm 221
Cooling rates 58, 68
Copper disc 80
Correlation integrals 221
Cracking 27
Critical point drying 202
Cross correlation function 219
Cross links 9
Cross linking, additives 4
— linking, agents 6, 15
— linking, epoxy resins 16
Cross power spectrum 220
Cryofracture 170

Cryoprotectants 36, 54, 56, 61, 68, 89
— properties 41
Cryo-ultramicrotome 70
Crystal, intracellular ice 71
— plant virus 70
Current, induced specimen 163
Cuticle, insect 166
Cutting qualities 24, 26, 31
Cytoplasmic protein 41
Cytoxic properties 25

Dark field image 248
Data acquisition 223
— flow diagram 258
Deep etching 94
Deformation 194
— particle 100
— plastic 71
Dehydrated 2, 44, 226
Densitometer, flying-spot 226
Density, current 158
—, distribution 114, 116, 118, 120, 121, 122, 142, 149
— electron peaks 256
— grain 132—134, 136, 139, 140, 144, 149—151
— projected object 217
Depolymerization, heat 4
Depth of focus 194, 289
Dermatitis 20
Design, instrument 178
Diameter, collection 159, 160
— probe 160
Diepoxy compound 24
Differential interference microscopy 200
Diffraction, change of pattern 279, 280
Diffraction limit 158
— rate 196
Dimethyl formamide 60
— sulfoxide (DMSO) 4, 57, 60, 68, 88
Dioxane 60
Distribution, developed grains 115, 116, 149
— integrated 133, 134
— non-uniform 146
DNA 242, 246
— molecule 241
Drying, artifacts 172
—, critical point 202
— freeze 202
Durcupan 10

Eggs, marine 57, 58

Electron microprobe 196
—, back scattered 163
—, beam evaporation 89, 90
— density peaks 256
— penetrating power 275, 276
— secondary 156, 164
— secondary image 184
— mode 159, 202
—, secondary radiation 163
— secondary signal 187
— transmitted 156
Electron microscopes, scanning transmission 160, 161
— — ultra high voltage 197
— stereoscopy 277
Embedding medium, ideal 28
— radiation-resistant resin 13
— warer soluble 11
Emulsion, coating procedure 113
— Ilford L 4 129
— interference color 113
—, thickness 118
En-block stain 281
Endoplasmic reticulum 41
End plate, mouse 149
Epoxy equivalent 22
— groups 21
Erythrocytes, frozen 70
Escherichia coli 125
Ester bonds 9
Etching time 80
Ethylene glycol 44, 57
Eutectic 44
Evaporation, electron beam 89, 90
Extraction, lipids 41
Eye 166

Ferritin marker 94
Film measurement 81
Fine structure, bacteria 126
Flying spot densitometer 226
Focus, depth of 194, 289
— spatial 153
Focusing 183
Fourier analysis 245
— masking method 230
— methods 217
— plane 222
— space 222
— theory 223
— transformation 215, 218, 219, 224, 225, 226, 231, 233, 236, 237, 244, 249

Fourier-Bellel transformation 219
Fracture faces 73
Fracturing, brittle 71
Free-radical, chain 8
— mechanism 6, 16
Freeze drying 202
— etching 61, 200
— fracture 41, 61
— device 74
Freon 12 74
Freon 22 54
Friction 30
Friedel's law 219
Frozen specimens 292

Gap junction 98—101, 284, 285
Gaussian probe diameter 158
Gelatin 2, 26
Geometric transformation 195
Giant chromosomes 287, 288
Glutaraldehyde, prefixation 88
Glycerol 44, 57
— impregnate 88
Glycocalyx 41
Glycogen 41
Glycol 44, 45
— ethylene 44, 57
— methacrylate 10, 60
— propylene 44
—, trimethylene 39, 44
Glycoprotein 41, 101
Goniometer 185
Grain 115
— density 116, 118, 121, 128, 132—134,
    136, 139, 140, 149, 150, 151
— distribution 115, 119—123
— histograms 144
— integrated distribution 116-121
Granulation 175

Half, membrane 96
Hardness 4, 8
Heart muscle 167
Helical diffraction pattern 257
— object 257, 258
Helium, liquid 54
Helium II 56
High intensity guns 202, 203
— pass filter 232
— resolution 26
Histochemistry 6, 11
— analysis 203
Hydration, degree of 40

Hydrogen bonding 61
— water 62
Hydrophobic bonding 36, 62, 101
Hydroxyethyl methacrylate 39
Hydroxypropyl methacrylate 39
Hypochlorite bleach 78

Ice, crystallization 58
—, damage 48, 49, 53
— formation, hexagonal 63
Ilford L 4 emulsion 129
Image, degradation 217
— formation 153, 215, 245
— interpretation of 177
— processing 216
Immunochemistry 6
Impulse response 218
Incident light interferometer 113
Induced specimen current 163
Inelastic scattering 293
Infiltration 2
— long 18
Information transfer 153, 193
Insect flight muscle 281
Instrument design 178
Integrated distribution, grains 116—121,
    133, 134
Interaction volume 179
Interference color 114
— color, emulsion 113
Interferometer, incident light 113
Ionic strength 38
Isobutyric acid 60
Iso-intensity area 188

Junction, gap 99—101, 284, 285
Junctional regions 94

Kidney, rat 124

Lanthinum hexaboride cathode 160
Leaf surface 166
Least squares method 263
Linear imaging process 215
— system theory 216
Lipids 41
—, membranes 96
Lipoprotein macromolecules 178
Liquid nitrogen cooled ultramicrotome 79
Liver 57
Living matter, observation of 281
Low pass filter 232

Macromolecules, lipoprotein 178
Magnification 192
— range of 156
Marker, ferritin 94
Mass loss 14
Mast cell, tissue 290
Medium, ideal embedding 28
Membranes 93
— artifical 105
— faces 97
— frozen 67
— half 96
— lipids 96
— natural 105
— plasma 90, 91
— protein 101
— structure 97, 98
— surface 92
Methacrylate, glycol 10
— hydroxyethyl 39
— hydroxypropyl 39
— thermoplastic 30
— water soluble 6
Methyl 2, 4
Microdissection 183
Microtome, liquid nitrogen cooled 79
Microvilli 284, 285
Microvision 154
Mitochondrion 142, 145
Mixing, turbulent 17
—, complete 18
Model calculations 266
Modeling techniques 190
Modulation transfer functions 216
Morphology, insect cuticle 166
Mouse trachea 290
Multiple microscope technique 201
Muscle, heart 167
— insect flight 281
— striated 41
Myelin sheath 13

Neuromuscular junction 133
Neuronal tissue 166
Newt, *Triturus* 126, 128, 141
Noise 217, 225, 228, 235, 238, 243
— considerations 180
— fluctuation 226
— function 248
—, image 249
Nonmetric properties 190
Nonoxidizing catalysts 15

Nonpolar 7
— solvent 173
Nonuniform distribution 146
Nucleation, crystal 62
Nucleoprotein 41

Opal 178
Optical, densities 257
— diffraction 240, 256, 257
— filtering technique 216, 228
—, sectioning 194
Osmotic shock 38
Overshot effect 232

Paleobiological samples 166
Pancreas 47, 69
— tissue 46, 50
—, tissue, rat 57
Paraffin 1, 2
Parseval's theorum 220
Particle deformation 100
Pattern recognition 188, 249
Penetration, rapid 3
Phase contrast 238, 240
— equation 245
— image 235, 242
— mechanism 216
— transfer function 243
— microscopy 200
— shift 236, 244
Phosphotungstic acid 39
Photographic granularity 228
— superposition technique 228
— system, automatic 179
Plasma membranes 90, 91
Plastics, fiber reinforced 30
Platinum-carbon 73
—, evaporation 76
— replica 67
Point spread function 218
Polycondensation 6
Polyesters, unsaturated 6
Polymer formation 8
Polymerization 3, 12, 13
— damage 4, 5, 18, 19
— free-radical mechanism 16
— methacrylates 18
—, ultraviolet 55
Polysaccharide 41
Power spectrum 220
Prepolymer, resorcinol-formaldehyde 27
Preservation, optimal 40
Probability circle method 129, 132

Probe, diameter 158, 160
— electron microprobe 196
— system 154, 155
Projected object density 217
Projections 263
Propylene glycol 44
— oxide 60
Protein, coagulants 35
— cytoplasmic 41
— glyco- 41, 101
— membrane 101
Protozoa 166, 174

Quenching 45, 48, 68
— tissue 43

Radiation damage 14, 15, 217, 230, 250,
    266, 277, 280, 292, 293
— electron 197
Radiation, mass loss 14
— resistant 13
— sensitivity 13
— spread 132, 134, 136, 138, 140, 144,
    146, 147, 150
Radiation-resistant, embedding resin 13
Radioactivity, shift in 135
Recrystallization 68
Refractive index 2
Replica, platinum-carbon 67
— technique, complementary 169
Resin, radiation resistant-embedding 13
— synthetic, urea-formaldehyde 26
— thermoset 30
— water soluble 11
Resolution 116, 118, 119, 129, 153, 180,
    187, 238, 240, 241, 247, 248, 253, 256,
    266, 275, 276, 277
— advances in 160
— anisotropic 266
— design 160
— high 26
— limit 216, 279
—, Z 188
Resorcinol-formaldehyde prepolymer 27
Ribosomes 41
— negatively stained 266
Risk factor 21
Rough endoplasmic reticulum 126

Scan rate 155
Scanning transmission 201, 203
— transmission, electron microscopy 160,
    161, 197

Secondary electron, coefficient 175
— image 184
— mode 159, 202
Secondary electron, signal 187
Sectioning, optical 194
— serial 3
Self-catalyzing 9
Sensitivity 149
— value 150
Shadow, cast 73
— platinum carbon 73, 77
— thickness 80
Shape transform 227
Shift, in radioactivity 135
Shrinkage 2, 16, 19, 20, 27, 57, 58
Signal, collection systems 201
— distortion 179
Signal-to-noise ratio 226, 228, 230, 238,
    249
Solid state properties 197
Solvent, extraction of tissues 24
— non-polar 173
—, structure breaking 63
— trimethylene glycol propylene 39
— water-miscible 36
Spatial, focusing 153
— mode 187
Specimen, chemical composition 197
—, damage 275
— frozen 292
— solid state properties 197
— thick 279
Spherical aberration 279, 287
Splits in cytomembranes 48
Spores, bacterial 292
Stain, block 48
— en-block 281
— materials 279
— phosphotungstic acid 39
— uranyl acetate 39
Statistical accuracy 139
— fluctuations 222
Stereo-pairs 165, 185, 202
Stereometric analysis 188
Stereoscopic viewing 162
Stereoscopy, electron 277
Structure-breaking 40
— solvent 63
Structured water 36
Sucrose 68
Supercooling 62
Superimposition 191

Superposition 217
— technique 192
— leaf 166
— tension effects 174
— texture 187
Symmetry 262
— cylindrical 261
— icosahedral 261
Synaptic regions, sub- 128
Sympathetic nerve terminal 147

T-4 phage tail 259
Television imaging 155
Temperature, controlled 82, 84
— specimen 84
Thermal advance 80
Thermoset resins 30
Thermoplastic methacrylate 30
Thickness, sections 197, 278, 289
— specimens 277, 288
Thin film measurement device 81
Three-dimensional, density function 259
— reconstruction 257, 258
Tilting angles 265
Tissue, brain 59
— contrast 14
— dehydrating, fresh 44
— kidney, rat 124
— liver 57
— mast cell 290
—, neuronal 166
— pancreas 46, 69
— quenching 43
— solvent extraction 24
Tobacco mosaic virus paracrystals 178
Total probe diameter 158

Toxicity 25
Transfer, function 221, 222
— of information 199
Transformation invariance to 199
Translational changes 193
Trimethylene glycol 39, 44
Turbulent mixing 17

Unsaturated, polyesters 6
Uranyl acetate 39
Urea 40
— -formaldehyde synthetic resins 26

Vapor pressure 21
Vestapal 39
Video signal 161
— —, visible light 163
Viscosity 10, 23, 25, 26, 30
— low 3
— resin 7
Vitreous state 68

Water-miscible solvent 36
Water-soluble, embedding 11
— methacrylate 6
— resins 11
Wiener spectrum 230, 238, 241, 248
Wiener-Khinchin theorum 220
Wound healing 167

X-rays 163
— characteristic 197
— diffraction 97
— mode 203

Yeast, Sacchromyces cerevisiae 98

Z resolution 188

Microautoradiography and Electron Probe Analysis

Their Application to Plant Physiology

Editor **U. Lüttge**, Technische Hochschule Darmstadt
With 78 figures. III, 242 pages. 1972. Soft cover DM 46,—;
US $ 17.10

Rather than a basic discussion of the principles of autoradiography and the photographic process, the authors adopt an empiric approach to the techniques of microautoradiography and electron-probe analysis, suitable for the laboratory bench, by describing actual experiments. The book is addressed primarily to the newcomer to the field and to the physiologist who wants to add the techniques described to the tools he already uses, but is not particularly interested in the method for its own sake.

Contents

Ulrich Lüttge: Botanical Applications of Microautoradiography.

Peter Dörmer: Photometric Methods in Quantitative Autoradiography.

John B. Passioura: Quantitative Autoradiography in the Presence of Crossfire.

Ulrich Lüttge: Microautoradiography of Water-Soluble Inorganic Ions.

Walter Eschrich and Eberhard Fritz: Microautoradiography of Water-Soluble Organic Compounds.

Reinhold G. Herrmann and Wolfgang O. Abel: Microautoradiography of Organic Compounds Insoluble in a Wide Range of Polar and Non-polar Solvents.

Jeremy D. Pickett-Heaps: Autoradiography with the Electron Microscope: Experimental Techniques and Considerations Using Plant Tissues.
André Läuchli: Electron Probe Analysis.

Springer-Verlag
Berlin
Heidelberg
New York
London München Paris
Sydney Tokyo Wien

Prices are subject to change without notice.

Elektronenmikroskopische Untersuchungs- und Präparationsmethoden

Von **L. Reimer**
Zweite, erweiterte Auflage
Mit 247 Abb. XII, 598 Seiten. 1967
Geb. DM 108,—; US $ 40.00

Inhaltsübersicht

Untersuchungsmethoden:
Elektronenoptische Grundlagen
des Durchstrahlungsmikroskopes.
Prinzipieller Aufbau des Durchstrahlungsmikroskopes. Andere
Abbildungsverfahren. Messung
wichtiger optischer Konstanten.
Elektronenbeugung. Bildkontrast
in amorphen Objekten. Phasenkontrast und verwandte Probleme.
Bildkontrast in kristallinen Objekten. Präparatveränderungen unter
Elektronenbestrahlung. Bildaufzeichnung und Intensitätsmessungen. Ermittlung der dritten
Dimension von elektronenmikroskopischen Präparaten. Abbildung
magnetischer und elektrischer
Objektfelder. – Präparationsmethoden: Objektblenden und
Trägernetze, Grundlagen der
Hochvakuum- und Aufdampftechnik. Herstellung und
Eigenschaften von Trägerfolien.
Oberflächenabdrücke. Herstellung
durchstrahlbarer Folien aus
Metallen und anderen Kristallen.
Präparation von Pulvern, Suspensionen, Stäuben und Aerosolen.
Gewebefixation. Kontrastierung
und histochemische Methoden. Einbettung. Ultramikrotomie.
Präparation organischer Teilchen
(Fraktionen, Homogenisate,
Bakterien, Phagen und Viren).
Autoradiographie. Monographien
über Elektromikroskopie. Bezugsquellen-Nachweis. – Namen und
Sachverzeichnis.

Elektronenmikroskopische Methodik

Von **G. Schimmel**
Mit 134 Abbildungen
in 249 Einzeldarstellungen
VIII, 243 Seiten. 1969
Geb. DM 78,—; US $ 28.90

Das Buch versucht, die Verbindung
von Objekteigenschaft — Präparateigenschaft — und Bild durch
Behandlung der physikalischen
Zusammenhänge und anhand von
ausgewählten Beispielen für den
Bereich von technisch-physikalischen Untersuchungen aufzuhellen.
Dazu wird nach einer Einführung
in die Theorie des Mikroskopes die
geometrische Theorie der Elektronenbeugung behandelt, aus der die
wichtigsten Erscheinungen des Beugungskontrastes abgeleitet werden
(kinematische Theorie). Nach einer
kritischen Betrachtung der Leistungsgrenze von Durchstrahlungsmikroskopen werden die Abdruckverfahren gebracht, die dann zu
Anwendungsbeispielen überleiten.
Hierbei wird größter Wert auf die
Kombination der verschiedenen
elektronenmikroskopischen Verfahren gelegt. Die Verbindungen
und Kontaktstellen zu anderen
physikalischen Meßverfahren werden hervorgehoben und gleichfalls
an Beispielen erläutert.

Springer-Verlag
Berlin
Heidelberg
New York

London München Paris
Sydney Tokyo Wien

Prices are subject to change without notice.